OD

OD

NALOXONE AND THE POLITICS OF OVERDOSE

NANCY D. CAMPBELL

THE MIT PRESS   CAMBRIDGE, MASSACHUSETTS   LONDON, ENGLAND

This book was set in Avenir by Westchester Publishing Services, Danbury, CT. Printed and bound in the United States of America.

Library of Congress Cataloging-in-Publication Data

Names: Campbell, Nancy D. (Nancy Dianne), 1963- author.
Title: OD : naloxone and the politics of overdose / Nancy Campbell.
Other titles: Inside technology.
Description: Cambridge, MA : The MIT Press, [2019] | Series: Inside
   technology | Includes bibliographical references and index.
Identifiers: LCCN 2019022684 | ISBN 9780262043663 (cloth)
Subjects: MESH: Drug Overdose--prevention & control | Opioid-Related
   Disorders | Naloxone--pharmacology | Naloxone--therapeutic use |
   United States | United Kingdom
Classification: LCC RC566 | NLM WM 284 | DDC 362.29/3--dc23
LC record available at https://lccn.loc.gov/2019022684

10 9 8 7 6 5 4 3 2 1

In solidarity with those who struggle, dead or alive

CONTENTS

ACKNOWLEDGMENTS

Writing a book about life and death means disentangling the actions of the dead from those who survived. Writing a book about a lively social movement makes one's first debt to the unnamed and unsung heroines and heroes who people the cause. The protagonists of this book survived the rigors of culture wars, political protest, and peer-reviewed research. Many more supported them in their roles, contributed to their causes, and hopefully will see their efforts reflected in this book despite going unnamed. My gratitude is also infinite for the people immediately beyond my closed study door, whose regular infusions of affection, food, tidings from the outer world, and marked-up manuscript pages sometimes continued round-the-clock; Ned Woodhouse, whose editorial sensibilities shaped both the abstract and concrete form this book ultimately assumed; and Isaac Campbell Eglash, Grace Campbell Woodhouse, Beans, and Cinnamon, who persisted in being and becoming themselves. My colleagues in the ever-shifting conceptual and material boundaries of the Department of Science and Technology Studies (STS) at Rensselaer Polytechnic Institute kindly weathered the time it took to move this project into print. Their patient congeniality rarely flagged; I especially thank those who survived, Atsushi Akera, Tamar Gordon, Abby Kinchy, Jim Malazita, Raquel Velho, and Langdon Winner, and the STS graduate students for maintaining intellectual community despite institutional adversity.

FRIAS, the Freiburg Institute of Advanced Study at the University of Freiburg in Germany, provided a convivial incubation period and gave me a room of my own. When in 2014 I spent my first two months at FRIAS, I began writing what I thought was an article or at most a chapter for an edited volume. Invited to FRIAS by Hermann Herlinghaus, whose friendship I cherish, I found long-term friendships and true scholarly support from Bernd Kortmann, Britta Küst, Petra Fischer, and Nikolaus Binder. Walking in the Schwarzwald, I made friends with Stephie Kirner and picnicked on Staufen's shifting grounds with Todd Meyers, Fanny, and Etta. Returning in 2017–2018, the FRIAS Fellows included Lorena Bachmaier, Oliver Braunling, Kate Burridge, Lawrence Chua, Leonie Cornips, Majid Daneshgar, Anne Harrington, Anne Holzmüller, Jacob Sider Jost, Henrike Laehneman, Veronika Lipphardt, Errol Lord, Evie Malaia, Cammie McBride, Azar Mir, Andreas Musolff, Jenny Reardon, and, far from least, Onur Yildrim. I am grateful to Stefan Pfänder for including me in the Synchronization in Embodied Interaction Research Group. My time at FRIAS was supported by a Marie S. Curie FRIAS Co-fund Senior Fellowship Award.

Humanists are often thrown onto their own resources when traveling to the archives and interviews comprising the empirical basis of their interpretive work. I gratefully acknowledge the School of Humanities, Arts, and Social Sciences at Rensselaer, from which I received a HASS FLASH Grant and subvention funds, and all of the people who made that possible.

For her precocious interest in this project, and her generosity in opening up her world to me, I am especially grateful to Maya Doe-Simkins. For sharing writing days of tea and friendship, I thank Hope Perlman. For particular help navigating the pleasures and occasional pitfalls of the journey, I thank Alice Bell, Dan Bigg, Sheila Bird, Heather Edney, Denise Holden, Carole Hunter, Amanda Laird, Hannah Lloyd, Brooke Lober, Rebecca McDonald, Lynn Paltrow, Iain D. [not Duncan] Smith, John Strang, and Brittany Trellis. The book would not look as it does without Heather Edney and Brooke Lober; would not be so well-documented without Hannah Lloyd and Rebecca McDonald; nor would the aesthetics and pragmatics have been so smooth without Rogan Zangari, who tracked down many assumed lost in the project's last months. I am grateful to

The MIT Press, and particularly to Katie Helke and Laura Keeler for seeing potential in this book.

My thinking in this book built on previous and concurrent articles, books, and talks; the conviviality of activity in adjacent fields; or my ongoing editorship of *Social History of Alcohol and Drugs*. For more than a decade of energetic and convivial ferment, I thank: Caroline Acker, Alex Bennett, Virginia Berridge, Elaine Carey, Anna Rose Childress, Claire Clarke, David Courtwright, Giovanna Di Chiro, Betsy Ettorre, Philippe Fontaine, Kim, Mike, Kora, and Lena Fortun, Suzanne Fraser, Will Garriott, Paul Gootenberg, Helena Hansen, Hermann Herlinghaus, David Herzberg, Geoff Hunt, Helen Keane, Kathryn Keller, Kelly Ray Knight, Howard Kushner, Anne Lovell, Zack Marshall, Todd Meyers, Alex Mold, David Moore, Jules Netherland, Dean Nieusma, J. P. Olsen, Jefferson Pooley, Lucas Richert, Eugene Raikhel, Sam Roberts, Marjorie Senechal, Susan Shaw, Christopher Smith, Joe Spillane, Laura Stark, Maia Szalavitz, Nathan Tauger, Trysh Travis, Luke Walden, Ian Walmsley, and Dan Wikler. Major losses along the way must be marked: Dan Bigg, Scott Christianson, Paul Copp, Ron Johnson, Herb Kleber, David Musto, Eric Schneider, Bob Schuster, Eddie Way, and Hayden White—none of whom envisioned inhabiting the same sentence and all of whom influenced my thoughts and feelings about the events of this book.

Finally, I thank my parents, David and Sandra Campbell; my siblings, in-laws, and out-laws, Connie Campbell and Tony Diehl; Dave and Amanda Campbell, who put me up when visiting southern Pennsylvania; Gary Campbell; Rebecca, Ed, Jasper, and Owen Allard, who put me up when visiting Seattle; my aunts, uncles, cousins, and assorted relations who keep me connected to Berwick, Beach Haven, and Fishing Creek, and keeping "something to crow about" in Hartman's Grove. Without all of these people and places, my world would be profoundly different.

INTRODUCTION: Making Overdose Matter—
Protagonists and Antagonists in the Social Lives of Naloxone

T-shirts sport the slogan "Keep Calm and Carry Naloxone." HERO bracelets signal that the wearer carries Take-Home Naloxone. Naloxone Ninjas (see figure I.1) tweet, tally lives saved, distribute naloxone kits, and wear silver ribbons to mark International Overdose Awareness Day.[1] Promotional campaigns highlight the role of naloxone in saving lives. Once a forensic matter almost out of sight, OD has evolved into a complex public problem demanding attention from governmental and nongovernmental organizations the world over—from Australia, Canada, Estonia, and the United Kingdom to Iran, Vietnam, and the United States. Widely understood as survivable, OD has moved from margin to mainstream. Naloxone, the opioid antagonist that can reverse even near-death experiences, has become represented as the special power of superheroes.[2] How did that come about?

With rare exceptions, the hidden arcana of overdose represented it as a lethal, acute encounter with mortality—one that was neither scientifically nor clinically interesting. For forensic pathologists and medical examiners, OD was a problem of categorization; for the popular press, an ugly death awaiting social deviants; for some moral registers, it was simply just deserts. Once "unmentionable," overdose prevention emerged in the last decade as the object of a social movement. Naloxone, a technology cast as "power in a bottle," was the movement's central artifact. Few

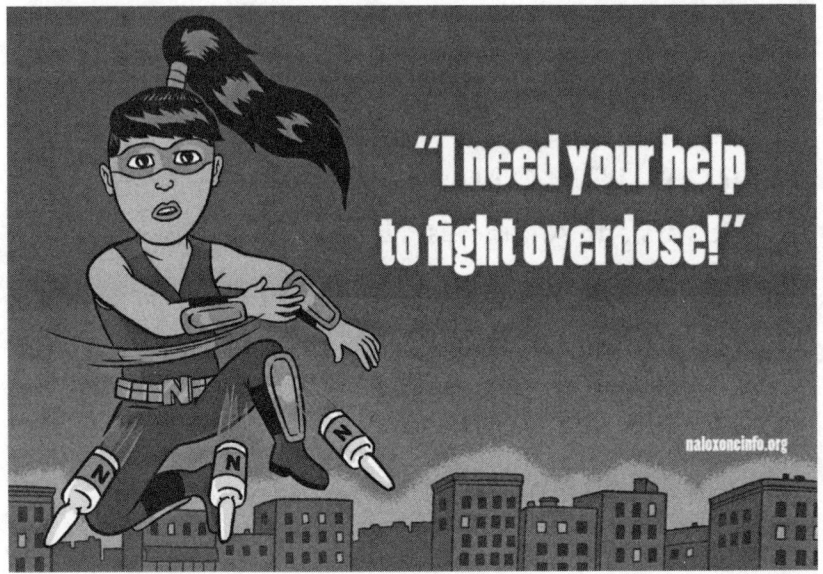

I.1   Naloxone Ninja.
Used with permission of Open Society Foundations.

outside emergency medicine had even heard of naloxone before 2010. Since then naloxone has exploded onto the scene. Thousands of kits have been distributed, "rescues" or "reversals" made, and "unnecessary deaths" prevented by the once-crazy idea of getting a naloxone kit into the hands of those likely to witness an overdose. Preparing communities to counteract overdose once would have been a pipe dream. Yet naloxone became a tool for shifting law, policy, clinical medicine, and science toward harm reduction. OD was remade as common, predictable, patterned, accessible, discernible, and above all preventable.

No longer set apart from daily routine, OD had regrettably become business as usual when I wrapped up this book: almost two hundred people were dying preventable deaths each *day* in the United States.[3] Close to one thousand people a day were treated in US emergency rooms for opioid-related issues. As death tolls rose, I had begun following naloxone as it was used increasingly to exhume questions about life, death, moral character, responsibility, identity, and the everyday quest for meaning. Harm reduction activists, public health officials, law enforcement, legislators, and clinicians have struggled over the terms on which opioids would be used in societies that depend upon them, but are deeply ambivalent about them. My curiosity about naloxone's peculiar antagonisms was piqued when I learned to think pharmacologically in the early 2000s. As a historian of science who studies addiction research and treatment, I sought to understand the concept of opioid receptors and the actions of agonism and antagonism. I wanted to interpret the dreams of molecular manipulation that have preoccupied the field of addiction therapeutics. But I also sensed a disjuncture between the official spheres where overdose was rarely mentioned and the back channels of a social movement where overdose prevention was being talked about. There was little imagination for overdose prevention or naloxone in the conversations about evidence-based practices (EBPs) that pervaded the endless official meetings that I observed in the early years of the current century. In retrospect, I wondered, "Why did it take so long for overdose prevention to be brought to life?"

By the early 2000s, both the United Kingdom and the United States aspired to place local-level drug treatment and prevention services on an evidence-based footing. Responsibility for treatment had migrated from the federal government to the states back in the 1970s. New York State's Office of Alcoholism and Substance Abuse Services was one of the largest state systems grappling with the imperative to infuse evidence-based practices (EBPs) into existing treatment programs. Listening for what counted as "evidence," I watched expert drama unfold around "bupe," the latest addition to the narrow repertoire of accepted evidence-based addiction treatments.[4] As with any drug, buprenorphine's safety and efficacy had been demonstrated via randomized, placebo-controlled clinical trials (RCTs). But what was behind the curtain—what were the politics of becoming an EBP? Could EBPs originate with clients, consumers, or harm reduction organizations? Did some practices meet resistance because they couldn't be subjected to RCTs? Did patients, publics, and politicians care about the rigors of the process of becoming certified as an EBP? Was naloxone—clearly a life-saving drug—an EBP?

During this time, Alan Leshner, then-director of the US National Institute on Drug Abuse, was preaching a new gospel that anointed addiction a "chronic, relapsing brain disease." A new priesthood—addiction neuroscientists—had arisen to save people from themselves by targeting their brains. Bupe was one of these molecularly targeted, high-tech drugs, having taken thirty years to wend its way to market. At the other end of the spectrum of accepted treatment approaches resided expansive notions of everything from friendly receptionists to the paint color on clinic walls. A "two-cultures problem" unfolded between science and practice, laboratory and clinic, scientist and clinician, each calling the other "ideological." To my ears the most important thing was what was not said and whose voices went unheard.

Drug users and overdose went unmentioned. Believe it or not, I never heard anything about overdose prevention with naloxone despite the uptick of deaths that we now know had begun in the late 1990s. The perception that there was a limited set of treatment modalities—and a "paucity" of EBPs—pervaded the meetings that I observed from 2000 to 2004. Anchoring this short list were pharmacotherapies validated by clinical trial: methadone, buprenorphine, and naltrexone. Naloxone showed up only as an ingredient of Suboxone (buprenorphine plus naloxone). The deafening quietude about overdose in gatherings of treatment providers, researchers, and bureaucrats felt to me like "undone science,"[5] a form of collective ignorance mixed in with historians' favorite quarry, amnesia. Yet the protagonists of this book were actively promoting naloxone for overdose prevention at the very time I was picking up radio silence in the official realm.

No doubt the lack of official attention to overdose and its reversal was due partly to the fact that it was difficult to imagine how to subject near-death experiences to clinical trials. Naloxone was an old drug—approved by the US Food and Drug Administration (FDA) for reversing opioid overdose in 1971, it had become a staple of emergency medicine and anesthesiology that did not leak into harm reduction communities for more than two decades.[6] As recently as 2010, when the Scottish government launched a National Naloxone Program, the story of naloxone remained stubbornly local. In 2012 my own Rensselaer County, New York, sheriff's department became one of the first in the country to require that all deputies carry naloxone. Just seven years later, naloxone was everywhere: repairing rock walls one summer after a Pennsylvania flood, I overheard four people explaining how naloxone "antagonized" overdose as they floated down the creek.

To my mind as an avowedly partisan researcher, naloxone is an essential medicine, and I was relieved to see it becoming common. Yet I was also keenly aware that naloxone is not nearly enough. Opioid death rates continued climbing despite harm reductionists' success at changing laws and building new channels to access naloxone. Turning the situation around will require multiple policy mechanisms, lots more naloxone, and cultural shifts in perspectives on drug use and treatment. Indeed, drug researchers, government officials, and the rest of us might actually have to confront the question of *why* so many people in postindustrial societies come to rely on opioids in the first place.

MOVING OVERDOSE FROM MARGIN TO MAINSTREAM

Designated an "epidemic" by the Centers for Disease Control and Prevention in the United States in 2010,[7] OD had by then become *the* leading cause of preventable death in many places. Outstripping traffic fatalities,[8] OD is an acute form of accident, injury, or poisoning (see chapter 1). With nine out of ten overdoses survived, it was recast as preventable, patterned, and reversible. Just as all "epidemics" overwrite the "endemic" conditions from which they spring, the "opioid abuse epidemic" in rural white and middle-class communities overwrote the urban drug economies where overdose had been endemic since the late 1950s in the United States. A focus on "white despair" deflected attention from endemic overdose deaths faced in urban communities of color. The new narrative seemed to fit places like the small Pennsylvania town where I grew up. But something in the white-despair narrative seemed off-key to my historian's ear.

Drug discourses have historically been used to make social order and to signal social disorder.[9] My hunch was that the epidemic was a symptom of something coming undone, a harbinger of new, sadder forms of social solidarity, and a way of talking about a hidden history of rare and unfortunate events. Resuscitation and reversal were not promoted as a public health strategy until this book's protagonists rescripted OD. Opioid use, often shorthanded as "addiction," "abuse," or "misuse," *was* about heroin, even though "polydrug" use—ingestion of other legal and illegal drugs along with opioids—has grown since the mid-twentieth century.

All these terms have been problematized both prior to and during the so-called opioid epidemic, a label that gathers together morphine, heroin, fentanyl, and pharmaceutical opioids such as hydrocodone, oxycodone, or codeine. Opioids are derived from opium poppies; classically, opium base has been necessary to produce "opiates." But contemporary "opioids" can now be synthesized without plant-based material. Compounds that occupy opioid receptors in the human brain are all considered "opioids" whether or not they derive from *Papaver somniferum*.

Opioids have bedeviled societies in the global North and South since the 1805 synthesis of morphine from opium. Medically necessary not only for pain, but for diarrheal diseases, dental problems, and respiratory illnesses, opioids were widely used in cough syrups. They were sleep aids during the period of industrial modernization. Wars, accidents, aging, cancer, and occupational injury from mining and railroads promoted opioid use—and their users hardly sat around getting high all day. The term "addiction" was little used in the United States until the early twentieth century; one might forgive a historian for thinking it overdetermined by the move to ban narcotics. Anti-narcotics organizing culminated in the Harrison Narcotics Act (1914), setting the scene for science and social hygiene to take the stage along with anti-narcotics policing.

The handful of clinicians and researchers who steered away from the word "addiction" on grounds it was morally stigmatizing got the World Health Organization to substitute "drug dependence" in 1965. This group defined drug dependence as a neurophysiological matter of central nervous system (CNS) stimulation, characterized by compulsion, drug-seeking, chronicity or tolerance, symptoms of abstinence upon withdrawal, and tendency to relapse. A cadre of clinicians, lawyers, sociologists, and scientists argued that "addicts" were "sick people" suffering a chronic disease:

[We should] treat addiction as we do other chronic disease, even cases which have already progressed to an apparently hopeless stage. Compared to the results which can be expected under present methods for treating chronic tuberculosis, chronic cancer, and chronic heart disease, the results achieved in the treatment of drug addiction are by no means discouraging.... [T]here is every reason to hope for a continued decrease in addiction as a chronic disease.[10]

Obviously, this prediction from the 1950s turned out to be overly optimistic. The label "addict" continued to be applied despite the implied and at times explicit pushback against criminalization. Those to whom that label stuck were deemed "deviant." The label slid right off the backs of the functional, professional, or otherwise respectable opiate user. People who left off using opioids without professional help made the news so rarely that they were thought not to exist. Although historians have concrete estimates of the massive amounts of "narcotics" (a collective noun for morphine, heroin, cocaine, and other psychoactives) that were produced, bought, sold, taxed, or consumed in previous eras, there is much we do not know about what people experienced. As addiction and drug dependence were painted as a "chronic, relapsing disease,"[11] the idea that "addicts" should be seen as "patients" was more than an intellectual debate. The reality was that "addicts" and "patients" struggled to maintain health and habit in the face of vigorous enforcement of the Harrison Act. Physicians were prosecuted under that act. They became not only reluctant to treat "addicts," but ignorant about how to do so. The prohibitionist policy context shaped the lived experience of "addicts" and "patients," doctors and police.

Overdose deaths were represented in this policy context as tragic casualties of acute reactions caused by ingesting too much or too potent an opioid agonist. Sometimes it has seemed to me that every term, knowledge claim, and finding that appears in this text was unstable; each has been redefined in relation to the social, political, and economic conditions in which claims about drug use and overdose are generated, situated, and received. The very technologies central to this book—naloxone and the agonists it antagonizes—operate at the molecular level but have multiple effects conditioned by social, economic, and political conditions that are in turn structured by drug policy and how the state enforces policy.[12] Yet how could this be? Isn't pharmacology *real*? Aren't these molecules that pleasure and pester brains and bodies just powders and solutions, their complex alchemy knowable and controllable through science, medicine, social norms, or practices?

Far more than pharmacology goes on in this book. Nevertheless, it is useful to know something about the micro-level interactions that distinguish opioid "agonists" from their "antagonists." Morphine, heroin, methadone, and fentanyl are agonists that can fully occupy the opioid receptors in human and animal brains. By contrast, antagonists interfere with the fit between agonists and their receptors, thereby preventing an agonist from exerting full actions and effects. Naloxone, one of the best-known antagonists, is an extremely short-acting drug that can dramatically reverse overdose and save lives if administered in time. Naltrexone, which is often confused with naloxone, is a longer-acting antagonist that does not rapidly reverse overdose and is instead used to treat people who are highly motivated to stop using agonists. Although there are many lesser-known antagonists, the two most significant for this book are nalorphine, central player in chapters 1 and 2, and naloxone, waiting for its humble entrance in chapter 3. Intricate interconnections between opioid agonists and antagonists in the brain can be schematically presented to render overdose exquisitely simple (see figures I.2 and I.3).[13]

Breathing is an orchestrated interaction between the heart and lungs, mediated by the brain. Although breathing and respiratory depression are complex matters, reversing an overdose may be broken down into steps: Put the person in recovery position. Manage the airway. Perform rescue breathing. Administer a pure antidote, the opiate antagonist named

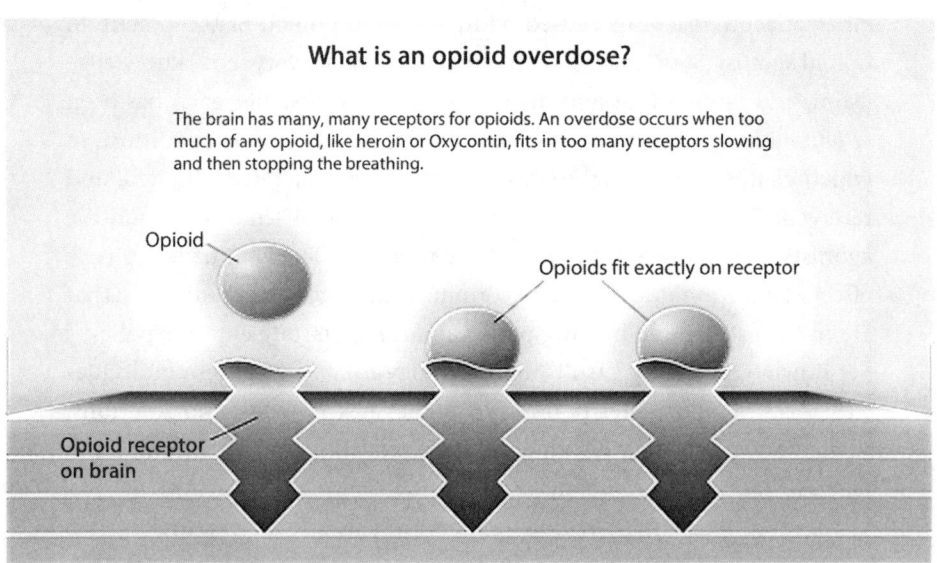

I.2   A spherical key lands on a receptor lock, depicting how an opioid agonist occupies a receptor.
Used with permission of Maya Doe-Simkins.

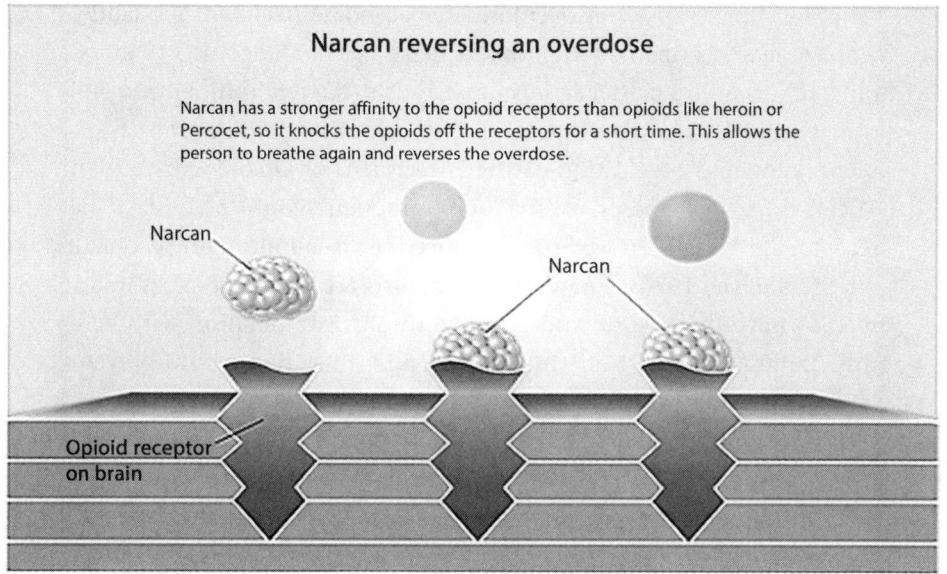

Narcan reversing an overdose

Narcan has a stronger affinity to the opioid receptors than opioids like heroin or Percocet, so it knocks the opioids off the receptors for a short time. This allows the person to breathe again and reverses the overdose.

Narcan

Narcan

Opioid receptor on brain

I.3 Antagonists have stronger affinity for receptors than do the opioids they "antagonize" by kicking them off the receptor. Their presence restores breathing, but the duration of their effects varies.
Used with permission of Maya Doe-Simkins.

naloxone. Check to see if breathing is restored. Call emergency services. Enough said. For the curious, details of dosage, duration, and delivery mechanism remain to be explored. While scientists know much about pharmacological interaction between brain receptors, their agonists, and their antagonists, the intricacies of respiratory depression remain unknown. Overdose is hard to study when it is happening—except in the surgical suite, where anesthesiologists use naloxone as an everyday tool of the trade. Despite the social, political, and economic complexities that shape each individual overdose, the straightforward solution is obvious: overdose is reversed when a molecular "antagonist" with stronger affinity for the receptor comes along and knocks the opioid "agonist" off the receptor just as a golf club knocks a ball off a tee.

Synthesized in 1960, naloxone became a tool for probing how opiate receptors worked. Thinking in molecular terms,[14] or what we might call "thinking with the molecule" or "thinking pharmacologically," naloxone also made it possible to imagine a treatment regime focused on preventing relapse and overdose. Indeed, some treatment modalities employ agonists (methadone); others make use of antagonists (naltrexone); others utilize partial agonist-antagonists (buprenorphine); and still others rely only upon abstinence. For the past fifty years, agonist treatment, known as Medication-Assisted Treatment or Opioid Substitution Therapy, has centered on methadone. Although agonist actions at the receptor site have been well studied since the visualization of the opiate receptors in the 1970s,[15] narcotic antagonists are a class of drugs imagined to have therapeutic and punitive uses in combination with agonists. Might there be some magic molecular ratio that would prevent harm, but enable people to enjoy the benefits of opioids? Might antagonists be useful for those seeking to control, police, or even punish those using agonists for purposes other than the medicinal? Such questions recur in ongoing conversations about opioids, which are often focused on the signal teaching moments provided by celebrity overdose deaths.

HOW DID PHILIP SEYMOUR HOFFMAN DIE?

Ask anyone, "How did actor Philip Seymour Hoffman die?" OD will be the answer. Substitute the celebrity of your choice—Kurt Cobain, Jimi Hendrix, Janis Joplin, Prince, Michael Jackson, Anna Nicole Smith, Jean-Michel Basquiat, Layne Staley, Paula Yates. No matter how contradictory the toxicology report, how multiple the substances knocking about, the answer will still be OD. Forensic detail does not matter when we "know" the answer is OD. Oddly enough, however, the term "overdose" has been clinically and scientifically puzzling. What do we think we know about overdose? How do we think we know it? To what purposes has that knowledge been put? And the perennial question that must always be asked: Who counts as "we"?

Who counts as "we" changes often when looking into the knowledge questions at the core of overdose orthodoxies. Far from straightforward, the term "overdose" is widely used not only in science, forensics, and medicine, but also among the general public, government officials, and participants in the harm reduction movements that emerged in Australia, North America, Europe, and the United Kingdom. In North America, Canada was exemplary in putting drug-user-led harm reduction into place, while the United States lagged behind and only gained traction in response to HIV transmission.[16] As with HIV/AIDS activism, harm reductionists moved from questioning science to producing it, a shift that gained "resources," including the social, economic, human, and above all political capital of research worlds, but produced new problems borne of political negotiations over what counts as expertise and who counts as an expert.

Some argue that the very term "overdose" misleads. "Heroin" or "opioid" overdose is often the announced result, even when toxicology screens show polydrug use, most often the copresence of alcohol, benzodiazepines, or other respiratory depressants. Simplistic accounts that do not acknowledge these complexities continue not only in popular media, but also among those whose pharmacological curiosities remain limited by profession, paradigm, or lexicon. The majority of those dying of overdose in the first two decades of the twenty-first century were not "pure" heroin addicts. It has become difficult to categorize these deaths; all of the categories intersect and overlap. To my mind, the problems involved in writing a history of overdose are not primarily sensationalism or even stigma, but the ways in which the cultural frames for making sense of drug use amplify harm, while failing to consider the needs that propel it. Writing the history of the overdose prevention movement within the context of harm reduction requires more fully documenting the history of needs and harms, and the ways in which markets were created to coordinate meeting demand.

Drug policy has long been criticized for pushing users out of legal markets, which historian David Herzberg dubs "white markets," and into "black markets" of illicit trafficking and "gray markets" composed of illegally diverted pharmaceuticals.[17] Dwelling on the national level obscures the shaping force exerted by overdose "endemics" in drug-using communities such as Edinburgh and Glasgow, New York City, Chicago, and San Francisco, as well as distinctly nonurban areas like the Española Valley high in the hills of New Mexico, or rural Rhode Island, western Massachusetts, and the Scottish Highlands. As pharmaceutical opioids, most spectacularly OxyContin, wormed their way into rural areas in the Midwest, the American South and Appalachia, West Virginia, Ohio, Pennsylvania, Maryland, Virginia, and North Carolina, people began to consider adopting harm reduction practices once confined to larger cities. The relentless focus on the United States is deserved, as it accounts for one out of every four drug-related deaths (DRDs) in the world. However, Estonia, too, has a very high DRD rate after more than two decades of experience with fentanyl. The United Kingdom, and specifically Scotland, has the highest DRD rates in Europe.

In the wake of Hoffman's demise, Australian researchers cautioned that few areas were as "replete with myths as heroin overdose."[18] Due to the needle in his arm and "nearly fifty envelopes branded 'Ace of Spades' filled with what [police] believe was heroin" strewn about, the cause of Hoffman's death was reported as the obvious—heroin OD. Days before official release of the cause of death, the Aussies accurately predicted the culprit would not be heroin but polydrug use. Indeed, Hoffman's death was later attributed to "acute mixed drug reaction" to benzodiazepines, amphetamines, opiates, and cocaine.[19] Medical correspondent Sanjay Gupta fit tolerance into the overdose puzzle, explaining how varying degrees of tolerance to different drugs used concurrently interacted: "As addicts take mixtures of drugs more chronically, they may not necessarily feel the effects of the narcotics, which still suppress the respiratory system. They're not feeling it, but it's still having an impact on their ability to breathe, and that's the real problem.... It's called stacking. You can stack the same drug too close together, or you can start to stack other drugs, one on top of the other. That's how people get into trouble."[20]

"Acute mixed drug intoxication"—casually referred to as "overdose"—has been clinically and scientifically puzzling, even downright misleading. Although stacking can lead to death, it is also quite purposeful as people attempt to achieve desired subjective states—sleep, pain relief, or lower anxiety, fear, and worry. There is something almost mystical about what drugs do in interaction with each other in the user's body, brain, and social milieu. Once in the body, all opioids are metabolized into morphine. Oddly, morphine concentrations in overdose fatalities are often found to be no higher than in those who survive. Motivated to dispel persistent myth-making around *over*dose, Darke and Farrell isolated three myths that exercised a stranglehold over media representations:

Myth 1. Youthful users: it is the young, inexperienced user who overdoses;

Myth 2. Functional injectors: there is no hidden group of functional or "recreational" users;

Myth 3. Death from impurities: variations in the purity of illicit opioids are the major cause of overdose deaths.[21]

Overdose represents a potent mix of purity and danger. Drugs themselves are attributed agency when assumptions about drug purity and potency take center stage; the result, presumably intended at times, is to deflect attention away from the social, economic, and political contexts in which drugs are produced, bought, sold, and consumed. All of these factors shape who dies what sorts of deaths—but because individuals overdose and individuals die, tolerance levels and neurophysiological conditions are taken as explanatory. Social context is mystified in almost every public reckoning, a profound deflection from the social worlds that structure and produce overdose death as a likely or unlikely "outcome."

Public health researchers and harm reduction advocates seek to undo mysticism with practical attitudes and evidence. But unraveling overdose myths only goes partway to undo the edifice on which they were established. "Overdose myths" remain commonly available despite the work of the protagonists of this book to identify a viable and simple solution to the problem of overdose deaths and to generate evidence that would definitively establish the effectiveness of their solution. Evidence alone would not do, so those less constrained by professional mores transformed into "Naloxone Ninjas," "experts by experience," "Momvocates," or "Heroines of Heroin." They adopted tactics running the gamut from distributing naloxone via reddit to ordinary lobbying for changes in law and policy to creating and sustaining new ways of knowing, organizing, and educating, and even transforming institutions. Would-be change agents met resistance, but also achieved success. Articulating a grand narrative of drug user health as a human right, activists molded their artifacts, practices, and protocols into an international movement. One naloxone kit at a time, one training at a time, one "save" at a time.

Outnumbered and poorly financed, social movements must rely on imagination and emotional connection, all the more so when drug users are involved. Harm reductionists rewrote the tired plot of overdose scripts to remap the outcome—from death to another chance at life. The prosaic details of how, why, where, when, and under what circumstances drug use becomes overdose, and overdose becomes death, matter for forensics, epidemiology, and statistical portraits of the many locations worldwide where drug-related deaths are occurring with rising frequency. Overdose orthodoxies provide such powerful and compelling stories that it is tempting to rely on them. But making sense of OD requires more than asking what the numbers are; also crucial are the who, when, where, and why. Why are these preventable deaths occurring in such numbers now? Who is seeking to reshape social contexts so as to prevent them? In other words, who has cared if drug users live or die, and why should we care about exhuming social histories of overdose?

People are dying these preventable deaths in much greater numbers than they should in a humane society. No causal narrative can fully encompass the wide range of people, places, and practices involved in different enactments of these most social of deaths. Yet more complex are the coordinating structures of legal and illegal drug markets and interactions within them. These larger and much more important factors are also seen as "too big" or "too complex." And indeed they are. But instead of considering these more audacious questions, the "simple" matters of drug purity and potency are perceived to be all that matters in making the difference between a drug user living or dying. These two factors continue to be widely interpreted as "causing" death, distracting from the social, political, and economic conditions that make overdose deaths more likely in some situations and less likely in others. For instance, the presence of alcohol, benzodiazepines, or other "downers" makes it much more likely for heroin to kill. Relentlessly centering the drug obscures how the drugs came to kill within social circumstances that either augment or undermine the likelihood of death. Shouldn't those who care about drug deaths try to change the social, political, and economic circumstances that make them more likely?

People matter. Place matters. Things matter. Situational plasticity can be put to use. Familiarity stabilizes. Novelty destabilizes: drugs kill in unfamiliar situations. People become "tolerant" to people, places, and situations just as concretely as they become tolerant to substance, dosage, and route of administration. Yet we know far more about the latter set of variables than we do about the former; the undone science of socially situated overdose has just begun. Why did it take so long to unfold? Following naloxone, I thought that was my question—why did it take so long for naloxone to be put into the social circumstances where it was needed to prevent people from dying? That turned out to be the question on the clichéd tip of the iceberg.

PLACE MATTERS: THE UNDONE SCIENCE OF PLACE

Undergirded by beliefs and commitments, pharmacology is a molecular science that understands interaction within brain and body, but not interaction beyond brain and body. Behavioral scientists have evolved research practices capable of dealing with some social nuance, designing experiments that show how familiarity with place, time, and route of administration matters for degrees of tolerance, intensity of withdrawal, and even drug effects themselves. Experimental conditions take place not only within laboratories, but also in the clinic, on the streets, and behind closed doors.

Overdose is surprisingly situation specific. Consider a 2015 story of an ill-fated, heroin-addicted couple who traveled from Alabama to Ohio for their infant daughter's surgery. The morning after the surgery found the mother dead in the child's hospital room, and the father alive but unconscious in the bathroom. The scenario recurs: using a drug to which one is tolerant under familiar circumstances may lead to death in strange places.[22] This has been noted since the 1970s. Experimental psychologist Shepard Siegel has patiently studied the relatively undone science of drug effects' "situational specificity," and how it affects who survives and who dies. In 1984, he interviewed ten methadone patients whose medical records indicated they had recently survived heroin overdoses, seven after injecting in unfamiliar settings.[23] Siegel argued that the places where drugs are used—and the social and environmental cues with which they are associated—mattered for the degree of tolerance formed, the intensity of effects, and intensity of withdrawal symptoms. Similar to studies on placebo effects, he posited that these associations were partly learned and conditioned in powerful ways.

An "enigmatic failure of tolerance," overdose death occurs in opioid-tolerant individuals when they are absent familiar environmental cues.

Siegel's repertoire of place-specific "retrospective reports" included a pancreatic cancer patient who died when his son gave him morphine in a brightly-lit living room, instead of the dimly lit bedroom where the father typically received exactly the same dose.[24] Why would an experienced opiate user who should have been tolerant die? Revisiting this question decades later, Siegel presented the enigmatic death of K. J., who died in Hungary in 1999 in what Siegel described as a "perplexing fatality." K. J. was an experienced heroin user who typically used with his wife at home. The day of his death, he and his wife had decided to quit. Unable to follow through but not wanting to diminish his wife's resolve, K. J. slunk off and shot up alone in what was for him a "novel environment"—a public restroom. He was found dead from a dose that should not have killed him—and did not kill anyone else who had purchased from the same source.[25] The public restroom has become "a pitiful sanctuary," in the words of Jessie M. Gaeta, a physician at Boston Health Care for the Homeless.[26]

Risk of overdose death increases when people use drugs—at dosages to which they are tolerant—in unfamiliar places and unfamiliar ways. Heroin users die at higher rates than their non-using contemporaries. But why do they die despite carrying a protective tolerance level?[27] Why don't most heroin users die? Tolerance is a form of protective adaptation in which "potential threats to survival are detected in their early stages and initiate homeostatic counter-responses that diminish the effect of physiologic alterations."[28] Tolerance was a functional adaptation that should have worked and didn't—a mystery little studied using modern methods in human beings. The science of how analgesics work in terms of tolerance, overdose, and abstinence has been subject to an institutionalized forgetting that has left no one prepared for current conditions. My purpose is not to reexamine overdose forensics, but to cast reasonable doubt on mainstream knowledge about it and suggest listening better to those who do know something about it—expert and experiential knowers. Under-researched aspects of overdose pertain not only to the complexities of tolerance and individual variation, but also to social factors such as drug control, disruption in drug markets, and the modifications users make in response.[29] The variety of enigmatic, unexplained, and under-researched phenomena runs from economic to neurophysiological—and yet a toxicology screen is all the media needs to tell us a convincing story about what happened when Philip Seymour Hoffman died.

Preoccupation with drugs themselves and routes of administration leaves little room for attending to complex social situations and conditions. In multiple drug toxicity deaths, each substance may well be "present in subtoxic concentrations."[30] This, too, has been known for a long time, but not really incorporated into practical knowledge. Warnings about the dangers of mixing opiates and alcohol date to the nineteenth century, yet until the harm reduction movement of recent years, drug users rarely got the message or passed it along to others. Harm reductionists looked and listened to what drug users actually did—as drug users themselves, in many cases, they didn't have to look far afield to create ways of living that reduced harmful consequences, introduced healthier lifestyles, or aimed for "any positive change" in their circumstances. They required little "evidence base" beyond the evidence of experience.

Toxicological designations such as "overdose" would need to be strongly situated within social context if better sense is to be made of the politics of who dies, where, when, and how. The sciences deemed relevant have neither concepts, methodology, or interest in social, economic, legal, and political dynamics necessary for a more complex story about the social situatedness of knowledge of overdose. Relegating place to background renders it inactive, weak, and inert, incapable of revealing how determinative setting can be for who lives and who dies. Place matters more for anthropologists, geographers, historians, and sociologists, all of whom have provided useful ways of knowing overdose. Yet in the scientific and forensic networks investigating overdose, place is enfeebled by a focus that zooms in on the immediate setting in which an individual does a drug and dies.

Formative aspects of pharmacology and toxicology remain contested in the overdose arena. Experts in this domain still hash out basic matters. Framed this way, the science of overdose still seems a new science; researchers perhaps cannot be blamed for not returning to earlier studies for fodder. Blame is not the point; my aim is to address amnesia and ignorance, ask different questions, and pose the goal of a better, more socially situated knowledge. Why do very low morphine concentrations sometimes kill? Why do people who use opioids in novel settings get into trouble they wouldn't have at home or in a dimly lit bedroom? At first blush, answering these questions seems to require some combination of molecular and epidemiological reasoning. On reflection, however, it becomes apparent that answering seemingly neurochemical questions requires bringing in many more kinds of expertise. Who can help situate drug users in relation to the biopolitics of legal and illegal drug markets, human rights, cultural connections, and social deprivation and dislocation? What fields can provide enough insight into the true depth of the fix we are in?

This book follows the contrails of an emerging harm reduction science by offering a glimpse into how situated knowledges about overdose and its remedies have been glancingly produced from the mid-nineteenth century to the present. In an opioid-using career, every injection is a risk: "Heroin users have an excess mortality rate approximately 13 times higher than their peers. We would expect half to be deceased by age 50."[31] Readers may be unsurprised that heroin users die at rates ten times that of non-drug users, but what about the fact that older, more experienced opiate users die at rates far exceeding younger novices? Media influences help create a misconception that those who die from overdose are young and inexperienced, when in fact the death rates are higher for older users supposedly protected by tolerance. Yet recycled depictions of youthful victims in the tragic vein regularly reappear. The trope of wasted youth, promising futures eroded by drug use, and need for redemption is belied by the numbers dying in their forties and fifties after a decade or more of heroin use.[32] Global variation in death rates indicates that heroin users in Asia are at far higher risk of death than those in western Europe, North America, and Australasia (in that order).[33] Place matters. How can we better take place into account in science, policy, and treatment?

PEOPLE MATTER: PROTAGONISTS IN AN UNNATURAL DISASTER

People always matter in terms of what science is done—what is thought to be a research question worth asking—and what science is left undone. Although respiratory depression or "drowning" has long been understood as the opiates' lethal consequence, overdose was not problematized as a significant issue for public health education and communication until recently. The sciences of overdose were considered of minor consequence, rarely appearing in the medical or popular press. Overdose was first framed as a problem of poisoning, a story told in chapter 1, "Poison Murders and Natural Accidents: Antidotes and Antagonists." Naloxone's predecessor, nalorphine, is discussed in chapter 2, "The 'Chemical Superego': Police Science, Social Antagonism, and Artificial Will," and in chapter 3, "Deaths from 'Narcotism' in the Mid-Twentieth-Century United States." Naloxone itself resulted from attempts to reduce one of the least savory effects of the opiates, constipation; chapter 4, "Bringing Out the Dead: Naloxone's Nine Lives Begin," and chapter 5, "Unnatural Accidents: The Science and Politics of 'Reanimatology,'" reveal naloxone's checkered adolescence.

Beginning in the 1990s, overdose was remade by this book's protagonists, who began working to reduce harm, reform policy, and extend human rights and health care to drug users. Getting preventable deaths onto the agenda of governmental, philanthropic, and community-based organizations was slow. Efforts took place alongside expansion of public health prevention programs centering on HIV/AIDS, hepatitis, and other blood-borne diseases. Efforts to make naloxone more available are addressed in chapter 6, "Adopting Harm Reduction: Early Democratizations of Naloxone," which describes direct provision of naloxone in Italy, diversion in rural New Mexico, and organized responses in California and Illinois; also recounted are discussions among early adopters in Australia and the United Kingdom that urged what came to be called Take-Home Naloxone. Local naloxone stories continue in chapter 7, "'Any Positive Change': Naloxone as a Tool of Harm Reduction in the United States," and chapter 8, "Harm Reduction Research and Social Justice: Working Naloxone into Public Health USA."

Three chapters tell naloxone's story in the United Kingdom: chapter 9, "Resuscitating Society: Overdose in Post-Thatcherite Britain," chapter 10, "'Growing Arms and Legs': The Scottish National Naloxone Program," and chapter 11, "Evidence from Pillar to Post: Researching the Varieties of Overdose Experience." Chapter 12, "Overdose and the Cultural Politics of Redemption," considers how different forms of cultural production such as econometric studies, popular films, and YouTube videos participate in making naloxone an artifact of moral redemption. The conclusion circles back to the question of people, places, and things central to the concept of situated knowledges, raising the question of what harm reduction infrastructure might look like in the future.

Renegotiating boundaries between ordinary and extraordinary, natural and supernatural, and expert and nonexpert, naloxone access activists created "peer-to-peer" or "bystander" training programs,[34] working to take naloxone out of the hands of medicine where it was regulated as a prescription-only, controlled substance, and to instead put it in the social circuits of active drug users and close kin.[35] Their efforts join a social transition in which people were becoming "experts of their own lives." Despite the rational world of evidence-based professionalism, the clipped tones of clinical guidelines, and a policy arena governed by the FDA, the Drug Enforcement Administration, and Her Majesty's Home Office, overdose deaths are emotional terrain. When people learn that they might have saved someone's life if they'd had naloxone, many become ready to take harm reduction into their own hands.

None of the sixty or so advocates, drug users, former users, friends, family, and witnesses, clinicians, or scientists interviewed for this book lacked good conscience or sense. I do not refer to the protagonists of this book as *subjects*, *informants*, *historical actors*, or *participants*. To me they are *protagonists*—the primary agents propelling the story. Because harm reduction affinities cross racial-ethnic formations, social classes, genders, ages, and professions, the protagonists are members of a distributed social movement that is about saving people's lives by making sense of overdose together. They brought harm reduction into a new affective domain of action, intellect, politics—and science. The work to displace the undone science of overdose with a robust new science grounded in harm reduction responded to the high expectations that social movements now face when advocating new practices within evidence-based policy environments.

THINGS MATTER: NALOXONE AS A TECHNOLOGY

But I came to wonder if the real question that should animate people is: Why did the goal narrow in on naloxone? Was it not an expansion of human rights to health and the pursuit of happiness, health equity, and social justice that motivated the protagonists? How did naloxone become the tool by which such goals were made concrete? Although the occasional researcher was attuned to rising incidence of overdose and regional differences in standards of care in the early 1980s,[36] such differentials were assumed to be ordinary risks in drug users' everyday lives. Undoing naturalized disparities requires forceful activism by social movements, advocates within relevant professions, legislative finesse, and political will. In an evidence-based age, undoing the knots that bind our knowledge of a densely materialized object like naloxone requires translating and resituating knowledge from multiple perspectives. The movement to prevent preventable drug-related deaths provides entrée into the social and political processes by which sciences move from "undone" to "done," from "raw" to "cooked," and from "questioning" to "unquestioned." Forthcoming chapters reveal how solidly overdose is now lodged as a phenomenon in contemporary life in many parts of the world.

Naloxone, the material technology I follow in this book, was once a tool of emergency medicine, its administration restricted to authorized medical professionals. Gradually extricated by the harm reduction movement, naloxone is now subject to myriad new policies, laws, and directives.[37] Changes in legislation governing naloxone access and Good Samaritan laws were crucial to this story, but also lie beyond it. In this social history of naloxone, I document the drug's role as a "technology of solidarity" in movements for social change, health equity, and human rights in Anglo-American countries by relying on concepts from STS, an interdisciplinary field that makes sense of how policy, people, and practices shape science and technology, and vice versa. The normative end of the STS spectrum produces studies aimed at equitably allocating the benefits and harms of technoscience. Such scholarship often takes the form of "hidden history" and shows where, when, how, and, above all, why the contentious politics through which new technologies—or new uses for old technologies as in the case of naloxone—arise.[38] Naloxone became a tool of empowerment, credited with helping drug users "change the way the world sees them—no longer as just an addict, but as a hero too" (see figure I.4).

The naloxone case offers rich data for thinking about the social organization of domains such as research ethics and the use of metrics, models,

I.4 HERO/IN ADDICT. "Help change the way the world sees them—no longer as just an addict, but as a hero too."
Used with permission of Martindale Pharm/Open LEC.

and outcomes to inform policy and reshape practice.[39] It enables inquiry into cultural ambivalence about pleasure and pain, and the intensely social forms of subjectivity and suffering[40] that drug users experience. STS enjoys a fertile conjunction with critical drug studies, wherein historians have sought to understand the social lives, cultural histories, and shifting boundaries of licit and illicit drugs and medical technologies.[41] Such domains catalyze unique conversations about the social movement to reduce drug-related harm, the "undone science" of opioid overdose, and what counts as evidence in highly politicized regulatory arenas.

Risk societies implement regulatory regimes and install resource-intensive infrastructures for other mortality risks such as automobiles, airplanes, alcohol, cardiovascular disease, diabetes, a host of other chronic and acute conditions, and even anaphylactic shock (the latter annually kills just about the number who die daily from overdose in the United States).[42] But opioid use—and nonmedical drug use broadly construed—is not just another mortality risk. Freighted with moral significance, it is framed within social contexts that police moral distinctions between medical and nonmedical opioid use. Overdose has become culturally central as urgency, pathos, hand-wringing, and blame directed toward pharmaceutical companies and their products lead to high-profile lawsuits and settlements. For all the fanfare, the endemic aspects of overdose still go understudied; the central myths and questions concerning overdose rarely get aired.

Overdose is not a contagious, infectious, chronic disease. Nor is it a natural disaster. It is a human-made disaster exacerbated by denial and disavowal of responsibility by those who structured and maintained distinctions between legal and illegal drugs, "patients" and "addicts," and physiological and existential pain. Those responsible are not simply pharmaceutical executives—they are policymakers and well-meaning clinicians, regulators and treatment providers, prison guards and administrators. Neither overdose nor its remedy is simple or straightforward. The history of the science of overdose is not just one of molecular agonists and antagonists, but of social antagonisms and political-economic agonistics. Although the agonistics of overdose are typically reduced to risk factors, risky environments, individual behavior, emergency response times, or the lumbering dynamics of health care systems, they reflect complex social and material conditions local to situations where they unfold. Knowing these well enough to design and build workable solutions that save lives has not been easy, and may even deserve portrayal as heroic, but it also requires that many of us give up the idea that we are innocent of this society-wide travesty if we are to build more livable worlds.

Naloxone encapsulates many questions about interactions between science and its publics. This book records a saga from the conceptualization of naloxone as a techno-fix to its current status as a harm reduction commodity. Naloxone may save lives but, like any commodity, it is limited in its capacity for changing lives or building more livable worlds. One of the protagonists viewed it as a

sad indictment of our times [that] countries like yours or mine ... cannot get past the war on drugs, the war on people that use drugs. Naloxone is driven by a continuation of the drug war that necessitates people needing access to life-saving interventions. So it's a sad indictment, yet at the same time I'm hopeful that as a result of more corporate entities and state interests spending money, that you do end up with hopefully less lives lost unnecessarily. But it's an uncomfortable place to exist, if I'm honest. It shouldn't be the case that our ambition is to offer the antidote when we could actually reduce the numbers by changing our policies, by changing some legislation.[43]

Naloxone has had multiple social lives. As such it encapsulates contradictory meanings and generates intense ambivalence. It promises an uncomfortable redemption, beckoning toward lives unnecessarily lost, and imparting hope that they might be found. But it is not an antidote that can heal—all it can do is give us a bit of breathing room and another chance to get it together. Perhaps we can use it to enable the reconsideration of the people, places, and things that have been hurt by the war on drugs and the war on drug users.

1

POISON MURDERS AND NATURAL
ACCIDENTS: Antidotes and Antagonists

Accidental poisoning, intentional poisoning, and suicide—in all of which laudanum, morphine, strychnine, or other preparations regularly appeared—were the categories into which nineteenth- and early twentieth-century deaths from overdose fell. Newspapers attributed such deaths to the regrettable errors of druggists and doctors, or to misbegotten children running errands for mothers ignorant of the power of the drug that they so often used to soothe infants and themselves. This chapter tells a surprising protohistory of overdose response, which varied by region: hot peppers in the American Southwest, for example, cold showers or ice in the groin further north.[1] An array of treatments, some dangerous, had traditionally been used in response to the toxic effects of opium or morphine, including alcohol, ammonia, amyl nitrite, atropine (once mistakenly considered an antagonist to opium), boiling water, caffeine, camphor, "capsicum per rectum," carbon dioxide, catharsis, cocaine, cold, coramine, digitalis, dilatation of the anal sphincter, electricity, emetics, epinephrine, ether, exercise, flagellation, gastric lavage, gelsemium, hot baths, iodine, lemon juice, lobeline, "milk, intravenously," nitroglycerine, Nux vomica, olive oil, oxygen, permanganate, picrotoxin, saline enema, saline infusion, stramonium, strychnine, tannin, tea, tickling, tobacco infusion, tracheotomy, turpentine, venesection, veratrum, vinegar, and inhalation of water vapor.[2]

Pain itself was thought the true antidote for opium overdose. Wisconsin physician I. H. Stearns urged his colleagues to "remember that the real antidote to opium is pain and, as far as known, there is no other."[3] He designed a thumbscrew device modeled on one used during the Inquisition to apply a "simple and exquisite torture that is unceasing in its agonizing character," and credited it with saving the life of a young woman who had intentionally taken ten doses of morphine. Although unresponsive for ten hours, the patient revived. Stearns also warned his fellows away from fashionable drugs for treating overdose, confident that twine and clothespins cobbled together would save many patients who would otherwise be lost. Assumptions that pain made people immune to addiction later played a part in casting chronic pain patients and the terminally ill as non-addicts who might depend on an opiate to which they were not addicted. Commonsense beliefs that pain offered immunity from addiction circulated long before pain was remodeled as the "fifth vital sign" in the clinical guidelines and pharmaceutical marketing campaigns of the late twentieth century. The very notion of the antidote prepared the ground; where there are poisons, there are antidotes.

APOTHECARIES' BLUNDERS AND LAMENTABLE ACCIDENTS: A CENTURY OF UNFORTUNATE EVENTS, 1850–1950

A scathing editorial appeared in the *New York Times* in 1854.[4] Lamenting the apparently eternally illegible handwriting of physicians, the writer called for stringent laws to mitigate "risks incurred from taking the wrong medicine." Pharmacists had become nonchalant about the "vending of all poisons, or medicines which in overdose may become such," allowing mere "boys" and "ignorant clerks" free hand in translating the "mystic language of the recipes" scrawled by doctors in "no language that is recognized by the living or dead." On occasion, pharmacists met with more than disdain. Criminal charges were made in the October 15, 1867, death of Mrs. Matilda Webster, a Gowanus wife and mother who "suffered very keenly from neuralgic affection." An invalid, she dispatched one of her nine children to a South Brooklyn drugstore, where the hapless child reported that her mother was in "great bodily pain."[5] Familiar with Mrs. Webster's usual dose, the druggist, Robert M. Kennedy, a Scot with thirty years' experience, packaged up two grains of morphine. After taking the full supply, Mrs. Webster "lingered in a comatose state, and expired" the next day.

After postmortem revealed that an overdose of morphia[6] was the undoubted cause of Mrs. Webster's death, Kennedy was arraigned on criminal charges. This was rare; accidental overdoses were typically reported as deaths from natural causes.[7] Newspapers recorded many notices of inquests and similar deaths from accident and suicide. Conflicting sympathies made the Brooklyn Court of Sessions jury unable to render a verdict when it convened on February 28, 1868. Most jurors wanted to acquit Kennedy; only two to convict after hearing "medical evidence"[8] that two grains was not an overdose in a tolerant individual. The pharmacist appealed to Mrs. Webster's "extreme suffering," insisting that he had intended for her to take just one-sixth to one-third of a grain at a time until she obtained relief. The hung jury did not yield satisfaction for Mr. Webster. Left to provide for nine young children, he was later awarded the maximum amount of damages allowed in such instances—five thousand dollars.

Questions about whether deaths were deemed intentional suicides or unintentional accidents arise when reading yesteryear's newspaper coverage. In 1866, the New York City Coroner's Office reported eleven deaths from accidental overdose of laudanum, four from accidental overdose of morphine, one from "overdose of strong tea," and another from poisoning by a decayed lemon. Couched in a list of "deaths from other causes," the office reported suicides in six men and two to seven women involving laudanum and morphine, but gave no clue as to how such categories were determined and laconically recorded.[9] Coroners held inquests in cases of accidental poisoning; these were rarely reported if the victim had been known to regularly take laudanum or morphine. Attempts to revive often failed. In one such case, the sixteen-year-old daughter of a stone cutter showed up at a Manhattan drugstore asking for an antidote for opium poisoning because she had taken too much for pain. Transported to Bellevue Hospital, she "became insensible" and died despite "the most heroic and scientific treatment."[10]

The need to sleep and suicide were the main reasons individuals used excessive amounts of morphine or heroin, as published coronial inquests revealed. As early as the 1880s, scorn was heaped on dangerous folk remedies used as antidotes for poisoning and other toxic events. "Mistakes as to quantity of opium preparations are by no means uncommon accidents," opined a newspaper columnist advising what should be done in the case of household accidents involving poisons.[11] Castigating mothers who gave poison to babies without doctor's advice, the column drew particular attention to breathing, which, if it should fall beneath ten breaths per minute, "will require all the skill that can be obtained to combat [death]." Although sleep was understood to be the enemy and pain the antidote, walking the person around or slapping them with wet towels were already cast as outdated customs in the 1880s. Yet the more up-to-date recommendations would raise eyebrows today: "If physical stimulant is to be used, lay the patient on a bed or lounge, and slap with the back of a hair brush or with a slipper. This is all the nursing necessary, so long as breathing keeps above ten to the minute." Artificial respiration was advised if breathing slipped below that threshold. A rapid diagnostic test for opium toxicity was to observe the "narrowing of the pupil of the eye to a small circle, which does not enlarge in the dark." Finally, practical advice for preventing accidental poisonings was to keep medicines in a locked cabinet. Similarities to today's campaigns are striking in both content and tone.

A panoply of poisons followed the term "overdose" in newspaper stories from the 1850s through the 1870s. Although mentions of laudanum were overtaken by mentions of morphine by the late nineteenth century, there were overdoses from aconite, arsenic, belladonna, strychnine, iodine, potassium cyanide, chloroform, ether, gelsemium, oil of tansy, valerian, carbolic acid, anodyne, mercury, quinine, tartar emetic, and "diarrhea medicine." Such nostrums were self-administered for bronchial and gastric afflictions, pain, insomnia, toothache, and delirium tremens, blurring lines between self-medication and suicide. Administered by friend or family member, they obscured the line between medication and murder. Disputes occurred regularly among prescribing physicians, dispensing druggists, and families of the deceased. Human causes ranged from "temporary insanity" or "mental derangement" due to lack of success in business to "pecuniary problems," marital disputes, or termination of love affairs. The overriding coronial question was whether actions were by accident or design.

Police officials and physicians were occasionally summoned to the scene during an "overdose," and were sometimes able to counteract it.

Such successes were often reported in cases of attempted suicides by young women. Typically, individuals were "relieved" of the poison by pumping out the stomach, inducing vomiting, or administering emetics. In one case, police reportedly administered a lamp oil emetic to good effect! Infant poisonings were not infrequent. Typically labeled "lamentable accidents," these were caused by ignorant mothers and the occasional father. Bank presidents, school teachers, businessmen, accountants, physicians, and nurses, many suffering from neuralgia or other nervous disorders, were among those who died of overdose. News of overdose deaths even appeared in a column titled "By Mail and Telegraph," an eclectic listing of notices of everything from upcoming frosts to saloon brawls to occupational injuries. Deaths from overdoses of chloral hydrate, laudanum, and morphine were typically listed with little comment. Occasionally, reasons for drug-taking were mentioned—"taken to quiet his nerves after a spree."[12] Overdoses of brandy or whiskey were listed; one death was even attributed to an overdose of water.

Overdoses resulted from "mistakes" such as using a bottle of laudanum instead of cough syrup; mislabeling (the dilute opium preparations available in the United States differed from more potent preparations manufactured elsewhere); and reusing a bottle that once held aconite, belladonna, hellebore, or more potent poisons. For example, seventeen-year-old Joel C. Steinhardt of Madison Avenue, New York, died after drinking a bottle of morphine that he thought was quinine prescribed to him for malaria.[13]

Despite the rarity of deliberate lethal poisonings in the United States, the 1890s experienced an "epidemic of poison murders."[14] Writing about an 1893 poison murder trial, Essig argued that public distrust in expert assurances sprang not only from the adversarial dynamics of the American legal system,[15] but from public disclosure of the uncertain nature of chemistry and toxicology within the medical sciences. Turn-of-the-century courts were treated to contradictory displays of expertise, leaving judge and jurors to make decisions based upon doubtful scientific testimony; indeed, physicians and pharmacists were often implicated in poison murder trials. Relative to the larger number of unremarked reports of overdoses appearing in the newspapers, poison murders were rare and sensational. Popular use of morphine and laudanum was rampant.

TOXIC EVENTS INVOLVING MORPHINE OR HEROIN

Heroin was first used "recreationally" in the United States during the 1910s, when it was generally sniffed rather than injected.[16] Sniffing was a safer route that reduced risks ranging from needle-borne diseases to the complications of nonsterile injections to overdose—all of which increased with the uptake of the needle.[17] Although the hypodermic was no longer a strictly "medical" tool, few people owned their own injection equipment and thus the practice remained uncommon prior to the 1950s.[18] A smattering of associations between "heroin" and "overdose" appeared in the *New York Times* before 1950. Among the earliest was a cluster of deaths among white female vaudeville performers. On January 31, 1922, Dorothy Wardell, a chorus girl said to be part of a drug smuggling ring operating out of the Montreal Actor's Club, was reported to have died a few days earlier of a "heroin overdose."[19] The ring used chorus girls in burlesque companies to convey small satchels of drugs into Canada and to bring large trunks of liquor back to New York City. "It is not suggested by the authorities that the theatre managers [in Montreal] or in New York were acting in collusion with or had any knowledge of this drug ring, but the fact remains that trunks have been endorsed by American customs officials at this end of the line and gone through to New York unexamined which were known to have contained a large quantity of intoxicating liquors." A second woman, Dixie Dixon, was also linked to the ring when she died in a New York City taxicab from "heroin poisoning."[20]

Deaths attributed to "morphine or heroin" were reported in well-off white women such as the granddaughter of department store magnate John Wanamaker, Miss Mary Brown Warburton, among whose Park Avenue effects were found "large quantities of an opium derivative" that was either morphine or heroin, along with the "reducing pills" she regularly took.[21] Following that report, few heroin overdoses appeared in the *Times* until January 1, 1950, when one death out of the eleven occurring on New Year's Eve in New York City was so attributed.[22] This death heralded those to come as the technological means for intravenous injection were becoming more widely available after the end of World War II. Dealers owned them and sharing was a necessity, as depicted in the first Hollywood portrayal of heroin injection, *The Man with the Golden Arm* (1955), when Frankie Machine, the Frank Sinatra character, ducks into his dealer's apartment for a fix.

"Mainlining," as intravenous injection was called in the United States, was promoted to a primary route of administration only after World War II, when market conditions made heroin less pure, more adulterated, and more available. When users first started injecting heroin directly into their veins, heroin's harms and pleasures increased, as explained in *Addicts Who Survived*:

An unusually strong bag could trigger a fatal overdose. This was a constant risk, since the addict usually had no idea of the purity of the drug he was sending directly into the bloodstream and brain. The survivors we interviewed were wary of this danger, and tried to minimize it by regularizing their intake, being careful about their suppliers, or by procuring medical narcotics of known strength.[23]

Reducing harm by limiting use was evident throughout the oral histories in *Addicts Who Survived*; those who survived overdose, infection, and disease took measures to minimize risk. Those who did not survive may have possessed such means, but their antidotes and consumption practices proved ineffective in the face of cumulative insult. Still, the relegation of all former antidotes to the dustbin of the "folk remedy" was a gradual process that began with the first proof of concept for narcotic antagonism.

AN IMPURE ANTAGONIST: N-ALLYLNORMORPHINE/ NALORPHINE/NALLINE

Naloxone's predecessor was a drug called nalorphine, produced and marketed by Merck under the trade name Nalline. Chemical characterization of N-allylnormorphine built on the concept of antagonism as refined by Polish pharmacologist Julius Pohl, who in 1915 demonstrated that a codeine derivative to which an allyl group had been added (N-allylnorcodeine) antagonized respiratory depression resulting from codeine overdose. Among the world's most widely used opium derivatives was the codeine in nonprescription cough syrups. Pohl's discovery was ignored until the 1940s, and the story of the narcotic antagonists is more generally one of fits and starts; dramatic delays due to social, technical, and linguistic barriers; and cycles of remembering and forgetting.

Narcotic antagonists were intimately bound to an ambitious search for the Holy Grail of a nonaddicting analgesic.[24]

Launched in the late 1920s, the quest for a nonaddicting analgesic was based in the ambition to tease apart morphine's "indispensable uses" from its dangerous effects. The goal was to get all opiates except those deemed absolutely necessary out of international commerce.[25] Exploration of narcotic antagonists was subordinate to this quest until a 1966 lecture at Kings College London, where the eminent American medicinal chemist Everette L. May acknowledged that the hunt had been "following a false scent [by] looking for an analgesic which would not depress respiration."[26] "In an attempt to eliminate respiratory depression, von Braun (1916) synthesized a new opiate derivative, N-allylnorcodeine (Fig. 1A), a compound that stimulated respiration when given alone and reversed morphine-induced respiratory depression (Pohl, 1915).... The next attempt to use an allyl nitrogen substitution to make an opiate free from respiratory depression came in the 1940s."[27] That attempt was central to the American quest to find a safer analgesic—one that would neither addict nor kill.

Nalorphine's initial chemical synthesis in the 1940s was a bicoastal contest. Pharmacologists Chauncey and Elizabeth Leake suggested reading Pohl to their doctoral students at the University of California, San Francisco, among them E. Ross Hart, Elton McCawley, I. Young Chen, and E. (Eddie) Leong Way.[28] By January 1941, Hart and McCawley *thought* they had synthesized nalorphine, and published results under the title "The Preparation of N-allylnormorphine."[29] An April 1941 abstract for the thirty-second annual meeting of the American Society for Pharmacology and Experimental Therapeutics stated that Pohl's findings had been validated. In a longer September 1941 article, McCawley, Hart, and Marsh realized that N-allylnormorphine modified morphine's actions. While their molecule would not reduce morphine's propensity to addict, it might counter respiratory depression in clinical situations where morphine overdose was routine—surgery and obstetrics.

As E. Leong Way explained in a 1983 consensus conference on agonist-antagonists, Leake's students "did not have in mind developing an antagonist."[30] Aiming at the Holy Grail—an analgesic that would not depress respiration—they turned to Pohl's work on antagonism only *after* identifying both antagonistic and analgesic properties in their compound. But it soon became apparent their compound was not quite nalorphine when Leake requested that the research director at Merck, his former Princeton colleague Randolph Major, prepare five grams of material for use in the experiments later reported by Hart and McCawley.[31] Merck chemists John Weijlard and Arthur E. Erickson reportedly failed using Hart, McCawley, and Marsh's method; they succeeded only by returning to the von Braun method used by Pohl. They included two proofs that their substance was actually N-allylnormorphine. They applied for a patent for "N-allylnormorphine and processes for its production" on December 6, 1941, which was granted on December 12, 1944, by the US Patent Office.

The next chapter of nalorphine's biography was written by Klaus Unna, senior pharmacologist at Merck from his emigration to the United States in 1937 until 1944. While Hart and McCawley were preparing their 1944 report, Unna published a definitive demonstration of nalorphine's antagonistic effects,[32] speculating that it "would prevent morphine from being effective if given prior to the drug."[33] When he moved to the University of Illinois, Unna took his Nalline with him and made it the linchpin of a form of pharmacological reasoning that led to wider circulation of the concept of narcotic antagonism; the conceptual basis for the existence of multiple opiate receptors; and clinical applications of antagonists such as naltrexone and naloxone. As an "impure" antagonist, nalorphine exerted interestingly mixed effects, acting as an agonist *and* an antagonist, leading to speculation as to whether its dualistic action could be used in mixtures that would have less abuse potential than narcotic agonists did on their own.[34] Nalorphine's more immediate use was as a "specific opiate antidote" for severe opiate poisoning. Eager for Nalline to acquire a clinical evidence base, Merck sent large supplies to investigators in hopes they would find opportunities to use it.

MAGICAL MIXTURES: THE WORK OF NALORPHINE IN THE CLINIC

Where could overdose be found so that studies could be accomplished? Despite its seeming rarity outside hospital walls, moderate overdosage was common during surgery.[35] Prompted by the lacuna of clinical studies on the treatment of acute morphinism,[36] anesthesiologist James Eckenhoff conducted the first critical experiment on possible clinical uses of nalorphine. Intrigued by nalorphine's variable effects depending on individual health status, his team found it had the uniformly dramatic effect of stimulating respiration in "heavily narcotized patients."[37] Speculating that it was acting in a "competitive fashion with the opiate for certain receptors"—an indication that theories concerning the actions of drugs at receptor sites were somewhat more widespread in the early 1950s than historians of science have believed—Eckenhoff and colleagues suggested that nalorphine did not completely block or replace the opiate with which it competed. Thus it exerted incomplete or nonspecific actions. It was also unclear why multiple doses of nalorphine had successively decreasing effects—a question that could only be settled empirically by clinical trial.

Although clinical trials were not yet required when pharmaceutical companies sought US Food and Drug Administration approval, tests were conducted in institutional settings such as mental hospitals and prisons where human subjects could reliably be observed. For instance, Abraham Wikler reported that large doses of nalorphine (30–75 mg) in twelve postaddicts caused "lethargy, mild drowsiness, vivid daydreams and dysphoria varying in intensity from vague anxiety to acute panic."[38] All the test subjects complained they could not repress such daydreams; morphine reduced them only modestly. Wikler's study was conducted in a unique epistemic niche—the laboratory within the US Narcotic Farm in Lexington, Kentucky.[39] The US Congress mandated research be part of the effort to "cure" drug addiction at the narcotic farms. The laboratory had opened in 1935 under direction of Clifton K. Himmelsbach. Named the Addiction Research Center (ARC) after 1948, it was the world's sole laboratory dedicated to studying drug addiction and was the key clinical testing site in the broader effort to identify safer analgesics.

A "morphine substitution technique" was Himmelsbach's way of measuring the addictive potential of a drug, based on the principle that "a substance which will support and maintain the 'addicted state' is essentially addictive in and of itself."[40] The Morphine Abstinence Syndrome was his signal clinical contribution—he mapped the predictable pattern of symptoms occurring after a morphine-dependent subject was deprived of the drug. The substitution technique involved administering another compound to see if it relieved withdrawal signs.[41] A drug's "abuse liability" could thus be compared to the gold standard—morphine.[42] Himmelsbach's scale subjected the messy processes of tolerance and withdrawal to a predictable, rational calculus.

Research at the ARC relied on tight control over material conditions. The research situation was designed to provide investigators ongoing access to human subjects who were former morphine addicts.[43] Himmelsbach played a pioneering role in thinking about withdrawal as a defining symptom of addiction. We might think of the transition as a shift in how the subject was framed: the thinking migrated from the chronically poisoned body suffering from the presence of poison, to the addicted body suffering from its absence.[44] Historian Ian Walmsley has shown that "leaving off" [withdrawal] was not medicalized during the nineteenth century. Twentieth-century clinical pharmacologists required tools for separating "subjective" from "objective" ways of knowing. The Abstinence Syndrome Intensity (ASI) scale was such a tool:

Instead of relying on the words of the patient, withdrawal was established in an "objective quantitative manner" (Himmelsbach and Andrews, 1944:288). The ASI allowed symptoms to be visualised, differentiated, broken down, classified and then compared to the normal functioning of the body. Abstinence deviations included respiratory rate; systolic blood pressure; temperature; blood sugar; diastolic blood pressure; sleep; caloric intake; and basal metabolic rate (Himmelsbach, 1941). The ASI allowed withdrawal to be reproduced and explained according to the rules of psychiatric discourse.[45]

Framed as dopesickness,[46] withdrawal came to signify and amplify an intensely individualized agonism.[47] Morphine itself transitioned from poison to chief among the brain's "mechanisms for the maintenance of homeostasis."[48] This notion that the brain was organically involved in tolerance, dependence, and withdrawal, in ways that could be measured, quantified, plotted, and visualized, was Himmelsbach's chief conceptual contribution to the history of addiction science.

The Addiction Severity Index functioned as a "technology for seeing and knowing withdrawal" that "allowed for differences in the intensity of addiction to be seen and known."[49] Dosage affected each of the component processes contributing to dependence. The steps could be plotted in an orderly way from the first moments of "experimental read-diction" to abrupt cessation. Each step had a parallel in the social world, but the primary way that withdrawal was handled in the post-Harrison Act United States was "cold turkey" in jails (i.e., without medical super-vision or assistance). Clinicians at the Narcotic Farm advocated keeping patients comfortable as they went through withdrawal. Concerned that the abstinence syndrome "may become so intense as to cause death,"[50] Himmelsbach erred on the side of caution by easing detox with sedative-hypnotics, aspirin, and soothing baths. Along with the comparatively com-passionate and affordable dental and health care available, these measures were the primary reason that addicted persons voluntarily presented themselves for treatment at the narcotic farm. People with addictions often claimed to have witnessed deaths from withdrawal in jails and pris-ons, and sought treatment at the narcotic farms to avoid this fate.

Growing significance was attributed to withdrawal in twentieth-century scientific accounts, accompanied by a growing popular dread floridly dramatized in *The Man with the Golden Arm*. Threatening to jump out a window, brandishing a chair at his lover, and crying like a baby, Sina-tra's character clamored for a fix. Shivering, twitching, and trembling, doubled over and fumbling with his clothes, he fell to the floor mutter-ing, "Molly, if you love me, kill me please." Sinatra observed people going through withdrawal in a California hospital in order to prepare for the role. This cautionary tale demonstrated not only withdrawal's physical effects, but its capacity to turn love to hate and life to death in the Ameri-can imagination.

Despite previous acquaintance with the agonies of withdrawal, ARC subjects volunteered to pass through their throes for the sake of science. Knowledgeable about the effects and risks of drugs they had encountered on the streets, they were drawn by promises of drugs. More than 125 volunteered for a large trial of methadone after it was brought to the United States from Germany following World War II.[51] By this time Harris Isbell had replaced Himmelsbach as research director. In these studies, the largest conducted there, most subjects tolerated cumulative, low doses of methadone. However, two African American males suffered "methadone repeated unless respiratory depression recurs. One should not be alarmed if the patient does not become fully awakened. Though Nalline may convert a comatose patient into one easily arousable by manual stimulation, such patients may remain sedated and drowsy for many hours. This causes no harm as long as respiratory function is well maintained. In the event of recurrence of respiratory depression, re-administration of Nalline — usually using a smaller dose — will promptly restore respiratory volume.

much more alert. He was no longer co-operate in making spirograms, arose and walked to the bathroom. Four hours and 45 minutes after the methadone had been given, an additional 5 mg. of Nalline was injected subcutaneously.

After the first dose of Nalline, the respiratory rate never fell below 10 per minute and respiration never became periodic. Recovery was uneventful, the patient being alert 4 hours (Fig. 2, *B*), and completely normal 24 hours, after Nalline had been administered.

Fig. 2. EFFECT OF NALLINE ON METHADONE POISONING.

A. *Comatose state prior to Nalline.* B. *Four hours after administration of Nalline.*

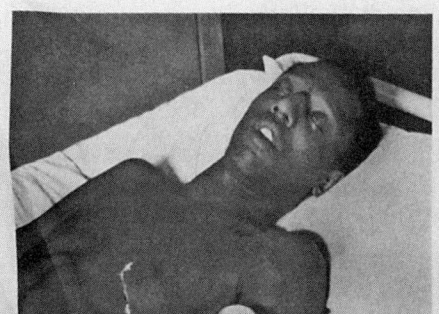 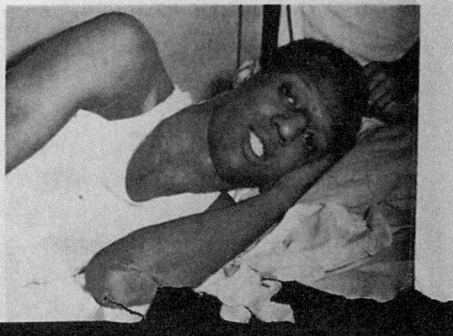

1.1 Effect of Nalline on methadone poisoning.
Source: Havelock Franklin Fraser, Abraham Wikler, Anna J. Eisenman, and Harris Isbell, "Use of N-Allylnormorphine in Treatment of Methadone Poisoning in Man: Report of Two Cases," *Journal of the American Medical Association* 148 (1952): 1205–1207.

poisoning." After standard responses failed to revive, someone recalled that Merck's large bottle of Nalline was sitting on a shelf and "induced spectacular and, possibly, life-saving effects" (see figure 1.1).[52] Subjects and researchers breathed a collective sigh of relief; the latter wrote: "Unless N-allyl-normorphine had been given, one would have expected that both patients would have remained in coma, with depressed respiration, for at least several hours. In fact, if N-allyl-normorphine had not been available, both patients might have died."[53]

The rigors of research were profound when the ARC explored nalorphine, which was found to precipitate a "violent abstinence syndrome that could not be antidoted by morphine."[54] "Postaddicts," as ARC researchers called their human subjects, intensely disliked nalorphine.[55] But what would "normal" subjects think of it? Henry K. Beecher and Louis Lasagna decided to find out,[56] the latter recalling,

I think it was Harris Isbell who first planted the notion in our minds. [W]e began, at the Mass General, to wonder whether we could come up with some magical ratio ([a] magical mixture of Nalorphine and morphine) which would give us without any significant loss the good things that morphine could provide and would allow us to drop out of the scene, one or more of the undesirable things that morphine did ... by the concomitant administration of an antagonist and agonist.[57]

No magic appeared, but to everyone's surprise there was a serendipitous finding: although a narcotic antagonist, nalorphine also acted like an agonist. It was an effective analgesic.[58] Lasagna later spoke of this time as if Beecher had come up with the idea of a "magical concoction":

We were responsible for the development of a whole class of analgesics. There was a drug that came along called nalorphine, which was supposed to be a pure antagonist, and so Beecher wondered if we could come up with a magical concoction of an antagonist and an agonist, and maybe lose some of the bad effects. We didn't care what bad effects we lost, just as long as we made progress [on the non-addicting analgesic project]. So I, to make sure that we could come to firm conclusions, said we can't just study morphine alone and morphine with one or another ratio of the antagonist, we have to study the antagonist itself. To our great surprise, it turned out to be an analgesic and that led to drugs like pentazocine and a whole class of agonist/antagonist analgesics, so that was one bit of unanticipated benefit.[59]

Strangely, Beecher and Lasagna also encountered a pharmacological paradox that went unreported. Called "hyperalgesia,"[60] the paradox is that pain relievers make some people *more* sensitive to pain. Nalorphine was considered exciting because it could be used to unmask the obscure processes of physical dependence. For instance, "spinal" dogs exhibited signs of withdrawal after just one dose of morphine and nalorphine—long before they were presumed tolerant.[61] Such findings led to the realization that addictive processes started earlier than had been thought, and that pain sometimes increased with the use of so-called painkillers.

Despite pharmacologists' fascination with the pharmacological "tells" of nalorphine, its clinical importance lay in its seemingly miraculous effects on respiratory depression in dogs and human beings. "Hitherto clinically untried," it was given to four hundred surgical and obstetrical patients in early 1951. Just about every woman who delivered at the University of Pennsylvania Hospital received either the new drug or a placebo (saline solution) just before delivery.[62] This double-blind, placebo-controlled study showed nalorphine reduced the time between delivery and the infant's first breath. Prophetically, study leaders Eckenhoff and colleagues noted that they had also successfully treated four overdoses apart from the study, but that it would "take time to determine the real place the drug is to take as a therapeutic agent in the treatment of narcotic poisoning."[63] Two Navy physicians considered the "advent of an effective antidote … of signal importance," and they urged all physicians to "become familiar with the pharmacology, dosage, and method of administering" Nalline.[64] They also precociously urged it be made available in "antidote kits."

Nalline found its way into large urban hospitals where heroin use was becoming more common. Clinical studies in such settings were often conducted by physicians treating indigent patients. In Chicago nalorphine appeared regularly in medical school demonstrations, where Klaus Unna's graduate students used it to illustrate the concept of antagonism by overdosing dogs on morphine and then waking them with nalorphine.[65] Heroin overdose occurred often enough at Cook County Hospital that cases regularly overflowed into corridors when Edward F. Domino was a medical resident. He recalled that the usual treatment was to "ventilate them periodically and give them picrotoxin or some other convulsive to try and stimulate respiration. It really didn't work. Sometimes there were so many of the overdose heroin cases around that there weren't enough people to ventilate them, and you'd have to make a decision of triage, who are you going to let die and who are you going to let live."[66]

Decades later, with tears in his eyes Domino told of accidentally overdosing a terminally ill, elderly woman with disseminated breast cancer and no family. His attending physician had been conducting trials of an opioid called levorphanol (on the US market since 1953 under the trade name Dromoran) for severe cancer pain, and Domino was treating her pain with the drug. After giving her enough of the drug to relieve her severe pain, he attended to another patient; when he returned, he found a nurse giving his first patient artificial respiration.

Her pupils are constricted. She's not breathing. I've overdosed her. Then the issue was, what am I going to do? I knew nalorphine was in the dog lab in Illinois. The nurse was continuing to ventilate the patient. It's called "bagging," giving her oxygen. She was doing okay, but now she's going to be that way for hours, and we're just bagging her.[67]

Seeking to avoid what he viewed as a futile effort at manual ventilation, Domino obtained permission from the attending physician to try the "new miracle drug" he had been using to revive dogs. He asked one of his colleagues to get it from the lab and bring it to the hospital.

I gave it to her intravenously. And you wouldn't believe what happened. First thing you know, she's breathing like a train. Then she wakes up and opens her eyes. Before, her pupils were constricted, but now they're dilated. She looks at me and she says, "Hi, Dr. Domino." A few minutes later, she says, "Oh, I'm hurting so much." Can you believe it? ... She lived for a while longer, and then she was transferred to a nursing home, where she passed away. After that the nurses thought I was god. The attending couldn't believe it. I was known as The Professor. That's when I became a pharmacologist. Then and there.[68]

Although Nalline was primarily pitched to surgeons through advertisements in the medical press (see figure 1.2), such ads consistently mentioned its use for heroin overdoses, buried in long lists among morphine, methadone, Demerol, Dromoran, levo-Dromoran, and Dilaudid. Heroin, of course, was never allowed into medical use in the United States, so this mention indicates that overdose did occur and was treated in mid-twentieth-century hospitals.

The first clinical case report of nalorphine being used to treat a heroin overdose appeared in early 1953.[69] The Brooklyn case—here I adopt the historian's convention of using the terms used at the time—involved a "well-developed, well-nourished, light-skinned mulatto woman" who was comatose when police brought her to Kings County Hospital at 10:35 p.m. on May 1, 1953. Earlier that day, she had apparently engaged in sexual contact with several known heroin addicts, one of whom had injected her with heroin at 6 p.m. that evening. When her breathing became labored and she became cyanotic, the men had become frightened enough that she might die that they called the police, who conveyed the victim to the hospital in an ambulance. She was placed on a ventilator, but removed when she "appeared to be lifeless." Manual resuscitation was continued and 1 cc (5 mg) of nalorphine was administered intravenously. "Within one minute, the patient started to breathe, her pupils dilated, and she rapidly became conscious and sat up. She then became belligerent."[70]

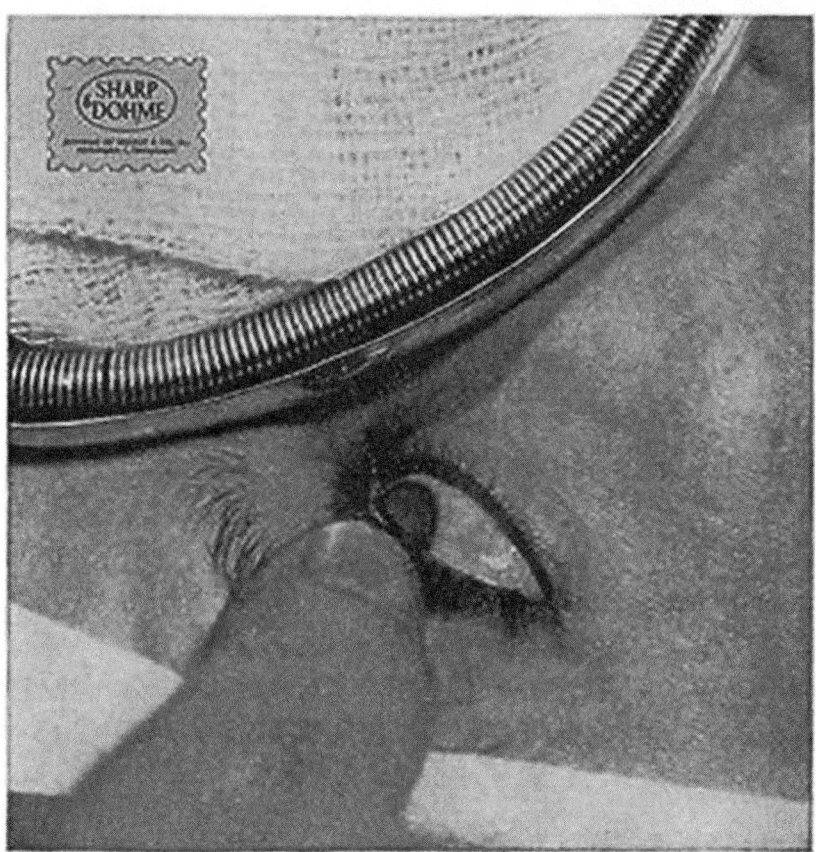

Dramatic benefits in heavily narcotized patients

NALLINE ®

HYDROCHLORIDE
(NALORPHINE HYDROCHLORIDE, MERCK)

CLINICAL RESPONSE: NALLINE is the most effective agent available for reversing narcotic-induced respiratory depression and any consequent circulatory depression. It promptly increases minute volume and respiratory rate 200 to 300 per cent. In surgical patients adversely affected by narcotics, NALLINE helps overcome hypotension secondary to respiratory depression and facilitates induction of anesthesia by increasing pulmonary ventilation. It has been reported to curb nausea, vomiting, and retching due to morphine.[1]

1. Adriani, J. and Kerr, M.: *Surgery* 33:731, May 1953.

INDICATIONS: Respiratory depression due to morphine and its derivatives including heroin, methadone, Dromoran®, Levo-Dromoran®, Demerol®, Nisentil®, Dilaudid®, and Pantopon®.

SUPPLIED: In 1-cc. and 2-cc. ampuls (5 mg./cc.). For parenteral use. NALLINE comes within the scope of the Federal Narcotic Law.

1.2 Medical advertisements for Nalline emphasized nalorphine's "dramatic benefits in heavily narcotized patients," making it the first choice for reversing "narcotic-induced respiratory depression" in medical settings. 1955–1956.

NALLINE materials/images reproduced with permission of Merck & Co., Inc., a subsidiary of Merck & Co., Inc., Kenilworth, New Jersey, U.S.A. All rights reserved. Used with permission of Merck, Inc.

Described as "quarrelsome, confused, and somewhat somnolent," the patient related that this episode was her first heroin experience.

As heroin overdose settled into the urban black neighborhoods where street-drug markets developed, hospitals stocked nalorphine and it was also sometimes available in prehospital settings. The range of individuals for whom nalorphine worked fell into two different patterns: experienced users who overdosed despite high tolerance, and opiate-naïve users who overdosed while also using other drugs. Overdose mirrored the extensive range of individual variation akin to that seen in opiate users themselves. Nalorphine, in other words, was not simply a research material, but saw clinical use well into the 1970s—even after naloxone had been approved.

MIXED PERFORMANCES: NALORPHINE'S ROLE IN THE HUNT FOR MULTIPLE OPIATE RECEPTORS

In addition to clinical uses, Nalline proved a good drug to think with pharmacologically about the conceptual problems posed by narcotic antagonism. Nalorphine's "mixed kind of pharmacological performance"[71] gave new impetus to the hunt for a nonaddicting pain reliever, a search propelled further by synthesis of two other antagonists more in use today, naloxone (n-allylnoroxymorphione) and naltrexone between 1957 and 1965.[72] The narcotic antagonists proved of "inestimable value"[73] when tested by the Committee on Drug Addiction and Narcotics (CDAN), which became the College on Problems of Drug Dependence (CPDD) in 1965.[74] The ARC found that "patients disliked the [morphine/nalorphine] mixtures intensely from the very beginning of the experiment and complained that the drug had 'no drive,' that it did not 'pick them up.' At times, they insisted they were being given water."[75] Yet they showed signs of tolerance—constricted pupils, depressed respiration, constipation, alternating sleeplessness and sleepiness, bad dreams, and involuntary twitching. They passed through an abstinence syndrome when the mixtures were withdrawn abruptly. These mixtures were the fruit of pharmacological thinking that increasingly took place at the molecular level.

Nalorphine in a sense became the ARC's Rosetta Stone. When Isbell and Wikler retired in 1963, William R. Martin (Unna's former student) took the reins. Intuiting that multiple forms of abstinence and tolerance must be in play, Martin realized that opioid tolerance was a change in homeostatic equilibrium—to which some organisms adapted by becoming hypersensitive, and others hyposensitive.[76] A neuropharmacologist, Martin was interested in interactions between drugs and their specific receptors which, he reasoned through a sophisticated blend of clinical observation and experimental evidence, had to be distributed throughout the brain.

For those interested in the early pharmacology of opioids, Krueger, Eddy, et al. (1941) prepared an extensive and comprehensive two volume overview of opiate action prior to World War II. What is impressive about this treatise is how much was known about opioid pharmacology before the current "molecular era." Indeed, it is humbling to see how much we forgot and what we have "rediscovered." Overall, one is struck with the complexity of the mu-opioid system, confirming the observation that "the only simple things in science are scientists."[77]

Martin's brand of pharmacological optimism aligned with the search for a nonaddicting analgesic. Those engaged in this project had great faith that science would relieve what they saw as great suffering. As the first narcotic antagonist, nalorphine was central to thinking pharmacologically about opioid receptors, which Martin named *mu*, *kappa*, and *sigma*.[78] Others embarked on a quest to visualize the multiplicity of opioid receptors, a search that unfolded in dynamic fashion during the early 1970s.

Scientific controversies and credit disputes are classic stuff in sociology of science. The simultaneous discoveries involved in visualizing opiate receptors in the human brain were high-profile disputes on unsettled scientific terrain, to which the differential effects of agonist/antagonist pairs were crucial. In "You've Come a Long Way, Baby!," Solomon H. Snyder captured the spirit of the time with an anecdote about a May 1974 Neurosciences Research Program meeting in which one of the progenitors of methadone maintenance therapy, Vincent Dole, questioned "each word of 'the opiate receptor,'" including the *the*. Acknowledging that "addiction was the bête noir of all psychoactives," Snyder described his one-time belief that "conquest of the opiate receptor would resolve the riddles of addiction."[79] Many joined him in assuming that addiction would become decipherable when the distribution, concentration, and function of the opiate receptors were mapped. "Receptor fever"[80] raged for decades, only ebbing with Snyder's recent admission that, "We must plead failure. There have been no major breakthroughs in the development of opiate drugs that can be attributed to our enhanced molecular understanding."[81]

Useful for demonstrating the heterogeneity of opiate receptor subtypes, nalorphine was truly a mixed bag: acting as a narcotic antagonist at low doses, it proved an analgesic agonist in high doses. Its unpleasant effects made it universally disliked; no one asked for more, or clamored for it "on the outside." As a pharmacological tell, its use endured; a nalorphine study concluded in the 1990s that correlations between the pharmacology of the classical opiate agonists and current understandings of receptor subtypes might mean that the "latest discoveries may prove to be extensions of theories proposed decades earlier."[82] However, nalorphine also had a legal career as a tool of "police science" and as a "tell" of a different kind. We turn to that career, which went all the way to the US Supreme Court, in the next chapter.

2

THE "CHEMICAL SUPEREGO":
Police Science, Social Antagonism, and Artificial Will

How can one tell who is a narcotic addict, and who is not? Are all users "addicts"? Without reliable indicators, those seeking to police narcotics had to determine who was a narcotic user with nothing more than circumstantial evidence. Harris Isbell, director of the Addiction Research Center (ARC) in Lexington, Kentucky, offered police a solution in a January 30, 1954, editorial: they could use nalorphine to detect who was an addict.[1] Nalorphine was not innocuous—too large a dose might precipitate abrupt withdrawal that was strong enough to endanger life. Isbell warned that the Nalline test protocol should never be carried out "without full explanation to the patient," nor should it be used on those suffering serious disease. Unusual for the time, the editorial urged full written consent be obtained from the person being diagnosed.[2] The Nalline test was the first diagnostic test for opioid addiction, accomplished by measuring changes in pupil size after a subcutaneous shot of nalorphine.

DETECTING "ADDICTION": THE NALLINE TEST IN THE POLICE PRECINCT

Law enforcement officers used the Nalline test (see figure 2.1) to surveil a broad population of potential addicts, deter casual drug users from becoming regulars, and control parolees and probationers. By the mid-1960s,

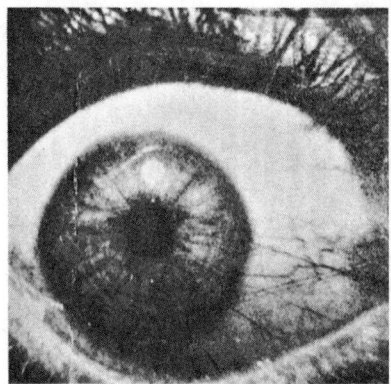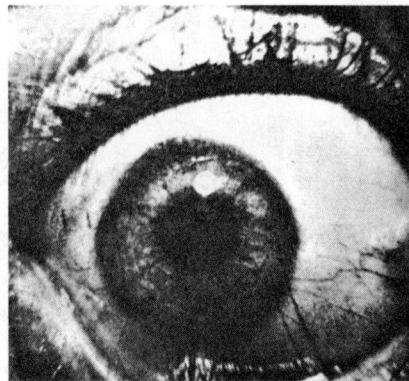

2.1   Pupils before and after administration of nalorphine, the so-called Nalline test. *Source*: Thorvald Brown, *The Enigma of Drug Addiction* (Springfield, IL: Charles C. Thomas, 1961), 288.

law enforcement in Hong Kong, Singapore,[3] and several US states used the test for a purpose not foreseen by scientists: to prevent people from falsely claiming to be addicted in order to obtain lenient treatment under state civil commitment laws. California Narcotics Bureau Inspector Fred Braumoeller and Dr. James G. Terry, an Alameda County medical officer, succinctly described Nalline's telltale effect: "while it cause[d] a non-addict's pupils to constrict, it cause[d an] addict's pupils to dilate."[4]

Although the Nalline test had been born in the laboratory, implementing a simplified version of it in the Oakland police precinct was Terry's idea. He claimed to have been inspired by an encounter with an "unknown pretty girl" whose pupils he had observed dilate after her first shot.[5] A physician working in criminal justice, Terry standardized dosage and a technique for measuring pupil size, as described in *The Enigma of Drug Addiction* (1961), a book authored by another of the Nalline program's chief architects and defenders, Thorvald T. [Ted] Brown, Commanding Officer of the Vice and Narcotic Division. Captain Brown was one of the primary advocates in the national press for adopting Nalline as a tool of antinarcotics policing.[6] He was nothing if not energetic about public relations; newspaper coverage of the program reached throughout the country.

Nalorphine intensified the unpleasant symptoms of abstinence.[7] Its accuracy depended on the timing, dosage, and level of the subject's tolerance. The laboratory was not the field; in the laboratory, morphine and nalorphine dosages could be precisely calibrated.[8] Negative tests some-times occurred if other drugs were present, a "consideration ... of practical importance since undoubtedly addicts have tried numerous drugs to beat the test."[9] Yet law enforcement considered the test definitive; there are hints as well that it was used in a punitive way to end heroin's "kick"—or in more scientific terminology, to "interfere[] with euphoria."[10]

No simple antagonist, Nalline's subjective effects varied: it made some people feel drowsy like whiskey, barbiturates, or small doses of morphine. Others were plagued with vividly disturbing daydreams or visual hallucina-tions.[11] Higher doses produced unpleasant anxiety or thoughts that raced uncontrollably. Administered soon after morphine, subjects nicknamed it "Climalene," named after a commercial cleaning fluid that "cleans you out of dope and makes you climb the walls."[12] Nalorphine's discontents were represented as the stark opposite of heroin. How could such a motile technology have found a decades-long niche as a supposedly standard-ized assay in police hands?

Police detailed to narcotics investigations were enthusiastic about the Nalline test. The Federal Bureau of Narcotics urged local law enforcement to adopt this tool of "scientific policing," and California cities including Eureka, Fresno, Los Angeles, Oakland, Sacramento, San Francisco, and Stockton had done so by the early 1960s. Increasing heroin use in the 1950s led California legislators to criminalize both the act of consuming narcotics and the condition of being addicted to them.[13] Those convicted under the 1954 amendments to the Health and Safety Code could be sen-tenced to the county jail for ninety days to one year. "One may be guilty under the act at any time or place he is found, so long as his condition or status is that of a drug addict, even though at the time he is arrested, he is then and there innocent of the act of using narcotics."[14]

Police in Alameda County instituted an elaborate Nalline testing program in 1956. The former chief of the Oakland Police Department, Wyman W. Vernon, was, according to Brown, the first law enforcement administrator to foresee the value of this test for moving drug users toward treatment.[15] Police were thus positioned on the side of treatment, and civil liberties laws and legal safeguards appeared as barriers to it! The Oakland Vice Squad formed a Narcotic Detail that worked closely with Terry,[16] who was the medical officer of a facility located in the Oakland foothills and variously known as the Alameda County Jail, a county prison farm, the Alameda County Sheriff's Rehabilitation Center, and the Santa Rita Rehabilitation Center—all names referring to the same facility.

The first in the nation, the Oakland Nalline Program served as a model for other cities, counties, and states desiring to use nalorphine as a detection device.[17] Program administrators fielded inquiries from Chicago, Detroit, New Hampshire, Connecticut, Pennsylvania, and Texas, and received official delegations from "Arabia," India, Iran, Korea, Japan, and Vietnam. Nalorphine was inexpensive and easily obtained, but its symbolic politics were also important: control over an unruly population was mentioned as an advantage of the technology. Nalline offered police and probation officers a "much firmer and more directive form of supervision."[18] The goal, according to Brown, was to "detect and isolate those who have been using narcotics so that enforced abstinence and treatment can be provided and control measures instigated."[19] His book included sample forms associated with authorization and release of results. Numerous photographs showed him examining suspected users for track marks. Also pictured were the equipment and setup used to position the "testee" (see figure 2.2).[20]

Depictions of track marks, sclerotic veins, and crude tattoos used "in a feeble attempt to hide the tell-tale scars" indicated that the Vice Squad was apparently in the habit of photographing injection sites—be they arms, shoulders, hands, or other body parts—and labelling each with a typed identification number and date affixed to the body part with transparent tape.[21] Such photographs testify to the extent of police scrutiny into addict embodiment in the course of the project to "rehabilitate" the addicts of Alameda County. Santa Rita, the "county prison farm," was the site of a mandatory ninety-day "drying-out" period. Considered a model in the 1940s and 1950s, Santa Rita originated as a military barracks and dairy farm.[22] In a bid to save taxpayers the cost of housing prisoners, the institution was intended to be self-sustaining, providing pork, lamb, milk, and fresh vegetables not only for those housed in the barracks, but also for many nearby hospitals, county juvenile detention centers, and women's facilities.

In the early days of the carceral state, this "health farm" was represented in glowing terms: "While at the farm, the emaciated addict is given proper nourishment, rest, and out-of-doors work; he gains weight, becomes physically clean and mentally healthy. It is at this point, his release back into society, that is the trying period in his rehabilitation and 'cure.'"[23] Enforced abstinence and "physical restoration of the body; proper food, rest, fresh air, and exercise" were believed necessary for rehabilitation. During the "trying period," police could use Nalline to identify addicts who had "slip[ped] back into their old habits."[24] That this should be presumed an "undeniable" police responsibility arose from the moral framing of vice, rather than from concern about public disorder. Heroin had scarcely any public face in Oakland. California habits were "fairly mild," according to Brown; addicted persons rarely showed signs of "excessive torpor" in public; indeed, "as a rule, he will be going through his suffering in the privacy of his hovel."[25] Believing that an unnatural taste for euphoria was the "principle cause of drug enslavement," Brown saw it as the major motivation for drug use: "Intensity of the euphoria may diminish but it is always present. There would be no point in continuing the use of narcotics if the injection simply made the addict feel 'normal'; it is the 'normalcy' to which he objects."[26]

As with other criminological models considered progressive at the time (including the US Narcotic Farm), changes in law, policy, economics, and enforcement patterns affected the demographic profiles of those incarcerated. The changes were attributed by staff to the social movements of the 1960s. A former sheriff quoted in the mid-1970s grumbled that the sources of prisoners "who could be worked" on the farm had dried up because of changes in prisoners' rights and laws regarding public drunkenness. The "revolving door drunks" that were once a "mainstay in the prison farm" were now placed in medical settings. Changes in probation subsidies incentivized counties to supervise more "hardened criminals," including heroin users, who were diverted from the state prison system and into county jails for the first time.[27] The resulting conditions fundamentally altered the character of Santa Rita.

In *Living for the City*, Donna Murch noted that Oakland was an important site for the rise of "preventive policing," an alliance between police and probation officers that reached into the lives of black youth through schools and recreation programs.[28] "Changes in East Bay law enforcement reflected a national trend toward 'legalistic policing,' characterized by modern equipment, formalized systems, and greater emphasis on juvenile

Fig. 34. Testee gives the officer his narcotic history.

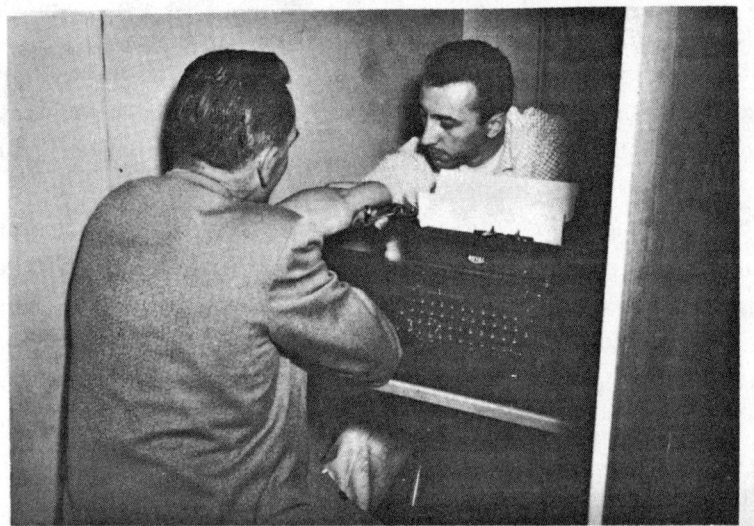

Fig. 35. Officer checks the arms of testee.

2.2 "Testee" gives a history of narcotics use and is physically examined prior to being positioned in a special chair for the Nalline injection. After injection, which is not pictured, the pupils are measured at a distance standardized by the apparatus pictured here.

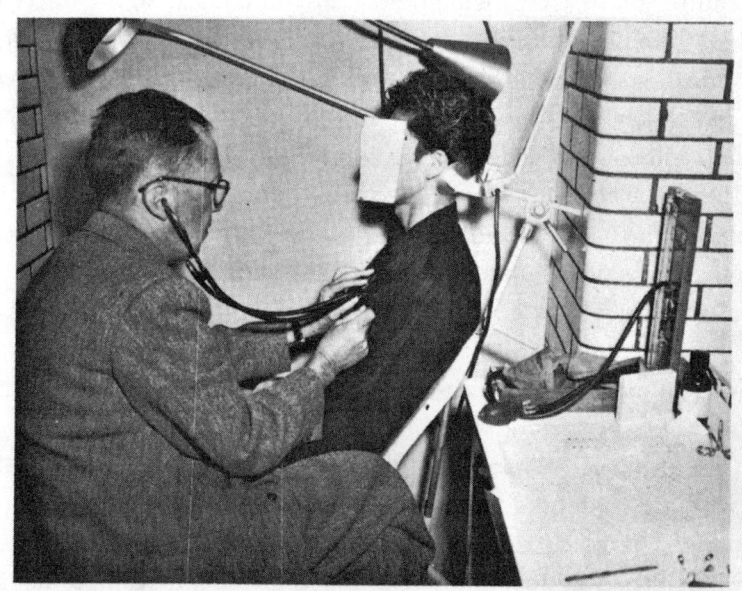

Fig. 36. Physical examination prior to injection of Nalline.

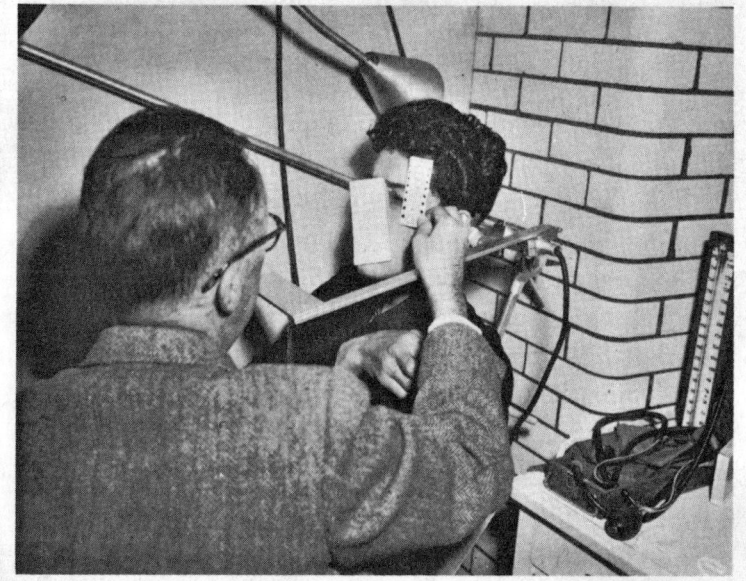

Fig. 37. Pupils are measured with pupillometer.

2.2 (continued)

Source: Thorvald Brown, *The Enigma of Drug Addiction* (Springfield, IL: Charles C. Thomas, 1961), 298–299.

detention."[29] There was perhaps no clearer instance than the direct violation of bodily integrity involved in the Nalline test, which collapsed the line that Murch identified between service and surveillance. Carceral places forcibly instituted "normal" habits, routines, and social expectations. Nalline was adopted as a tool to compel drug users to adopt conventional social norms. Brown's perspective displayed striking similarities with how methadone maintenance was later spoken of as a tool to reduce crime: "The addict who of necessity (Nalline) has cut down his habit, has in many cases been able to support himself and his family; he has been able to keep a job, stay out of jail and out of trouble on a more or less permanent basis so long as he knows he has to live with Nalline. This is his crutch and society's control."[30]

Consider Carl S., a 45-year-old parolee from San Quentin who had been an addict all his adult life except when incarcerated. Hardly a good treatment prospect, his parole was conditioned upon passing weekly Nalline tests. After forty-two weeks, Carl S. reportedly realized that Nalline "would catch me and I wasn't going to go back to San Quentin this time."[31] His "one-year cure" served as Brown's success story and as the basis for a recommendation to widen the scope of the Nalline test to physicians in private practice, who "seldom see addicts and probably would not recognize one if he were to walk into the office." To protect themselves from confidence schemes, doctor-shopping, and patients whose questionable pains could not be verified by examination, doctors were evidently supposed to inject Nalline to confirm or disconfirm their suspicions that a potential patient was or was not a drug addict.

A pharmacological optimist at heart, Brown hoped that Nalline would have rehabilitative benefits for addicted persons by strengthening their will or "superego." Police departments all over California were urged—by no less a personage than State Attorney General Stanley Mosk in a Town Hall address in Los Angeles on March 31, 1959—to adopt the "miracle drug" to solve one of "law enforcement's worst problems."[32] That year, in the midst of rising heroin use in southern California, Governor Edmund G. (Pat) Brown funded a state-level Nalline pilot to prevent a group of southern California parolees from becoming readdicted in their first two months after release. Newspapers reported that the program sent "sickly shivers down the spines of the Southland's needle-and-dope set."[33] Two

Nalline control units were set up in East Los Angeles and in Huntington Park (Long Beach), eventually expanding to five sites. Nalline was heralded as "prelude to a breakthrough to an effective statewide narcotics addiction control program."[34]

Police, parole officers, and probation officers sought control over relapse, a stance they repeatedly billed as pro-rehabilitation. Nalline's powers as a technological fix were regularly associated with clarity, strength, and stability, contrasting with the portrayal of addicts as weak-willed, unstable, and in need of a "chemical superego" to prop them up. Captain Brown presented Nalline as giving police the leverage of a "therapeutic crutch."[35] Police participation in "therapeutic" procedures imparted the status of science and medicine to police and probation officers, which many appeared to embrace as a form of professionalization. Later critics of the program pointed out that a "rhetoric of rehabilitation" affected Oakland police to such a degree that officers felt it "part of their business to change and 'cure' drug users as well as to merely catch them."[36]

The Nalline test was said to produce a definitive form of evidence that augmented police expertise. In Pasadena, the first southern California community to adopt the technology, it was presented to the public as "a 'hound dog' drug that can sniff out the narcotics user."[37] Nalline was personified as a detective poised to "pinpoint dope addict violators" at the second southern California pilot site, the State Adult Parole Department in Huntington Park.[38] According to one journalist, those who failed the test were arrested, but those who passed were allowed to walk out the door; the "antinarcotic" injection meant they were "clean." Nalline was not presented as a cure-all: "there are no cures, except within the addict himself," according to Dr. Charles T. Hurley, one of the first southern California physicians to administer Nalline. His program began with 100 Nalline tests in October 1959 supplied under the state pilot program. By the early 1960s, this program had administered ten thousand tests but detected only five hundred users. Administrators like Hurley claimed that drug users had a "healthy respect" for the test and even referred potential employers or family members to the parole department to prove they were clean. Parole officers who carried seventy to seventy-five individuals in their caseloads might also be said to have seen a crutch in Nalline.

Probation officers also played key roles in the Nalline programs, participating with judges and the district attorney in planning the Oakland program. Donald O. Thompson, Director of the Division of Adult Probation in Alameda County, emphasized mutual respect and cooperation that resulted between police and probation officers in the program. These agencies found in Nalline a convergent technology that reinforced existing lines of authority, and gave police a tool to combat "crafty propaganda" from addicts who promoted "vicious fabrications" about officers involved in narcotics control.[39] Tales of corruption were said to have been devised to sully police reputations—Brown reported a "ridiculously simple bit of gossip connecting a prominent and fearless narcotic officer in a relationship with an addict-prostitute of another race and naming him as the father of her unborn baby."[40] As noted in *Using Women,* fears about racial mixing have been integral to antidrug discourse since nineteenth-century concerns that San Francisco's "opium dens" were attracting white women. This "dirt," as Brown called it, was produced at a time when police corruption was rarely called out; he dismissed the possibility that there might be any truth to this rumor beyond the "momentum" that it gained as it "buzzed around the skid-rows of several metropolitan centers."[41]

White males received one-third of the Nalline tests at first, but in short order African American men and women comprised slightly over half the population tested. The racial-ethnic breakdown of the first 1,060 people administered Nalline tests from April 12, 1956, to June 30, 1959, was majority black. An "other" category was "more than 90 percent Mexicans."[42] Both exceeded their proportion in the population. Only 100 white or "other" women were ever tested, indicating that this tool of police science was not aimed primarily at them. Women of color tested positive at a rate higher than did white women (the rate of positives was almost exactly the same in white men and women). Additional evidence that Nalline was applied in ways that produced racial disparities was that almost 40 percent of African American men had positive reactions—the highest positive rate of all of the racialized gender categories recorded. This was largely due to their overinvolvement in the criminal justice system: there were dramatic differences between arrestees testing positive for narcotics violations who were not on probation (43 percent) versus those on Nalline-supported probation (4.1 percent). These statistics reinforced the idea that Nalline was a useful deterrent.

Along with the "vigilance of local peace officers," the Nalline program was credited with heroically driving heroin traffic out of Alameda County, the "most efficacious development in the fight against addiction so far devised."[43] Objections to the test issued from a few quarters, most notably drug users themselves. Although few consulted users about their perspectives, one quoted in the news dismissed the test as easy to beat—"Look, man, you don't fix before you go in for your test, you go out and fix AFTER the test." Unlike the Oakland program, the Long Beach program found that 47 percent of enrollees reverted to narcotics use during their first six months. Legislators in New York State declined to use the test because they saw it as a "chemical superego" or "artificial will," talk that Brown found preposterous because an addict was "likely to need as much in the way of his own resources as he can bring to any problem, rather than a chemical conscience."[44] In his view, users displayed "weakness of will, emotional deficiency, [and] inability to cope with problems," and furthermore, if they had possessed any problem-solving "resources" of their own, they would not have turned to chemicals in order to escape their realities.[45] The debate over the social utility of inducing or strengthening an "artificial will" by "substituting" a drug for a conscience or superego later surfaced in struggles over methadone maintenance.

The Nalline clinic made assumptions about the moral purpose of law enforcement, constant surveillance as a basis for rehabilitation, and the advantages of quasi-scientific production of evidence for use in criminal proceedings. The ultimate goal of the Nalline clinic was to reduce drug-related crime, but it enabled police to constitute themselves as scientific experts who could be useful in courts of law, where rules of evidence played a regulatory function.

PROOF POSITIVE: THE FRYE TEST AND THE FORENSICS OF NALORPHINE

Despite proponents' widely publicized enthusiasms about the rehabilitative powers of the Nalline test, the evidence produced by this technological means was by no means incontrovertible. Terry's ability to read small differentiations in pupil size was challenged in 1957. An entire courtroom decamped to his modified barber chair in the basement of the Oakland police station, where Terry demonstrated how he measured changes in pupil size precisely one half hour after nalorphine injection.[46] Explaining that changes in pupil diameter were not the sole indicator by which he determined someone was an addict, he spoke of a combination of visual markers such as track marks and tattoos, and a history of narcotics use. These were later codified by the California Department of Public Health.[47] Although Terry was presented as an expert witness for the prosecution, defense attorneys played him for a fool by asking him if he had ever tested any other chemicals in the opiate "family" to see how they reacted in the Nalline test. When Terry inquired about the meaning of the term "family," the defense introduced the technical term "alkaloids," an attempted humiliation that Terry shrugged off by saying that he had not taken a chemistry course in a quarter century.[48]

Trusted in courtrooms as "proof positive" of recent opiate use, the Nalline test's appeal lay in its capacity to produce seemingly unassailable, "objective" results. Legal questions did arise concerning whether proof of the use of an opiate constituted sufficient evidence that an individual was addicted. Accordingly, the state of California amended its code so that probationers and parolees had to submit to mandatory tests of a "synthetic opiate antinarcotic" (i.e., Nalline). Prior to the advent of nalorphine, evidence that a person was addicted to narcotics was largely circumstantial—riding upon lay knowledge of the effects of narcotics due to past experience; visual signs such as needle marks, red eyes, or contracted pupils; or the evidence of the senses, such as smells. Any witness in these instances was a lay witness, police included. Having the results of the Nalline test on hand helped police shift their status to that of expert witnesses. Faith in the definitive nature of results was reinforced when judges accepted the test under rules governing admissibility of scientific evidence in court.

Although administered by doctors, knowledge of how to conduct the Nalline test was not widespread among medical professionals. Thus did the Nalline test fall short of the Frye test, which had been established in 1923 to govern admissibility of evidence based on a standard of "general acceptance" among experts in a given scientific field. The Frye test was characterized as a "judicial straitjacket on the field of scientific evidence [that is] very unrealistic in view of modern developments, and fails to recognize the vast specialization which is part of our modern society."[49] The state of California skirted the Frye test by recognizing Nalline's validity regardless of whether or not most members of the medical profession knew how to administer and interpret it.

Not all authorities agreed that Nalline provided much leverage for the law. Probation officers and city health officials in Portland, Oregon, did not think it would work to deter most addicts because they had family ties and would not pack up and move because a Nalline clinic opened. Other critics opposed Nalline as police harassment. Noticing in the mid-1960s that no one had asked drug users themselves what effects the deployment of the Nalline test had on their mobility, Stanley E. Grupp, a sociologist at Illinois State University, administered questionnaires to people who had experienced the Nalline test in Chicago, the Oakland Santa Rita Rehabilitation Center, and the Lexington narcotics farm. As Grupp drily put it,

Addicts' views regarding the effect of the Nalline Test on the mobility of addicts are not in any substantial agreement with the views of those defenders of the test who feel it fosters the movement of addicts to non-Nalline areas. Neither does the estimated numbers of addicts known to have moved within the state or to other states suggest a high rate of mobility.[50]

He argued that the Nalline test had little effect on addict mobility, despite claims to the contrary and the framing of it as a "reasonably tight form of narcotic control."[51] Even if a given narcotic control measure could be shown to impact addict mobility, "this can hardly be viewed as an overall gain for the state or nation. Arguments advanced in favor of the Nalline Test or other devices as catalysts for addict mobility are clearly shortsighted since nothing is gained except to move the problem from one place to another."[52]

Sociological research on the effects of the Nalline test was conducted by Carey and Platt, who tallied eight thousand Nalline tests conducted by the Oakland Vice Squad up to 1963, with narcotics detected in only about 1.5 percent. They included a map of the clinic, an account of interactions within the physical setting, and exact transcripts of encounters there. Whereas the doctor remained in his office and interacted with "the testee," the Vice Squad dominated the setting; "their presence is acknowledged and felt by everyone."[53] Everyone in the situation—cops, doctors, social workers, and "testees" alike—was aware that Nalline served "only a limited practical purpose."[54] Concerned that any "ritual" involving weekly injections might encourage relapse rather than deter it, Carey and Platt saw the "ritual purpose" of the clinic as encouraging the very behavior it was supposed to eradicate.[55] The clinic, after all, made no treatment or health care available, and it forced attendees to center their lives around injection. The physical proximity of vice police and medical doctors symbolically signaled convergence between these authorities:

> The medical relationships are authoritarian and de-personalized. The combination of medical and custodial authority heightens the feeling of depersonalization on the part of the testee. The public health doctor, a benign figure in American mythology, emerges much more forcefully as a medical policeman. The ease with which he assumes this role suggests some elements of the public health doctor as a social type which have not yet been made explicit.[56]

The Nalline clinic thus enacted relationships that reinscribed assumptions about addicted persons as needing to be constantly surveilled, kept in line by the forces of order, disciplined by law and medicine. Although the drug itself obviously did not accomplish this, Nalline provided grounds for the convergence among law, medicine, and policing. In this sense, scientists and clinicians who developed and deployed the drug might be considered influential co-constructors of a heavy-handed technological approach to governing opiate usage.

BOON AND BANE: EXPANDING SPATIAL CONTROL OVER ADDICTED PERSONS

Law enforcement justified the Nalline test because they thought it deterred dealers from moving into an area and setting up shop, and motivated drug users to flee to districts where testing had not been instituted.[57] Restricting mobility was precisely the point. Proponents recognized that this "geographic cure" shifted drug problems rather than solving them, but this suited officials' limited objectives and possibly their parochial values. In 1960, for example, the city of Eureka in Humboldt County in northern California adopted the Nalline test. The purpose, according to

Matthew O'Connor, supervising agent for the State Bureau of Narcotics Enforcement, was to "keep addicts out of this area because they know they will be detected here."[58] Directed by jail physician Dr. James T. Collins, Eureka's program detected narcotics use in a quarter of those tested in its first months; triumphant pronouncements about the wisdom of adopting the program for its deterrent effects appeared in local newspapers for years.[59] The *Humboldt Times* practically crowed: "Narcotic addicts have met with more than their match in a clear, innocent-looking fluid identified in the medical profession as N-allylnormorphine. Better known to the layman as Nalline, the solution is fast becoming one of society's most potent weapons in the war against dope addiction" (see figure 2.3). County authorities—some of whom accompanied Alameda County officials to the first White House Conference on Drug Addiction convened in 1962—saw Nalline as their weapon in a "narcotics war."

Although many date the War on Drugs to the 1970s, some local officials clearly felt engaged in waging one much earlier.[60] Pictured enacting their official roles, they also acted out unofficial roles of their "patient and narcotic suspects," grimacing as if they were on the receiving end of Nalline injections and appearing to derive more than official gratification from their fight to restrict the movements of addicts into logging and lumber mills. When testing identified zero heroin users in 1964, a sergeant on the Eureka Police Narcotics Detail insisted that the first question potential convict jobseekers asked was whether there was a Nalline program—and when they found out there was, "they head in some other direction." Perhaps by coincidence the story appeared directly beside a report of two men using a fictitious name to purchase cough syrup containing codeine.[61]

California legislators also decided in this era to revoke driver's licenses from anyone testing positive for opiates until the offenders "voluntarily submit to and successfully pass a series of Nalline tests."[62] As Brown put it, "prior to this program, addicts drove from city to city making connections with pushers and peddlers and locally, used their cars for transporting narcotics and loot taken in shoplifting, thefts from parked cars, and other criminal activity indulged in to support their habits."[63] Nalline supposedly deterred such activities because anyone stopped and unable to produce a license supplied police with "reasonable cause" to search for

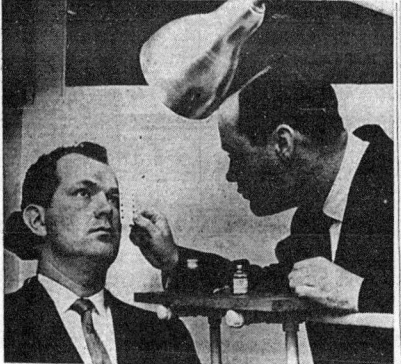
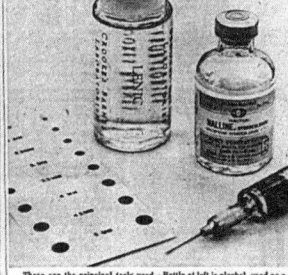
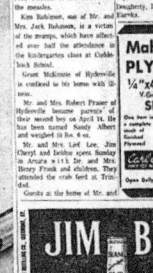

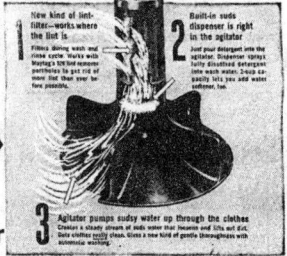

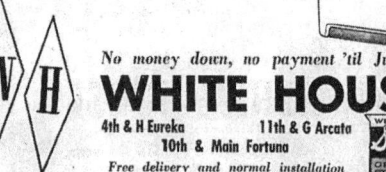

2.3 Nalline was a "scientifically developed solution," part of a "carefully controlled test." "Nalline New Weapon in Narcotics War," *Humboldt Times*, April 20, 1960. Used with permission of Marc Valles, *Eureka Times-Standard*.

"contraband narcotics, evidence of addiction, paraphernalia for injections, and stolen merchandise."[64]

Ever more vocal about restrictions placed upon them, such as the need for "reasonable cause,"[65] police complained vociferously about growing attention to the civil rights of criminals. Proponents of the Nalline test dismissed civil rights complaints on grounds that persons suspected of being addicts could simply decline to take the test. In a report to the United Nations Office of Drug Control (UNODC), one of the Federal Bureau of Narcotics's most trusted (and prolific) law enforcement assistants, Malachi L. Harney,[66] wrote:

The subject might well object to the administration of such a test as a violation of his civil rights. In practice, no person is required to take the test against his will. When the addict, showing indicia of addiction, such as old scars and fresh needle marks on his arms, nevertheless denies drug use, he generally will not refuse to take a Nalline test.... [T]he authorities in California, as well as those in Illinois and St. Louis, Mo., have found Nalline more than a detection device. Some former addicts, wishing to remain abstinent, were found to be relying on the prospect of frequent, and sometimes surprise tests for some support in keeping away from drugs. They seemed to find satisfaction in the interest and examination of the doctor; successful passing of the Nalline test was a source of pride and the accomplishment of the next test a good limited goal.[67]

Despite the multiple and conflicting meanings projected onto the Nalline test by police and policed, Nalline passed muster to be admissible as evidence in court, and remained so for many years.

Under the Harrison Act, possession of narcotics was a felony, whereas California made using narcotics or the state of being addicted to them a misdemeanor. California courts were aware that circumstantial evidence was the product of lay opinion. Circumstantial evidence did not meet the same evidentiary standard as a "sample of the narcotic taken from the body of the accused and properly analyzed and identified as a narcotic by an expert in the field."[68] Nor were there unequivocal visual signs. The Nalline clinic was designed to record such signs and to render them unassailable through photographic proof. Testimony from "experienced narcotic investigators" to the effect that an arrested person had "unusually red eyes," contracted pupils, was "somnolent," or was wearing clothing that smelled of narcotics could support a charge of narcotics *use* against a person in *possession* of narcotics after the amendment of California's

Health and Safety Code in 1954. All these "tell-tale signs" stood in the California courts. Yet after the 1945 case *People v. Moore*,[69] trained police officers who demonstrated substantial familiarity with "people under the influence of narcotics" could still render only "lay opinion."

The real use of the Nalline test was to convert "lay opinion" into evidence, reinforcing police qualifications as expert witnesses. In 1958 the California Department of Public Health released recommendations on procedures and dosage for administering Nalline, and California law enforcement officials promoted its merits. Police adopted it to combat the rising incidence of opiate addiction perceived in their localities. Results of the Nalline test were admitted as evidence for the first time in *People v. Williams*,[70] a case that came before the California Appellate Court in 1958. In conjunction with track marks and a prior history of admitted narcotics use, test results had been used to convict a heterosexual couple. The state's primary witness was Terry, who had administered the tests and confirmed the presence of an opiate in each person's system. However, on cross-examination Terry and two other medical personnel serving as witnesses admitted that most physicians did not know about the test and therefore would not necessarily accept results as evidence. The court, however, opined that rules of evidence should be liberalized—freed from the "straight-jacket" of the Frye test.[71] In 1956, in *People v. Jaurequi*, the California courts had differentiated vagrants, who commit crimes borne of idleness and "refusal to work and become productive," from persons who repeatedly performed wrongful acts as a "condition of existence" of being a "common drunkard, common thief, or common prostitute."[72] Narcotic addicts were considered the latter type of criminal: "Once established, addiction is not a crime of the will; rather, for the addict, it has become a condition of existence. His repetitive act of using the drug is determined by his addicted condition and its demand for the drug. The acts of the addict are not acts of the will. On the contrary, in some cases, the act may be in spite of his will."[73] Nalline was seen as providing a chemical crutch for the will. Masquerading as supporting public health and safety, the Code made being an addict a crime and a positive result on a Nalline test the evidence of that crime.

"EVEN ONE DAY IN PRISON WOULD BE A CRUEL AND UNUSUAL PUNISHMENT"

On a chilly Los Angeles night in early February 1960, two narcotics officers pulled over a car with an unilluminated license plate moving slowly on a darkened street. Officers Lawrence Brown and S. T. Wapato were not traffic police, but were from the Los Angeles Police Department's Wilshire Felony Unit. The officers inspected all four occupants' arms and pupils with flashlights. The green, four-door 1947 Nash was driven by its owner, Charles Banks, whose wife Norma occupied the passenger seat. A twenty-five-year-old Army veteran, Lawrence Robinson, was in the back with a "lady friend." Ordering Robinson to "bare his arms," the officers noticed marks upon them and arrested him.[74] They let the women go and took both men back to the station, where they were held without bail.

According to Los Angeles Deputy City Attorney William E. Doran, Robinson exhibited "a purported nervousness and perspiration" and, when asked, volunteered that he had used narcotics two weeks prior. At the time of the arrest, Robinson had no attorney and he appeared without counsel well into the proceedings.[75] Another narcotics officer confirmed the marks on his arms as needle marks, and Robinson was charged with suspicion of being a narcotics user under California's Health and Safety Code. Arraigned on February 4, 1960, a jury convicted him of this offense on June 9, 1960. The jury was supplied with a magnifying glass to assure itself that the marks on his arms resulted from narcotic injections.[76] The judge instructed the jury that it "could convict Mr. Robinson even if it found no proof of actual use of narcotics by him[,] so long as it found him an addict."[77] The jury was told that it did not need to disclose the basis of its decision to convict.

Placed on probation for two years, the first three months of which he served in county jail, Robinson was required to submit to Nalline tests at whatever intervals probation required. By the time he was released, Robinson's mother had engaged a lawyer, Samuel Carter McMorris, to argue her son's case. He did so—all the way up to the US Supreme Court. The case is considered a watershed precedent—not on Fourth Amendment rights to be free from illegal search and seizure, but on Eighth Amendment rights to be free of cruel and unusual punishment. Confused, prolix, and scattershot in oral arguments, McMorris's most coherent point was that "any statute which punishes a status as distinguished from an act or omission is unconstitutional."[78] He argued Robinson had been convicted on grounds of "suspicion itself"—the suspicion that he was an addict, a "wholly involuntary status."[79] He argued that

an addict no more intended to become that than does an alcoholic intend to become an alcoholic because he may have liked to drink socially.... [A] person intends to take a drink, intends perhaps to take [a] shot of heroin or smoke a marijuana cigarette or intends to have an illicit sex act. None of these intended to acquire venereal disease or to become an alcoholic or to become an addict.... No one, I submit, intends to become an addict."[80]

Advocating treatment and not punishment, McMorris dismissed the California Health and Safety Code as "vague, indefinite, and uncertain," because the term "addict" referred, firstly, to mere use of the drug; secondly, to being "under the influence"; and thirdly, to being "addicted to a narcotic."[81] Only those using narcotics out of compulsion and only those whose bodies and physiological functions were clearly "disorganized" by drug use should be considered addicts, according to McMorris. Throughout, he implied that this was not Lawrence Robinson's situation.

Pointing to the suffering entailed in "cold turkey California penal treatment," McMorris stated that "terrible violent withdrawal[s]" in jail had led to deaths of which he was personally aware. Claiming great experience with narcotics, he pointed to the dangers of collapse and death due to withdrawal.[82] "In every case, they suffer untold agony ... which could have been tapered off" with methadone or deep-sleep treatment.[83] Cold turkey, argued McMorris, was "cruel and unusual punishment for the condition of addiction."[84] The justices specifically asked McMorris whether he had experienced cases of "double jeopardy," where individuals were prosecuted multiple times for "the same addiction," to which McMorris exclaimed, "Every day. It's constant."[85] He accused police of punishing any "evidence of continued use."[86] Gaining little traction, he was directed to pursue the constitutional questions that his exhausted appeals at the state level had placed before the court. The very last question that Justice Earl Warren posed to the hapless lawyer was whether his client was still on probation and McMorris assured the Supreme Court that his client remained on probation and subject to the Nalline testing program.[87]

McMorris was an unlikely champion of the rights of drug users.[88] While preparing for the Robinson case, he took a night course in criminology. At one class, a guest speaker—an "expert from the LAPD narcotics squad"—mentioned that marijuana was not considered a narcotic, and in response McMorris wrote a paper titled "What Price Euphoria? The Case against Marijuana." He argued that if the US government did not ban marijuana, LSD, and other substances causing "temporary insanity," "we shall find ourselves at the threshold of that day in our history at which the American dream will evaporate into a nightmare of generally self-imposed narcotic insanity, from which there will be no return." Viewing drug use as a form of brain damage comparable to insanity, McMorris distinguished involuntary mental illness from voluntary acts under intentional control. Yet civil commitment of drug addicts was precisely what the State of California had inaugurated in 1961 (followed shortly thereafter by New York State's Narcotic Addict Control Commission and in 1966 by the federal government's Narcotic Addict Rehabilitation Act).[89] Under the guise of a therapeutic regime premised on civil commitment, the California program placed restrictions so severe they might be thought of as punishing.

The US Supreme Court's 1962 decision was heralded beneath such headlines as "Drug Addiction Ruled No Crime."[90] Commentators on the implications of *Robinson v. California* noted the Court's willingness to "set its face against those who are content to imprison the addict and not to cure the disease."[91] The goal, according to one law review article, was "a modern, civil and successful treatment of this sad, sick segment of our people."[92] The decision was understood as a watershed in the decriminalization of so-called victimless status crimes. One effect was to disallow California's unusual statutes and court decisions enabling conviction without proof that defendants had used, purchased, or possessed illegal narcotics at the time of the arrest. Although the Supreme Court invoked the Eighth Amendment's "cruel and unusual punishment" clause, Justice Harlan White's dissent chided his colleagues for making a strained attempt to write into the Constitution their "own abstract notions about how best to handle the narcotics problem."[93] The majority opinion itself, authored by Justice Potter Stewart, still reads as a liberal call to arms, despite its affirmation that states might confine addicts indefinitely for "compulsory treatment, involving quarantine, confinement, or sequestration" for therapeutic purposes.[94] The majority wrote, "It is unlikely that any state at this moment in history would attempt to make it a criminal offense for a person to be mentally ill, or a leper, or to be afflicted with a venereal disease.... Even one day in prison would be a cruel and unusual punishment for the 'crime' of having a common cold."[95]

But Lawrence Robinson had neither leprosy nor a common cold. Nor had he found Nalline tests either deterrent or therapeutic. He had been found dead in a Los Angeles alley of a "probable" overdose on August 5, 1961—fully ten months before the Supreme Court rendered its opinion.[96] His lawyer had either given a very credible performance of not knowing that his client was dead, or was genuinely unaware of that cold fact.[97] Once it came to light, California's attorney general filed a petition requesting that the decision be vacated and all proceedings rendered moot by Robinson's death. The Supreme Court denied the petition and thus Lawrence Robinson's name lived on, invoked as a liberal bellwether in the decriminalization of "status crimes" or "crimes of condition."

SELF-APPOINTED APOTHECARIES, GUINEA PIGS, AND THE SURVEILLANT CITY

Overdose deaths, suicides, murders, and "deaths listed as 'accidental,'" both "undetected" and "ascertained," were common in the "illicit drug realm."[98] Commentator Thorvald Brown attributed accidental deaths to unsafe practices and "ignorance—an unawareness of the ingredients of the dose and the potency of the drugs," complaining that addicts greedily used themselves as guinea pigs with little fear of danger. "Every dealer, pusher and addict is a self-appointed apothecary, mixing poisonous prescriptions to inject into his or another's arm."[99] Many of Oakland's overdose deaths followed a period of abstinence due to jail time, and the decedent's reversion to the same dosage used prior to abstinence. Brown cited overdose deaths that had occurred over a six-week period to illustrate this point. Two involved women with fairly mild habits. One twenty-five-year-old woman was "gallantly" injected by a "pusher" at a "family gathering of addicts." She immediately "fell to the floor in a semi-comatose condition" and died on the way to the hospital after family members tried to revive her with cold baths. Only when it became apparent that "medical treatment was of the utmost urgency" did they call an ambulance.[100] But most overdose deaths occurred not in new users, but in those whose heavier habits were interrupted by enforced abstinence.

Murders were sometimes disguised as overdoses, as when a police informant was found dead after failing to appear for an appointment.[101] When the "narc" was found slumped on his bed with a needle at his feet and paraphernalia nearby, Brown deduced that "an overdose [was] deliberately given to him by unknown members of the underworld in retribution for informing."[102] Fluctuations in the potency of "Mexican brown" [heroin] brought about deaths in "single-room skid row hovel[s]."[103] Brown described the death of Toby, a twenty-year-old man whose friends had redressed him after attempting to revive him from a deep coma in a cold bath, and had then dumped him outside the emergency entrance to a hospital. Revived briefly by administration of nalorphine and oxygen, "he sat up and tried to speak but fell back dead."[104] This death was considered "typical of the ignorant manner in which addicts indifferently dabble with drugs and ignore the death-dealing power of the unknown quantity and quality of the junk they shoot into their veins."[105] So potent was this dealing in death that the addict's life "must be truly charmed for the mystery is how do they survive."[106] Throughout the book, Brown marveled that overdose deaths were the exception and not the rule—that most drug users survived.

Why was the Nalline test adopted widely and used into the 1970s despite little or no evidence that it did the things proponents claimed? Playing to the ignorance of addicts and the powerful cultural myth that they lack self-control—as Brown and others regularly claimed—is one explanation; weekly Nalline testing was seen to instill control and even self-knowledge that buttressed the will to abstain. Reliance on Nalline as a deterrent preceded reliable urine testing (Endo Labs and the ARC explored such technologies in the 1960s, but these did not assume reliable form until the 1970s).[107] Critics linked emergence of amphetamine use to long-running Nalline programs, seeing in the turn to stimulants attempts to ensure drug use would go undetected.[108] Where drug users experienced the Nalline test, they became familiar with its surveillant and punitive aspects before realizing—if indeed they ever did realize—that nalorphine could also be used to reverse overdose. Although Nalline clinics phased out in the 1970s,[109] the test remained admissible in the most recent edition of Jefferson's *California Evidence Benchbook*. As a "chemical superego," nalorphine died a slow death. Positioned as both relapse prevention and deterrent, it was never used to reverse overdose outside of hospital settings such as operating rooms and emergency rooms.

3

DEATHS FROM "NARCOTISM" IN THE MID-TWENTIETH-CENTURY UNITED STATES

Clinicians working with people who used drugs began to recognize that nalorphine could save lives: "many lives can be saved by prompt and intelligent action on the part of the police, physicians, and other persons who may come into contact with deliberate or accidental poisoning from drugs."[1] Treatment of opiate poisoning by nalorphine was almost always successful unless patients were already comatose. For overdoses occurring beyond the limited reach of hospital-based nalorphine, remedies fallaciously thought to be efficacious included West Coast addicts believing that injecting instant coffee or milk was effective for opiate poisoning, and East Coast addicts injecting saline solution.[2] That such practices came to the authorities' attention indicates that overdose was not uncommon among postwar drug users.

Ray E. Trussell of the Columbia University School of Public Health surveyed 247 adolescents treated at Riverside Hospital in 1955. Eleven members of this youthful cohort had died from overdose in just two years, constituting a "very high death rate for the age group in question."[3] Once a proponent of Riverside's unique adolescent treatment program, Trussell became Commissioner of Hospitals for New York City in 1961 and thus was responsible for shutting it down. He acknowledged that "Any organized program dealing with the problem of drug addiction is faced with the historical fact that no therapeutic success of any significance has ever been recorded."[4]

Overdose deaths were taken as a signal that the progressive, treatment-oriented policies adopted in New York State before the 1973 shift to the Rockefeller Laws were failing just when heroin was becoming increasingly a "problem of black and Hispanic minorities."[5] Although Courtwright cites evidence that this racial and class shift had begun prior to the war in northern cities, it did not become evident to authorities, including those at the narcotic farm in Lexington, until the early 1950s. In New York City, those who used heroin regularly came not only from the black and Puerto Rican neighborhoods where heroin markets were located, but also from the outer boroughs. Postwar patterns of drug use would eventually make overdose business-as-usual in some places, but rare in others. The rarity of overdose deaths rendered them remarkable, an ambivalence that marked overdose both mundane and sensational during the mid-twentieth century.

THE HAUNTING ICONOGRAPHY OF "NARCOTISM"

Deaths from "narcotism" ticked steadily upward in Chicago in the late 1950s. Due to the drug control approach taken by the Federal Bureau of Narcotics (FBN), the problem was traced to nefarious "pushers." Historian Eric Schneider told the story of Chicago teenager Tommy Gordon, whose father demanded an explanation for his son's 1947 death from the FBN.[6] Realizing Tommy was "over and out," his friends put ice on his temples and groin and force-fed him milk, both "standard measures for handling overdoses" that had worked for them before.[7] Already mundane for those akin to Tommy, overdose retained its novelty for parental and governmental authorities. The Chicago FBN investigated in response to Tommy's father's inquiry, gathering details about the lives and deaths of a network of high school heroin injectors. Schneider found that "none of the Chicago cases conformed to the prevailing narrative in which a dealer supplied heroin to an unsuspecting youth in order to addict him."[8] Instead, heroin use was said to be influenced by "adolescent desire to be part of the hip, new happening thing."[9] Primary motivations for drug use centered on pleasure, peers, and knowledge of experiences within social networks. FBN agents had to tap into peer networks to explain why Tommy Gordon died what was even then a death both mundane and novel.

Deaths from narcotism also rose to public notice in New York City. According to the Office of the Medical Examiner, 3,200 such deaths occurred between 1950 and 1964, two-thirds of them after 1957. Such deaths were portrayed as a "Negro"[10] problem; twelve "Negroes" died for every "white," and the sex ratio was four men to every woman. The ratio of "male negroes" to "white males" was 10:1; the female ratio was 15:1. At the time Puerto Ricans were counted or classified as "white," comprising 35 percent of males and 20 percent of females. More than 90 percent of fatalities occurred before the decedent was fifty years of age.[11] Those dying tended to be unmarried, working class, "inversely proportional to the intelligence involved in the job, with unskilled workers providing the highest percentage of addicts."[12] Deaths were concentrated between 79th and 165th Streets in Harlem, "with a predominance of those deaths among the Negro population which resides heavily in that area."[13] The mid-twentieth-century wave of heroin use illustrated a number of relevant features of urban drug markets. More "addicts" died in that area than actually resided there because it provided "the narcotics and the opportunity to indulge in addiction to non-resident addicts."[14] Urban drug markets drew from all over; users in Baltimore, Philadelphia, and Washington, DC, had New York suppliers. Whites traveled to black neighborhoods if that was where the heroin supply was.

Location was as important to "medical detective work" as it is said to be in real estate, providing important clues for certifying deaths from narcotism. Placement and position of the body provided clues to the cause of death. Photographs were taken to supplement toxicology reports and autopsies. "Negroes" rarely appeared in such exhibits as shown in figure 3.1, possibly because black-and-white photography was technically ill-suited for capturing darker skin tone with the contrast necessary for documenting visual evidence of track marks, abscesses, tattoos, or "the works" dangling out of the dead person's forearm. Decontextualized photos of the dead are dehumanizing, evoking disgust and increased social distance between viewer and subject. The visual iconography of overdose portrayed the ephemera of narcotism: dead bodies awkwardly posed, surrounded by the paraphernalia of needles and syringes, cookers, bent and blackened spoons. No traces of alcohol bottles or other drugs clutter

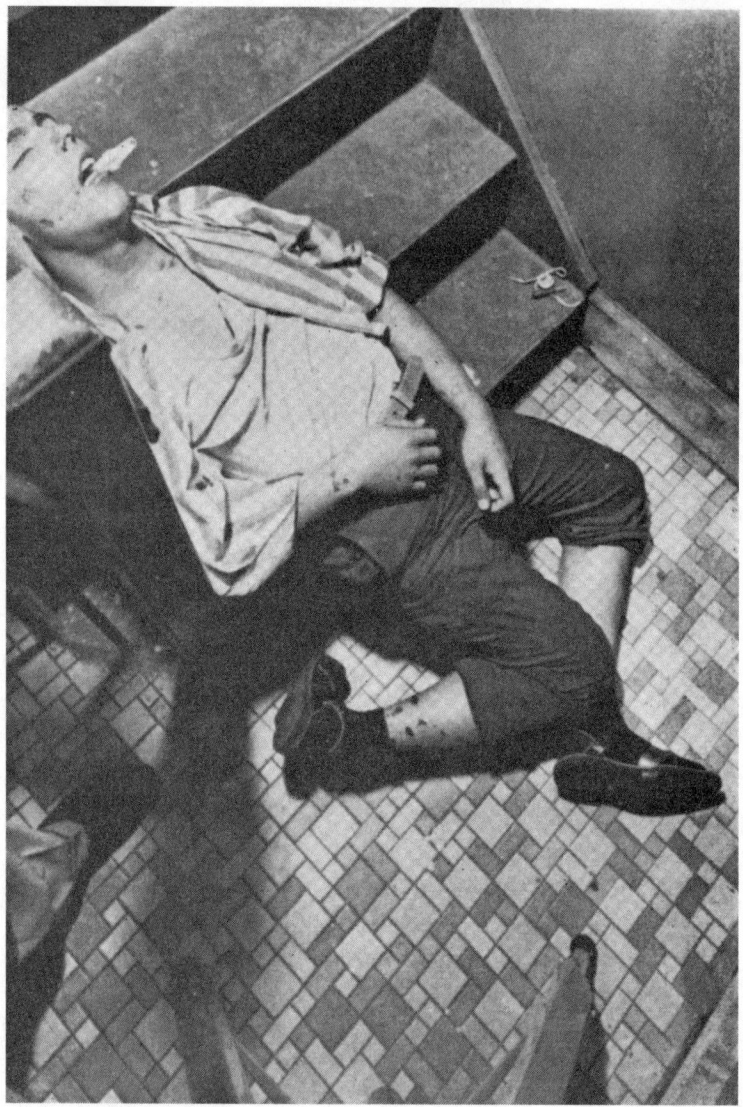

3.1 One of Milton Helpern's oft-used visual aids. Original caption read "Intravenous narcotic addict found dead on tenement house stairway. An absorbent tissue wad was found in the mouth and throat. The 'works' are on lower steps. Crude attempt was made to revive deceased after collapse."

Source: Milton E. Helpern, "Causes of Death from Drugs of Dependence," in *The Pharmacological and Epidemiological Aspects of Adolescent Drug Dependence: Proceedings of the Society for the Study of Addiction*, ed. Cedric William Malcolm Wilson (Oxford: Pergamon Press, 1968), 222C.

these scenes, which feature a visual repertoire as narrow as "the life." One might consider these to be autopsy photos of social death.

The science of overdose, too, narrowed to pharmacological essentialism about the properties of the opioid drugs, the molecules, and their physiological effects. Well into the 1960s, the science of pharmacology assumed universal actions and uniform effects despite the well-known and thoroughly documented range of individual variation. Two prominent explanations for overdose deaths emerged in forensic studies to exert a stranglehold over the field—one centered on potency, the other on purity. Researchers now insist that there is no accepted toxicological evidence for these explanations; yet the folk notions continue to enjoy wide circulation in media overdose events.

Overdose deaths occurred at or beyond the margins of respectability, with families like Tommy Gordon's experiencing their losses quietly. Tommy's father was a rare instance of someone who wrote to relevant authorities to learn more about the circumstances of his son's death. Urban drug markets shifted between the 1950s and the 1960s in ways recorded by anthropologists Edward Preble and John Casey in the case of New York.[15] For no other city was there such a range of material showing how heroin became an organizing force around which whole lives centered just after the end of World War II. The typical heroin addict of that era, according to Preble and Casey, was a "very busy man," a "hustler," full of the purposive activity of "taking care of business." Changes in illicit drug markets have distinct implications for social relations between users, law enforcement, and dealers. Disrupted by local and federal law enforcement, mid-twentieth-century drug markets were in the process of reorganizing in the late 1950s. Social relations within them deteriorated, as illustrated by Schneider's work on ethnicity and the market for Mexican "brown" or "black" tar heroin not only in Chicago, but in the Mexican American barrios of East Los Angeles, which began to experience rising heroin use in the late 1940s. In early-1950s Los Angeles, gangs were socially cohesive, their members sharing social bonds, philosophies, and activities that mitigated against ongoing involvement with heroin. But "peer-organized resistance" to use of drugs and alcohol declined in the face of market disruptions resulting from law enforcement anti-gang initiatives.[16] Although Chicanos were observed to be "faithful" users of heroin since its introduction in the 1950s, the drug was simply integral to patterns of sociability along with marijuana, barbiturates, and alcohol.[17] Few differences in life outcomes and cultural background appeared between gang members who had not been heroin users, and those who had used the drug regularly at some point in their lives.

Where social networks were broken, the illicit economy expanded. New York City's population of known heroin users doubled across the long 1950s, far exceeding growth in the adolescent male population. From 1947 to 1951, anthropologist Edward Preble and John Casey had encountered socially cohesive networks to which tight relationships and communal patterns of use were essential. There was little crime; the price of heroin was low; and those using it had some type of skill or training on which they could rely to generate disposable income. As the situation changed, younger people became less likely to develop skills before becoming regular heroin users. The racial-ethnic demographics of urban heroin use on the East Coast shifted from majority white to a substantial contingent of African Americans and Puerto Ricans.

Struggle ensued over what policy responses should be put into place during the 1950s—a struggle best visualized as a pendulum swinging between the poles of crime and disease. Defining drug use as a crime clearly won, an FBN victory underlined by the first adoption of mandatory minimum sentences for any crime in the 1951 Boggs Act. Juvenile delinquency became a national political issue.[18] Job opportunities in the legitimate economy began a long-term decline in inner cities. As structural conditions changed, policymakers perceived heroin as the culprit and attacked it with vigor.[19] Changing enforcement patterns and internal dynamics increased violence in urban drug markets. Organized crime withdrew (sometimes keeping a hand in by financing independent traffickers). The day-to-day risks of overdose death and violence increased during a heroin shortage or "panic" that waxed and waned from 1961 to 1968. Heroin prices skyrocketed, demand rose, and dealers responded with "short counts" and adulteration. Trust, needless to say, declined. No longer members of a subculture with shared rituals and tight relationships, heroin users became willing to commit crimes against one another. Preble and Casey noticed that "hustling" activities had displaced all other meaningful pursuits.

Changes in social relations and norms before, during, and after the heroin "panic" of 1961 concerned them as they observed these changes: "There is no longer any social cohesion among addicts. The competition and struggle necessary to support a habit has turned each one into an independent operator who looks out only for himself.... There is no longer a subculture of addicts based on social cohesion and emotional identification but rather a loose association of individuals and parallel couples."[20] Caught on camera in late 1964, one such "parallel couple" were Karen and Johnny, habitués of Needle Park at 71st and Broadway.[21] Looking like innocuous members of the white middle class, the two opiate users lived frenzied lives, as depicted in a photo of Karen "tying off," holding a tourniquet in her teeth with the fixity of an animal about to tear into its prey (see figure 3.2).

A competent "dope fiend" and sex worker, Karen was known to have a knack for reviving people from overdoses. In a riveting series of photographs, Eppridge caught her reviving a "trick" who had shot up heroin on top of taking five Doriden pills (see figures 3.3–3.5).[22] The photographs attested to the familiarity of overdose as Karen matter-of-factly went about a sequence of actions, one of the first visual characterizations of overdose and revival to appear in mainstream media.[23] Set in a single-room occupancy hotel in Manhattan, the overdose scene had an educational side effect in that it portrayed overdose as a survivable matter of know-how. Several accounts of this story appeared (in some the name "Helen" substituted for "Karen," and "Bobby" for "Johnny"; all pseudonyms).

Life's gritty realism was garnished with a few fantastical elements in a 1965 Signet paperback by journalist James Mills. One scene in *The Panic in Needle Park* positioned readers to overhear two addicts talking:

"But [heroin's] not the best high," one addict said. "Do you know what the best high *really* is?" The voice was serious and everyone turned and stayed very quiet to hear, maybe, of a new kind of high that was better than heroin, better than anything else. The best high—the voice was low and somber, heads leaned close to listen—was death. Silence. Man, that's outa sight, that's something else. Yeah, no feelin' at all. Everyone agreed. The best high of all was death.[24]

One of the story's main characters is Helen, whose "brilliance was widely recognized ... in the treatment of an OD, an overdose—a shot that unexpectedly contains more heroin than the body can survive. Helen always

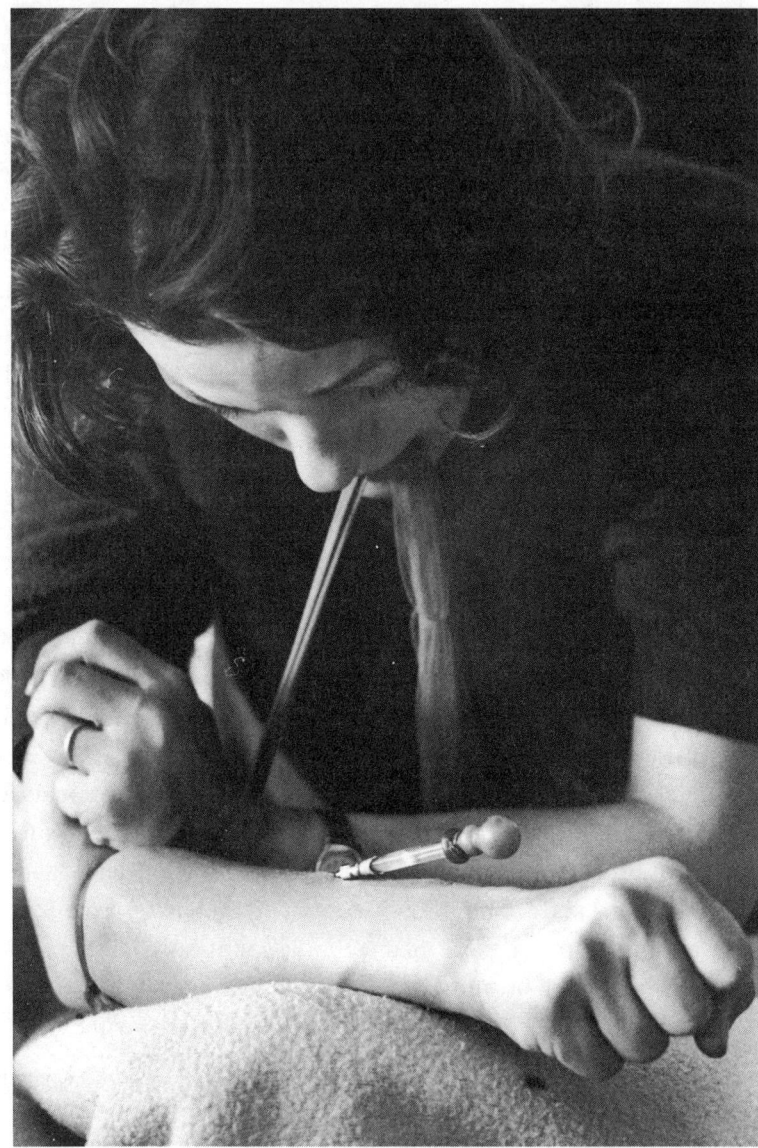

3.2   Karen tying off with the frenzy of an "animal with no name," as read the headline to the February 1965 *Life* magazine article. Photograph by Bill Eppridge, 1964. Used with permission of TIME, Inc.

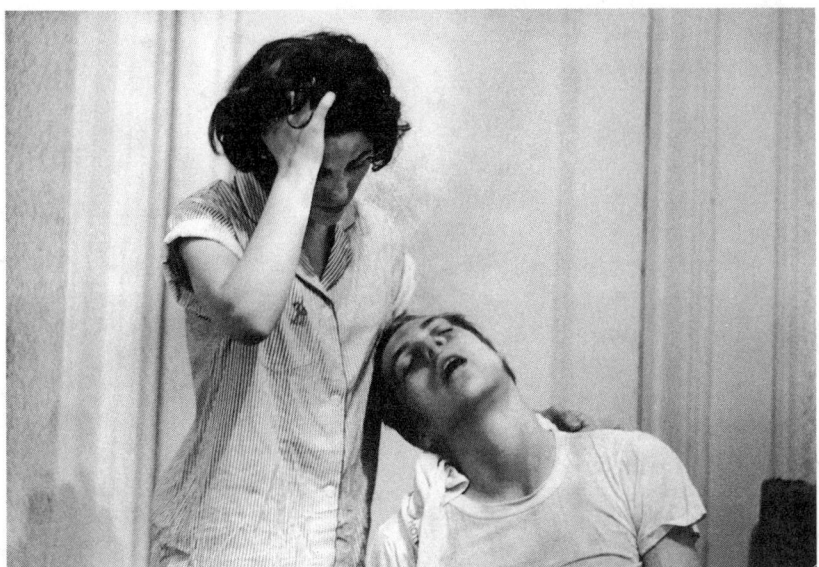

3.3 Karen readies herself to care for the overdose victim and begins by applying a cold compress. Photograph by Bill Eppridge, 1964.
Used with permission of Time, INC.

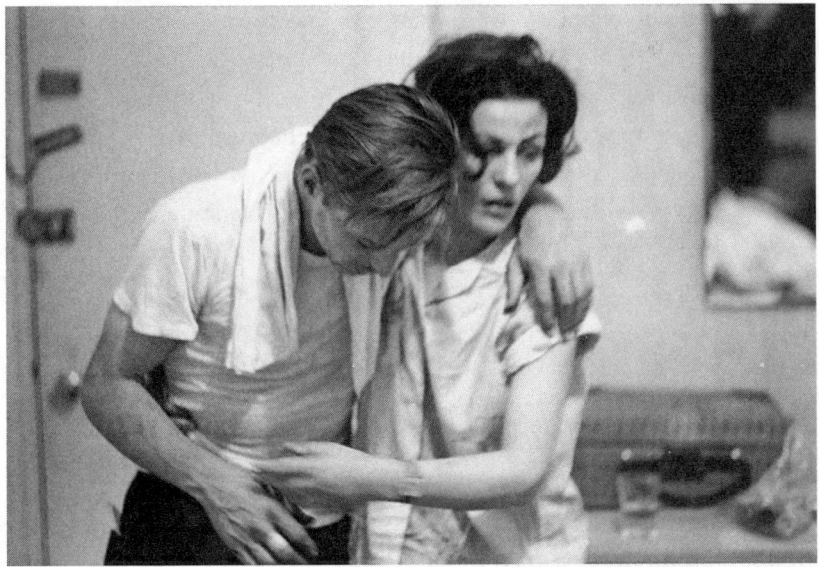

3.4 Karen steadies her "patient" as she walks him around the room. Photograph by Bill Eppridge, 1964.
Used with permission of TIME, Inc.

3.5 Karen's evident relief as her "patient" appeared to revive. Photograph by Bill Eppridge, 1964.
Used with permission of TIME, Inc.

knew what to do about an overdose."[25] She is compared to a doctor who "treated many ODs," or to the medical examiner who had "seen quite a few bodies disposed of—tossed out windows, hauled into alleys," or dragged down staircases. Helen was aware "that almost every day in New York City a junkie dies of an overdose."[26] Depicted with unmistakably maternal affect, she was a medic tending fallen comrades while others went about taking care of their own business.

The main overdose scene in the book has Bobby and Helen sneaking upstairs into the room of Marcie, a seventeen-year-old Puerto Rican woman. Marcie is preparing herself to meet with a paying customer, while her baby lies unattended on the bed, and her boyfriend, Frankie, is doubled over on a chair with his head almost touching the floor. Recognizing an OD in progress, Helen moves rapidly to her patient while dispatching Marcie for cold, wet towels to wrap his head. Helen props up Frankie, shouts in his ear to wake him, and then walks him around. These techniques will come to be maligned in later years as "folk remedies" for OD victims, but they worked in this case. The hapless Frankie rallies enough to exclaim, "Man, *that* was a good bag." The team, including the baby, sits the "stupefied" Frankie on the toilet; the others huddle in the cramped bathroom long enough for Marcie to conduct her transaction.

"Romeo and Juliet on junk," as a Signet tag line read, was a story "both fiction and fact: fiction in that none of its characters ever existed exactly as described; fact in that none of the characters or events is an impossibility in the junkie world. Everything that happens here has happened many times, to individuals not unlike the characters described in this book."[27] The story was also dramatized on the big screen as *Panic in Needle Park* (1971).[28]

These fictional social worlds offer entrée into a specific amnesia in which heroin use by white, relatively middle-class people frequently has been depicted as historically new.[29] Repeatedly spoken of as "novel"—despite whites having been the predominant racial-ethnic group using opioids in the United States since the nineteenth century, and it was African American and Spanish-speaking cultures that were actually new to it—such stories made surprise a recurrent staple of middle-class, white opiate use. Overdose there has often been framed as survivable; whereas overdose in communities of color has assumed the aura of inevitable fatality. Continually remade as "shocking" in white communities, overdose is blamed on circumstances beyond individual control—drug supply, deadly adulterants. By contrast in communities of color, overdose is depicted as endemic, an expected event occurring in a less individually differentiated "population." As popular awareness of overdose deaths spread, they were hushed and hidden, rarely appearing as the cause of death printed in obituaries in the United States until very recently.

THE "HEROIN OVERDOSE" MYSTERY: "MYTHS," "FACTS," AND AMNESIAS

Indeed, it is not stretching truth too far to opine that almost everything that we firmly believed to be the case regarding overdose was subsequently shown to be incorrect. As with many cherished beliefs, it was only in the harsh light of research that this was shown to be the case.
—Shane Darke, "Addiction Classics"

"Myth" is an unnerving term for historians, who study social processes by which "mythic facts" and "fact-ish myths" are produced and circulated to certify truths. Constrained to a world of "myths" and "facts," how do we explain category shifts, or the amnesia arising in moments when we are urged to forget not only past "myths," but also past "facts"? Such were my thoughts as I returned to the work of Edward M. Brecher, whose once familiar, grass-green paperback, *The Consumers Union Report on Licit and Illicit Drugs* (1972), was one of the most widely read sources on narcotics. I read it one hot summer in the late 1980s in the Mendocino County Library in Ukiah, California. No particular design guided my reading—I was a member of the general public taking the air-conditioned opportunity to better inform myself on the historical and social contexts of drug use, about which I thought I might someday write a dissertation.

My own historical amnesia was fueled by the misplaced specificity of national statistics on overdose deaths, given the intensely local character of opioid overdose and the numerous category mistakes at work in epidemiological profiles produced at the national level. Misplaced specificity is a pattern disorder in which cultural authorities, often policy elites, mistake local dynamics for a national "crisis" on which they might wage "war." From the midst of Nixon's War on Drugs, Brecher saw through the fact-ish mythology of overdose, which he designated "Syndrome X." His book was a bellwether text for the Consumers Union and patients' rights movements, which characterized drugs as technologies with implications for individual and social health. By questioning the "heroin overdose myth," Brecher did not deny that overdose deaths occurred, but sought a point of intervention into how they did so that was motivated by curiosity about how to prevent them. He implicitly acknowledged that drugs don't kill, people do—that the social, political, and economic contexts in which drugs are used matter for how deadly they are; who dies and who does not; and how significance is attached to these manners of death.

Characterized by the *New York Times* as a "towering work of scholarship" of comprehensive scope, the book contained a chapter on "The 'Heroin Overdose' Mystery and Other Hazards of Addiction." The list of hazards read like a research agenda for undone science: basic social and pharmacological dynamics of opiate overdose; copresence of depressants like alcohol, barbiturates, and benzodiazepines; and interaction between the protective effects of tolerance, dangers of abstinence, and frequency of relapse. Overdose was the most "widely misunderstood," if most publicized hazard of heroin injection.[30] Marshaling the evidence of his day, Brecher focused on what was then the epicenter of overdose, New York City, and on its chief medical examiner, Milton Helpern (see chapter 4). He considered the mortal event an "acute reaction" rather than an "overdose."[31] Having worked at the Office of the Medical Examiner since 1931, he questioned why massive pulmonary edema was found in most such deaths in the 1960s, but not in earlier cases. What accounted for the white froth taken as an indicator of heroin overdose in the 1960s and 1970s?[32] Why was it present even when heroin was either absent or present in such minute quantities that these deaths could not be characterized as "overdoses"? What about cases where toxicology turned up no "evidence of alkaloid in the tissues or body fluids"?[33] Were such deaths due to an "overwhelming shocklike process due to sensitivity to the injected material,"[34] or were they overdose deaths?

Yet more surprising than his willingness to question such matters was that Brecher urged preventing these deaths by "administering an effective antidote: a narcotic antagonist known as nalorphine (Nalline)."[35] His book came out more than a year after the Food and Drug Administration approved naloxone for overdose reversal, yet naloxone was unmentioned. Nalorphine, on the other hand, Brecher knew to be stocked in hospitals and pharmacies throughout the country, and he argued that it should have made "the death of anyone due to heroin overdose ... very rarely excusable."[36] Although Brecher warned persons at risk of dying this way that they could prevent such a consequence, he also represented this idea as "wholly ineffective in the current crisis, *for the thousands of deaths attributed to heroin overdose are not in fact due to heroin overdose at all.*"[37] One question was whether there was time for intervention during an acute reaction? Death from opiate overdose was ordinarily slow, during which Brecher noted there was plenty of time for prevention.[38] Failing to find overdose to be the major cause of death among American heroin addicts,[39] Brecher demanded an "intensive clinical and experimental search for what is in fact killing these addicts."[40] Others concurred that the "main cause of death in New York City narcotic addicts is an idiosyncratic reaction to an intravenous injection of unspecified material(s) and probably not a true pharmacologic overdose of narcotics."[41]

By contrast to specialists' definitional quagmires, mainstream scientific press data on opioid overdose—even where record-keeping was quite new in the early 1970s—was presented as completely reliable.[42] Although the District of Columbia did not turn to a medical examiner system until July 1971, Robert DuPont, director of the Special Action Office for Drug Abuse Prevention, and Centers for Disease Control (CDC) epidemiologist Mark H. Greene presented data to readers of *Science* on opiate overdose deaths as unquestionably reliable a few short years later. Eighty percent of deaths occurred in "young, black, inner-city males with a history of narcotics addiction"; toxicological evidence of morphine, quinine, or methadone in their systems was produced in a "modern, sophisticated" laboratory by expert forensic pathologists and toxicologists.

Designed to show that the heroin epidemic was ebbing, DuPont and Greene were realists, uninterested in the mysteries of overdose or fantastic notions that overdose was somehow not killing heroin users. As foil for Brecher's project, DuPont and Greene presented epidemiological claims and recognized the limits of their characterizations for asking why the heroin epidemic appeared to have been curbed in the nation's capital. "We must emphasize that these data describe very complicated events that have occurred in a very brief time span."[43] Nevertheless, DuPont and Greene went on to attribute the observed decline of heroin use in ways convergent with their interests—to a combination of methadone treatment availability, vigorous law enforcement, and "development of an antiheroin attitude in the community."[44]

Heroin is now "out." No longer is the pusher seen as a glamorous individual, a fabulously successful businessman. He has become a parasite in the community. Youngsters in the susceptible age group have become more aware of the dangers of addiction and, therefore, less willing to experiment with heroin. It is one thing to grow up knowing there is an older fellow on the next block who is addicted to heroin, and quite another to have many of your friends "strung out," in jail, or even dead from an overdose. As the appalling consequences of heroin addiction have become apparent, previously susceptible teen-agers are no longer willing to take the risk of experimenting with heroin.[45]

Couched as "lessons" within scientific and epidemiological morality tales, framings of overdose presented it as a cautionary tale. This was a major reason for the policy enshrinement of overdose deaths as a way to track heroin epidemics—and a clue to why statistical claims about them were generally accepted at face value. It was in no one's professional interest to question them.

"Why is the overdose myth almost universally accepted?" Brecher's own answer to this question lay in the peculiar practices of coroners and medical examiners—persons whose task it was to conduct autopsies and report on findings to family and to the press.[46] Pointing to the early 1940s, sometimes pinpointed at exactly 1943, Brecher wrote of a "strange new kind of death" that arose to bedevil coroners. Although customary for medical examiners to categorize all unexplained deaths in heroin users to overdose, a custom of convenience adopted by many coroners was to attribute users' deaths to tetanus,[47] endocarditis, trauma, tuberculosis, or suicide. Sudden deaths that could not be assigned to other causes were categorized as "overdose." As the nation's foremost expert on narcotic deaths and protocols by which coroners and medical examiners should certify them,[48] Helpern chose to raise thorny issues rather than neatly covering up the fact that sometimes addicts died from their "usual injection," and sometimes they did not. What made the difference?

Changes in route of administration were one plausible explanation; "mainline shooters" did not much appear on the urban American scene until 1926 and syringes did not become widely available until the aftermath of World War II. Before that most heroin users sniffed the drug; after 1926 they inserted medical droppers into hypodermic needles with a paper flange that completed the assemblage of "the works."[49] Delivering the drug into the vein, the works was a far more hygienic route of administration than the method used if needles were scarce—a safety pin was used to jab a hole in the skin into which a dropper was inserted.[50] "This crude technique often resulted in septic infection and scarring"— not to mention wasting heroin.[51] Differences between "mainlining" and "skinpopping" resulted in gendered health disparities—women in that era were rarely inoculated against tetanus, which manifested at high levels among female "skinpoppers." Chronic sepsis and cellulitis plagued those who relied on this route of administration, along with endocarditis, viral hepatitis, perforated ulcers, various pneumonias, and many forms of organ damage. The lungs were "very suggestive in the addict"—perhaps

the "most striking organs of all" as they were heavy and stiff with conges-
tion, sometimes containing "coagulated milk given to resuscitate."[52]

Helpern's postmortem presentations relied on graphic photographs
that were familiar to me because they were widely reproduced in newspapers and antidrug pamphlets in the 1960s and 1970s—and in Brecher's book. One depicting pulmonary edema was especially memorable
because the linoleum flooring was an exact replica of what my parents
chose for their new kitchen in 1974. The photograph depicted an unidentified white male with a wad of paper towels stuffed into his mouth, said
to be either an attempt to clear the airway or to administer milk, often
considered an antidote along with "all sorts of queer ideas."[53] Helpern
associated "queerness" with drug use, and often mentioned or illustrated
the "high incidence of homosexuality" in the addicted population. His
photographs were selected to accomplish cultural work.

For example, the body of a young woman dragged to a junk heap was
an obvious conflation with the term "junkie." Close-ups of forearms with
syringes carelessly hanging out of them, bodies amidst a litter of glass-
ine envelopes, or the detritus of bobby pins, matches, and spoons, established parallels between the impoverished nature of the domestic settings
where dead addicts were found and their state of social death. Lest his
audiences think such deaths occurred only among the young or poor,
Helpern regularly exhibited the cloisonné pillbox of a well-to-do addict
whose supplies were "seemingly inexhaustible," attracting a steady clien-
tele to his "hovel" on The Bowery.

THE LETHAL SYNERGIES OF POLYDRUG PROBLEMS

Polydrug use was a growing category in the 1970s, when the term was
coined to indicate the copresence of alcohol and barbiturates in so-called
heroin overdoses. Both the United Kingdom and the United States
witnessed barbiturate overdoses, as well as a "spectacular increase in the
number of suicides by barbiturates" from the 1930s into the early 1960s.[54]
Their popularity—many viewed them as legal alternatives to the opiates—
outlasted several cycles of concern. By the mid-1930s, the British medical
community was so acrimoniously embroiled in conflict over barbiturate
prescription and regulation that one commentator dubbed the period the

"battle of the barbiturates."[55] By the late 1950s, deaths from barbiturate overdose or poisoning outstripped those from all other poisons, including opioids, with rates increasing in major US cities such as New York, Los Angeles, and Chicago, as well as in England and Wales.[56] Celebrity suicides and/or accidents such as Marilyn Monroe's 1962 death from "acute barbiturate poisoning" focused public attention on *acute* events and overshadowed attention to *chronic* use.

Heroin's illegality made barbiturates seem somehow "safer," but much the opposite was the case. Controversy in the medical press contrasted with popular enthusiasm for barbiturates and an impressive market share commanded by their manufacturers. According to historian Nicholas Rasmussen, the United States experienced two separate "moral panics" centered on barbiturates. The first lasted from the mid-1940s to the late 1950s, by which time over half of US states regulated barbiturate sales through prescription-only laws. The second began in the early 1960s and led to federal regulation in 1965.[57] Regulation came about a full decade after ARC director Harris Isbell had testified before Congress advocating the control of barbiturates as "dangerous drugs."[58] Physicians had been undereducated about barbiturates and often prescribed the medicines with little concern for dependence. Isbell had found that abrupt withdrawal from chronic administration of high doses of barbiturates resulted in life-threatening seizures, convulsions, and temporary psychosis.[59] Often involved in celebrity overdoses, the pills were a mainstay of sleepless doctors, businessmen, and anxious wives. Brecher wrote, "You can get drunk on them; you can become addicted to them; and you can suffer delirium tremens when they are withdrawn."[60]

Alcohol, too, has made a perennial contribution to opioid overdose deaths, but the role of drinking in overdose has been repeatedly forgotten. The "hazard of death from shooting an opiate while drunk on alcohol or a barbiturate" was emphasized by Brecher in the 1970s and was "ignored by authorities on drug addiction—and by coroners and medical examiners— through the years."[61] Clashing modes of expertise led "addiction" experts to ignore alcohol, even as "alcohol" experts ignored narcotics. However, it has been reliably established that it was relatively rare for New York City heroin users to drink in the 1950s.[62] The Riverside Hospital drug treatment program found that drug users believed that "use of narcotics and alcohol in combination is dangerous and might possibly lead to the death of an individual."[63] Awareness of that lethal synergy had dissipated, however, by the time New York City toxicology reports revealed alcohol's role in a sample of 549 addicts who had died in 1967. Researchers noted "summer peaks" in "addiction activity" indicators such as hospitalizations for cellulitis, hepatitis, and nonfatal overdose.[64] Whereas, from 1964 to 1966, overdose had been a "stable" index of "addiction activity," the authors found it increasing for reasons other than heroin use per se. The knowledge about not mixing alcohol with opioids, and other ways of averting overdose, was no longer being transmitted. The origin, maintenance, and loss of this lay knowledge deserves further research.

During the late 1960s, San Francisco physicians working directly with addicted persons also drew attention to the hazards of mixing alcohol and drugs. At the first National Heroin Conference in 1971, Haight-Ashbury Medical Clinic physician George Gay reported that 37 percent of heroin addicts at the clinic used barbiturates "for sedation or sleep," and 24 percent reported using alcohol for similar reasons.[65] He reinterpreted the "overdose" deaths of Jimi Hendrix and Janis Joplin as caused by interactions between alcohol, barbiturates, and opiates. Divisions between clinicians and researchers in the alcohol field and those in the drug field became further entrenched when the US federal government created separate, competing institutes, the National Institutes on Alcohol Abuse and Alcoholism (1971) and the National Institute on Drug Abuse (1974). Alcohol and drugs were separated so clearly from one another that drug researchers considered alcohol largely uninteresting research terrain. One exception was a study of heroin-related deaths in the District of Columbia between 1979 and 1982.[66] CDC epidemiologist A. James Ruttenbar and James L. Luke of the Armed Forces Institute of Pathology analyzed autopsy protocols from the DC medical examiner. Copresence of ethanol and heroin (morphine) was statistically significant in those deaths, leading them to conclude that alcohol augmented heroin's toxicity, particularly when people used heroin "nonaddictively" (i.e., infrequently).

By the 1970s the scope of the heroin problem could be characterized by quantitative indicators that enabled better comparisons between local, state, and national levels. Heroin overdose deaths plateaued late in that decade but then began rising in the 1980s, during which between 1 and 3 percent of heroin users died annually. Deaths of young, inexperienced users did not propel the rise. Rather, older and more experienced opioid users began engaging in novel patterns of polydrug use that were killing them. Most "heroin-related deaths" occurred on spring and summer weekends, a pattern that Ruttenbar and Luke interpreted as evidence that "chronic ethanol abusers" were "casually" or "recreationally" turning to heroin as a "discount high."[67] This made them vulnerable to dying heroin-related deaths in summer weather, as if warm weather brought on a taste for the drug.

Another reason for missed connections between overdose and alcohol was that researchers recruited human subjects almost entirely from treatment-seeking populations. The protocols called for individuals who primarily used heroin. Drinkers were excluded from studies because they presented confounding variables. Smokers, on the other hand, were ubiquitous among human subjects of heroin studies. Accessing subjects unconnected with treatment proved challenging. Those working with non-treatment seeking populations tended to use more anthropological, qualitative, or ethnographic methods that required building long-term relationships with drug-using communities. As the drug research apparatus became ever more biomedical and epidemiological, all but the most stalwart qualitative and ethnographic drug researchers sought other fields.[68]

The social processes of biomedicalization reshaped drug and alcohol research just as surely as they restructured the political economy of health care.[69] In conjunction with generational and geographic differences, some aspects of substance use, abuse, and overdose were overstudied, while others were understudied. One that did get attention was the coincidence of overdose deaths with relapse subsequent to release from incarceration or hospitalization. In a study conducted at Bethlem Royal and Maudsley Hospitals on *all* deaths to addicted persons between 1965 and 1969,[70] Ramon Gardner searched UK coroners' records and identified 112 deaths in "non-therapeutic" opioid users, 75 percent of whom had died within a thirty-mile radius of London.[71] Assiduously tracing each death, Gardner found almost half involved both heroin and barbiturates; in all but one instance, the death occurred "shortly after a period of opioid abstinence" of at least ten days. Only 20 percent of deaths occurred in younger, inexperienced users who had not registered with the Home Office; the older, more established users who were registrants of the British system comprised 80 percent of deaths. "A major factor leading to accidental death from overdosage in established addicts was loss of tolerance to opioid drugs, usually after institutional admission."[72]

Like his US counterparts, Gardner noted that accidental overdose occurred suddenly, "the body being found in a lavatory, the street, or even a bath, often with the syringe inserted in the arm or clenched in the hand."[73] Finding that such deaths were increasing, Gardner emphasized the protective nature of tolerance, and pointed to increased vulnerability following release from prison or hospital, return from abroad, or outpatient withdrawal.[74] Methadone could prevent such deaths when properly prescribed and administered (though he opposed self-administration on grounds that methadone was just as hazardous as heroin).[75] While he did not mention narcotic antagonists, Gardner wondered, "Why, then, do so many addicts discharged [from institutions] … survive?" Speculating that some had managed to maintain themselves inside hospitals or prisons, he realized that some drug users knew about the hazards of abstinence and had knowledgeably avoided "coincidental misuse" of opioids, alcohol, and other central nervous system depressants.[76] His emphasis on individual levels of experience, tolerance, awareness, and knowledge was prescient, particularly as it concerned the mixing of heroin or other opioids with barbiturates and alcohol. The latter was by far the most significant drug dependence problem in the United Kingdom.[77]

Via different drug policy trajectories, then, both the United States and the United Kingdom ended the 1960s with increasing mortality from drug-related deaths.[78] In London and New York, mortality was read as indicating that the opiate-using population was growing.[79] "The 'Heroin Overdose' Mystery" noted sardonically that if eight hundred "respectable citizens" had suddenly dropped dead, public health researchers would have left no stone unturned to figure out what was killing them. While Brecher did not directly discuss the racial politics of this undone science, we can infer that his use of the term "respectable" implied that he saw the lack of alarm over overdose deaths as rooted in the continuing stigmatization of drug addicts, partially on the basis of race and ethnicity. The amnesia and ignorance regarding the contribution of alcohol and other drugs to "opioid overdose" has been and continues to be a dangerous form of ignorance. "Polydrug toxicity is *the* major factor in opioid overdose."[80] Yet such knowledge was not widely shared or acted upon even by clinical and research communities positioned to use it.

4

BRINGING OUT THE DEAD: Naloxone's Nine Lives Begin

Dubbed the "unsung hero of overdose prevention" and a "brilliant but humble man," Jack Fishman, the Polish émigré who synthesized naloxone in 1960, is said to have never become aware of its significance for prevention of overdose deaths beyond hospital walls.[1] He was adopted by the harm reduction movement, however, and his family has become active on the issue since his death in 2013. His career was spent studying steroids and hormones at Sloan-Kettering Institute for Cancer Research. When he synthesized naloxone, he was searching for a way to reduce one of the less savory aspects of chronic opiate use—constipation.

THE INGLORIOUS SYNTHESIS OF NALOXONE

In "The Antidote,"[2] Ian Frazier revealed that Fishman was moonlighting in a "small lab under the elevated tracks on Jamaica Avenue" in Queens. He was working on narcotic alkaloids in the private laboratory of Mozes J. Lewenstein, head of the Endo Labs Narcotics Division in Richmond Hills, New York. Founded in 1920 as an independent, family-run business, Endo enjoyed its first pharmaceutical success in 1948 with Coumadin, the anticoagulant still used today. Inspiration for naloxone came from Harold Blumberg, who headed Endo's biological science division, through a time-honored form of pharmacological reasoning by analogy.[3] If an allyl group was attached to oxymorphone (the same tactic Leake had suggested to Hart and McCawley with nalorphine), it might yield a more potent narcotic antagonist,[4] a goal Blumberg had been urging on his colleagues since 1956.[5] Once Fishman synthesized naloxone, Lewenstein licensed it to Endo so Blumberg could conduct in-house biological studies demonstrating that naloxone was a potent, rapid-acting, pure narcotic antagonist.

Headquartered in a purpose-built Brutalist building designed by modernist architect Paul Rudolph (see figure 4.1), Endo Labs was an up and coming company when William R. Martin assumed direction of the Addiction Research Center (ARC) in 1963. He enrolled Endo in the search for a "magical ratio" to treat addiction (a line of thinking that yielded buprenorphine).[6] Whereas many companies and researchers were looking for ways to ameliorate the most attention-grabbing effect of the analgesics—their capacity to produce dependence—far less glamor lay in the antagonists, which were treated as afterthoughts. And so it was with naloxone to such a degree that the initial development team did not even write up scientific results.

4.1 Endo Labs Headquarters in Garden City, Long Island, New York, 1964. Photograph by Robert Perron.
Used with permission of Andrew Perron and Sarah Perron.

But Lewenstein and Fishman did understand that they had synthe-sized a potent antagonist that might have potential use, so they applied for a patent in March 1961. They waited five years for it to be granted.[7] Meanwhile, the Japanese pharmaceutical company Daiichi Sankyo inde-pendently developed naloxone and applied for a British patent in March 1962; the ever-efficient British granted it in October 1963. By 1983, over 100 papers in the *Science Citation Index* cited the US patent or Blumberg's 1961 abstract; only one cited the 1962 Japanese paper. "Absolute priority is difficult to determine," stated citation studies founder Eugene Garfield, for whom the primary factor for determining priority was citation in English.

All origins are multiple and all priority disputes interested, and the social organization of science had implications for naloxone's clinical fate. Jack Fishman's day job at Sloan-Kettering was tracing estrogen's role in breast cancer. Although he later joined efforts to prolong naloxone's action by creating a "sustained naloxone delivery system," his focus lay elsewhere.[8] His name went unrecognized in alkaloid chemistry and other fields where naloxone became clinically and conceptually important—addiction medicine, anesthesiology, emergency medicine, obstetrics, and pain research.[9] Not only did most relevant professionals not recognize Fishman's contribution; he himself did not realize it. The incident testi-fied to the power of socially organized failures of scientific imagination.

Organizationally, something held Endo back from another Couma-din.[10] Endo supplied investigators with naloxone but never imagined a market for it.[11] Naloxone in research settings was most useful for refin-ing pharmacological concepts and studying interactions at receptor sites.[12] It became central to the quest to "discover"—or more accurately to visualize—opiate receptors in the human brain, and it was the "most useful single tool" for elucidating the endogenous opioid system (as pre-viously described at the end of chapter 1).[13] In other words, naloxone played a major role in scientific efforts to comprehend the basic workings of the human brain.

Researchers also speculated about naloxone's possible clinical con-tribution to a "magical ratio," an innovative combination of naloxone mixed with an agonist such as morphine or methadone (interestingly, mixing it with heroin was not suggested). Mixtures were proposed to prevent diversion of legal opioids to illegal markets, or to make opioids "tamper-proof" by negating agonist effects if injected or crushed instead of swallowed[14]—all tactics that were tried as means to subvert misuse in subsequent decades. In contrast there was little fanfare about naloxone's capacity to reverse opioid overdose, although it was approved by the Food and Drug Administration (FDA) for that purpose in 1971. Nevertheless, naloxone proved so well suited for overdose reversal that it gradually edged nalorphine out of clinical settings and emergency medicine altogether.

This process took surprisingly long, considering that naloxone has been portrayed in retrospect as quickly becoming the "antidote of choice,"[15] and considering nalorphine's toxic profile. Not until the mid-1980s did pharmacologist Louis Lasagna speak of naloxone as "now widely used in the management of certain kinds of untoward clinical events like accidental or purposeful poisoning with narcotics."[16] Within the addiction research enclave, naloxone was used experimentally to define and characterize the actions of antagonists as a class. Naloxone lacked abuse potential: it did not constrict pupils or produce dependence, tolerance, or subjective effects that typically accompanied these states.[17] As a "true" antagonist, naloxone was instead thought to possess "desirable therapeutic effects," and clinicians were urged not to shy from "administer[ing] repeated doses of naloxone to patients intoxicated with excessive doses of narcotic analgesics."[18]

EPICENTER OF OVERDOSE: THE NEW YORK CITY OFFICE OF CHIEF MEDICAL EXAMINER

Shortly after Milton Helpern started working at the Office of the Medical Examiner, he autopsied a heavily tattooed body of a heroin-addicted seaman and found, to his surprise, falciparum malaria parasites in the brain.[19] "Perhaps if it hadn't been a slack day that October 12 [1933]—Columbus Day, I recall it distinctly—the death might have been signed out rather perfunctorily as one from addiction, although such cases were uncommon then."[20] The young Helpern took to the Tombs (the city jail) and the corrections hospital on Welfare Island, collecting blood from "live addicts" and tracing their social and sexual interactions with "dead victims."[21] Some passing through what appeared to be florid opiate withdrawal turned out to have malaria and were treated with quinine. This bit of "medical detective work" ignited in Helpern a career-long interest in "narcotics addiction deaths." Between 1933 and 1943, needle-sharing practices among intravenous drug users were implicated in nearly 170 confirmed falciparum malaria deaths in New York City.

Eager to identify "less advanced cases walking around, awaiting dis-covery and diagnosis,"[22] Helpern imagined a role for medical examin-ers among the living. Within the queer social network he had happened into, "homosexuals" sported stage names like "Lady Astor" and "Panama Flo."[23] As if to justify passing from the austere world of autopsy into such colorful drama, Helpern cast himself as an anthropologist tapping into a "drug addict fraternity."[24] But the outbreak "vanished as mysteriously as [it] had come"; after 1943, there were no more malaria deaths in New York City. Opioid use declined during World War II because of a her-oin drought created by government stockpiling of opiates, which had restricted street supplies. As soon as stockpiling ended, intravenous opi-oid use spiked such that Helpern could later boast that he was the "chief medical examiner of the city with undoubtedly the greatest drug problem in the world."[25]

What was available on the illicit market was adulterated, most often with quinine. Helpern billed the addition of quinine to heroin as an instance of what in retrospect might be called harm reduction: "I like to believe that this practice began because addicts and their suppliers dis-covered that it prevented the development of malaria—a sort of built-in prophylactic."[26] The quinine theory later settled into the status of fact for him:

The use of quinine as a component of the heroin mixture appears to have origi-nated in the early 1930s, a period when malaria was prevalent among intrave-nous heroin addicts.... It appears that addicts, and those who prepared the drug mixtures, soon became aware of quinine as a cure and preventative of malaria and began to add it as a component diluent in the heroin mixtures. It is a remarkable fact that although there has been a considerable increase in addiction, no new cases of artificially transmitted malaria have been observed since 1943, when the last death of an addict from malaria in New York City took place.[27]

The "scores of sordid squalid scenes of death" that he had witnessed in the 1930s drew Helpern's eye to the "misery and degradation that so often accompanies this desperate craving for heroin, morphine, cocaine, and other drugs of dependence."[28] His careful reconstructions of scenes of death, dispositions of bodies, and "crude" paraphernalia surrounding them helped develop authoritative knowledge about the forensics of drug deaths.[29] As the world's foremost authority on the subject, Helpern argued against overreliance on toxicology in death certification; he argued that toxicology had to be triangulated with findings from autopsy and investigation of the social circumstances within which the death had occurred.[30]

Overdose deaths remained relatively rare prior to the 1960s, when the incidence of deaths from "acute narcotism" began to climb. Such cases were differentiated from deaths due to chronic health problems or infections due to faulty injection technique. By 1959 the annual toll reached 300, up from approximately 100 per year during most of the 1950s; the number then rose to an average of 425 per year between 1960 and 1969. Few other cities kept tabs on overdose deaths, making authoritative estimates or comparisons hard to come by. At that time only New York City had defined criteria for ascertaining overdose as a cause of death. A moral entrepreneur in the matter of deaths from acute narcotism, Helpern lectured colleagues on how to interpret relevant physical clues or "giveaway features" beyond obvious track marks or presence of "works."[31] A major question dogged him—why should one person drop dead and another not do so when using the same substance, from the same dealer, at the same dose?

In October 1969, narcotics-related deaths in children headlined *New York Times* reports about a conference on the Official Investigation of Medically Unattended, Unexpected, Sudden, Suspicious, and Violent Deaths, a forensic sciences gathering at the New York City Office of Chief Medical Examiner.[32] "Fresh from the horrors of the morgue autopsy room," Dr. Michael Baden, second in command, imparted a sense of urgency about getting coroners and medical examiners involved in solving the problems of the living that led to premature deaths, rather than merely dissecting bodies and accumulating data. The office used the sharp rise in heroin overdose deaths to solidify its own reputation in the forensic sciences but also with city government. Aptly titled *Autopsy*, Helpern's memoir described the office as fragmented and underfunded in the 1960s.[33]

The *New York Times* began to publish a running total of children who died of overdose. Two weeks after turning twelve years old, Walter Vandermeer died on December 14, 1969. His death transformed him from "just another kid running loose in the Harlem ghetto" to a "notable statistic"—he was the youngest person ever to die of overdose in the city of New York.[34] Depicted dualistically, Walter was said to have been capable of taking care of himself "just like a little hustler," but he was also portrayed as a light-skinned, four-foot-eleven, eighty-pound innocent wearing a Snoopy shirt. He died of an apparent heroin overdose in a bathroom on a Sunday morning with "equipment at his side."[35] Among the fascinating details was assistant medical examiner Michael Baden's insistence that Walter had used heroin for two years prior to his death, despite no forensic evidence of long-term injection. If Walter Vandermeer was indeed an "addict-pusher," his injection technique was impeccable: "Dr. Baden reported that the boy's body bore no marks from unclean needles but that it was still possible that he had been addicted to heroin for some time." Walter had a prior possession arrest and was said to be widely known as a dealer who "roamed" the streets of Harlem after school. Walter's mother, recently evicted after complaining about an unrepaired toilet, did not think Walter used drugs. Her knowledge was dismissed, while Baden's speculative claims were elevated to authoritative status thanks to the international stature cultivated by the New York City Office of Chief Medical Examiner.

Although white people comprised the majority of opioid drug users in the United States, the harms of drug trafficking and consumption have been disproportionately borne by communities of color since the 1950s. The Vandermeer case joined other reports on heroin in Harlem. There was no mention of the possibility of preventing Walter's death—he was represented as beyond redemption. Nor was there naloxone for Antoinette Dishman, a seventeen-year-old African American Barnard College student, who died on January 31, 1970, after sniffing heroin at a party in the Bronx.[36] Her obituary and articles discussing her death commented on race relations and expressed shock at events leading up to her death. A high school honor student from Chicago, she had lived in a dormitory with other African American women and was known to be disinterested in illegal drug use. One of her classmates intimated that, "Because we're black, we all know about heroin and its dangers and just because of this knowledge I always thought we were protected from it. It's not as if we're some white girl sniffing for the first time out of ignorance."[37]

Dishman's new friends at Barnard included "black militants" such as ABLE, the Academy for Black and Latin Education, and BOSS, the Black Organization of Soul Sisters. One friend commented on Dishman's social life, "All the parties we go to are black parties.... Our lives are black. Some people call us militant; we call ourselves nationalist. All the guys we date are black. Me, I've never had any social contact with whites. To me, Harlem is like home." Dishman herself was characterized as "rather friendly with some white people, but she preferred to be with blacks as we all do." The "white world of Barnard and neighboring Columbia" represented "the Establishment" to another classmate, who explained, "We've been fighting that all our lives." The fight had recently erupted into organized protest. Just a few weeks before her death, Dishman participated in a four-day sit-in at St. Luke's Hospital in Harlem. The goal, which succeeded, was to get a medically assisted detox and treatment program for adolescent heroin addicts.[38] These activities indicated the degree to which heroin use—and with it overdose—had become endemic to the point that the community demanded drug treatment.

Despite her role as a protester, Dishman was portrayed as an innocent by Barnard president Martha E. Peterson, who referred to the death as a "needless waste of a young life." Barely a month later, on February 26, 1970, a grand jury indicted Harold Burnell and Roy Garner on charges of homicide, second-degree manslaughter, criminal negligence, and six narcotic violations connected to Dishman's death.[39] This was the first time that drug suppliers were charged with homicide. An assistant district attorney called the apartment in which Dishman died a "'regular factory' for the packaging of narcotics for sale." The students renting the apartment, Wayne Brewington, 19, a student at City College, and Reggie Blackwell, 22, a student at Bronx Community College, were said to hold frequent parties. A teenage neighbor said, "I know that Reggie. He's like, you know, real cool." Reported around the nation with consistent emphasis on Dishman as pretty, smart, and militant, the case appeared beneath headlines: "Pretty, bright young militant dead at 17"[40] and "Pretty Antoinette is Morgue Case No. 483."[41] Another article titled "'Herd Instinct' Turns the Young to Heroin" quoted Helpern and Baden, for whom "proof of heroin's power" was on display every day at the New York City morgue.

Harlem organizing for drug treatment and against drugs was led by Mothers Against Drugs, which marched later that spring to expand treatment at Roosevelt Hospital.[42] The black community demanded control over drug treatment programs, aware that by the seventh week of 1970, thirty-four teenagers and 104 adults had died in ways attributable to heroin overdose.[43] Telling his building superintendent that he was feeling unwell, seventeen-year-old Peter James Stergios lapsed into a coma and died before police arrived. His parents claimed that he had been using heroin and buying pills—"four for a dollar"—since junior high school. The city was on track to outstrip the 1969 death toll, which was 224 teenagers (twenty under the age of 15) and 600 adults, an increase of 700 percent since 1966, when there were only 33 deaths of teenagers and 38 in 1964.

New York City public schools became a focus for drug law enforcement. Said to be available in almost all public high schools, heroin was supposedly sold by a veritable army of juvenile "pushers."[44] A front page *New York Times* story, "Heroin 'Epidemic' Hits Schools,"[45] ran not long after Dishman's death. The epidemic "enveloping" the schools was portrayed as a multiracial problem that could affect "a white girl of 14 ... or a black one of 17." Yet only African American boys and mothers adorned the front page as they awaited treatment at the Addicts Rehabilitation Center in Harlem.[46] Founded in 1957 by James Allen, a former heroin addict who had voluntarily taken the cure at the US Narcotic Farm, the ARC was Harlem's first drug-free treatment program.[47] The photographs illustrating the story are tellingly arrayed: two white teenage girls with long hair flanked very young-looking black schoolboys as if they were two sides of the same coin. Later that year, the overdose deaths of Jimi Hendrix and Janis Joplin would be similarly presented.[48] Despite their predominance in heroin use and trade, white males did not appear. The long-haired white women illustrated a section boldly titled "spreads to middle class," which indicated that white and Jewish youths "hooked inside the schools" predominated in treatment facilities. As with abortion, which was alternately medicalized and criminalized by race,[49] individual encounters with institutions of social control were racialized matters stirred into complex admixtures, such as the civil-commitment program administered by the New York State Narcotic Addiction Control Commission. Generally speaking, whites were sent to treatment, and blacks and Spanish-speaking people to the criminal justice system.

Although rarely explicit, comments on the racial patterning of heroin use—said to be moving from Harlem to the outer boroughs—were evident in articles like the one under the microscope here. Although initially considered as "primarily the scourge of Puerto Rican and Negro slums,"[50] the reporter traced the heroin epidemic beyond Harlem to a Queens street nicknamed "Sales City" and to a white high school dubbed "The Drugstore." Heroin-using teens were referred to as "addict-pushers"; officials opined on the "revolving-door" pattern with which teens who were arrested returned to school on probation and were treated as "folk heroes" who had beaten the system. Focused primarily on drug use and dealing by the hip children of the white middle classes, the article differentiated between the black "children of the slums" who learned about heroin early in their lives, and white youth who were just learning about it. "We are dealing with the third generation of drug addicts in Harlem, while whites are dealing with the first," Allen observed from the ARC. He cynically predicted that prevention money would be spent in primarily white areas, whereas "experiments, like methadone, will be pushed in black areas."[51] The article closed with a teacher's thought, "Now that heroin is getting out of the ghettos, everybody is beginning to see it as a problem."[52]

Portrayed as escaping the "ghetto" as early as the 1960s, heroin was represented as responsible for overdoses in Central Park, where eleven- and twelve-year-old children were said to be skin-popping and even mainlining.[53] Michael Baden's flamboyant then-wife, Judianne Densen-Gerber, founder of a controversial drug treatment program called Odyssey House, spoke to the press. The "therapeutic community" she directed assumed custody of children as a condition of enrollment. The medical examiner's office assumed a similarly activist stance, advancing its own fortunes in response to what leaders clearly viewed as a national or even an international crisis. Helpern flew to London to take part in the first international symposium on adolescent drug dependence in 1968.

Overdose had also surfaced as a concern in the United Kingdom. The president of the British Society for the Study of Addiction, F. E. Camps, noted that deaths associated with drug addiction had been rare in the United Kingdom until several Canadians visiting London in 1966 had overdosed following withdrawal.[54] Heroin was legally available as an oral tablet in Britain, whereas in Canada and the United States it was confined to the nonmedical market—and thus prone to injection, adulteration, and contamination. Helpern's presentation used fatalities as "indicators" of the extent of narcotics abuse in New York City (see figure 4.2). In "Causes of Death from Drugs of Dependence," he argued that social and economic changes had catalyzed the increase in heroin-related deaths among youth.[55] Between 1950 and 1959, overdose had become a more prevalent cause of death in urban communities of color in New York City, as illustrated by the overall ratio he had calculated, which stood at 12:1 "Negro" to white (10:1 among males and 15:1 among females). Overdose decedents were 4:1 men

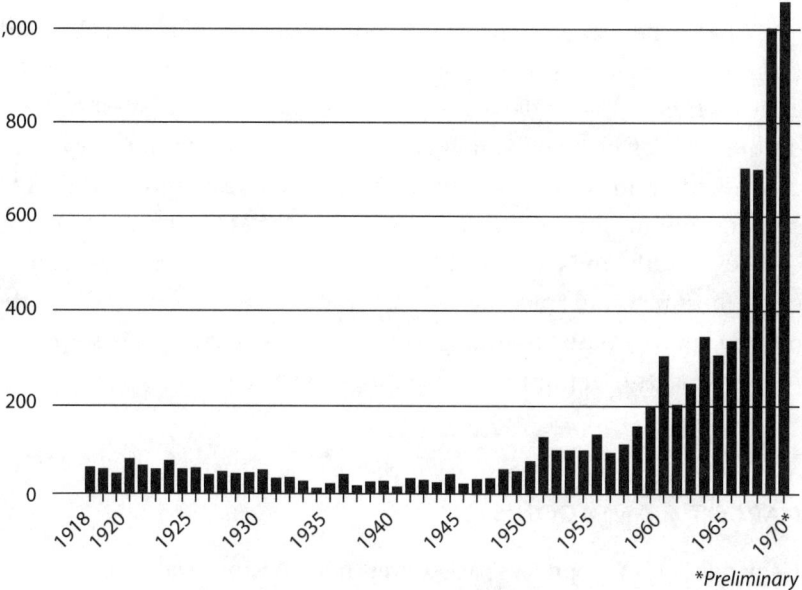

*Preliminary

4.2 Deaths from narcotics abuse in New York City, 1918–1970. From Milton Helpern, "Fatalities from Narcotic Addiction in New York City: Incidence, Pathologic Circumstances, and Pathologic Findings," *Human Pathology* 3, no. 1 (1972): 16. The New York City Office of Chief Medical Examiner.

to women because women were more often subcutaneous injectors who reached the hospital alive and could be revived with nalorphine.[56]

His reporting on racial-ethnic differentials to his British colleagues revealed that among "whites," 35 percent of men and 20 percent of women were "of Puerto Rican extraction."[57] Most heroin-related deaths occurred to unskilled workers who were younger than age thirty and who died between 79th and 165th Streets. Many of those who died of acute narcotism in Harlem were found not to reside there.[58] Another American attendee in London, Dean F. Markham,[59] noted the paucity of reliable epidemiology beyond his conclusion that adolescent heroin users were increasingly drawn from urban communities of color.[60] The calling out of persons of Puerto Rican descent from the category "white" was further evidence that adolescence was an important signal in the racialization of the drug problem. Despite the vast majority of drug use, drug traffic, and drug-related deaths continuing to come from established rather than emerging adults, adolescents—particularly black teens—were drawn as both threatened and threatening.

By 1970, there were about eight hundred overdose deaths in New York City each year, approximately one-quarter in teenagers, and this excited a media frenzy.[61] Drug deaths had become the "single greatest cause of death in adolescents and young people in the fifteen to thirty-five-year-old age group in New York City. This exceeded deaths from accidents, suicides, homicides, or even natural disease."[62] Yet only 20 percent of overdose deaths in New York City were from "pure" heroin or methadone; most involved "mixtures of everything under the sun."[63] Given the public clamor aroused by rising heroin overdose deaths, it is surprising how little attention the approval of naloxone received.

APPROVING NALOXONE: THE PURE NARCOTIC ANTAGONIST

Naloxone's FDA approval packet was not controversial. Endo sought approval for overdose reversal, and included studies showing favorable outcomes and effective response within seconds. Recommended adult dosage (0.1–0.4 mg) was calculated not to produce the withdrawal symptoms observed when dosage was higher. Dosage was set at 0.4 mg when naloxone was approved for "prevention, control, and treatment of narcotic-induced respiratory depression in general medicine, surgery, anesthesia, and obstetrics."[64] Naloxone's lack of efficacy against nonopioid drugs such as barbiturates, tranquilizers, nonnarcotic anesthetics, and sedatives was noted. As in most clinical studies, a maximum dose was specified,[65] and preclinical studies included the usual Lethal Dose (LD) studies in different animal species. However, investigators found it difficult to kill *any* experimental animals with naloxone, and had to estimate the LD50 dose. One LD study dosed rats with 200 mg/kg/day with no fatalities; the LD50 for rats was determined to be 640 mg/kg subcutaneously.[66] Comparisons to nalorphine revealed that the newer drug not only was safe and effective, but also lacked "side effects" in the largest clinical study of over two thousand pre- and postoperative patients or pregnant patients in labor.[67] Studies also showed that naloxone effectively reversed respiratory depression in newborns. The FDA insisted on warnings that "Narcan should be administered cautiously to persons including newborns of mothers who are known or suspected to be physically dependent on opioids. In such cases the abrupt and complete reversal of narcotic effects may precipitate an acute abstinence syndrome." Two pediatric studies were included in the application, one in which naloxone was administered to nine children who had accidentally ingested methadone, and one in which nine children who had ingested peripherally acting antidiarrheals were given naloxone. Most of the studies included were conducted in groups of thirty to forty patients in whom naloxone "adequately decreased respiration depression." No abuse potential was found in any of the eight to ten clinical studies, all of which were performed by outside investigators.

One of Endo's independent investigators was experimental psychiatrist Max Fink of New York Medical College, a well-known clinical trialist of psychoactive drugs such as chlorpromazine, widely used to quell patient behavior in mental hospitals. His studies used the electroencephalogram (EEG) to render drug effects discernible. Considered revelatory for brain function and dysfunction, the EEG was regarded as the "single most sensitive and recordable index of brain function in intact man."[68] EEGs offered experimentalists studying psychoactive drug response a noninvasive method to literally trace changes within the brain as they occurred.[69]

With National Institute of Mental Health (NIMH) funding, the EEG was adopted in state mental hospitals for studying mental pathology and for screening drugs to treat it. Fink was awarded an NIMH grant to set up EEG research at Hillside Hospital in Queens, New York, and at Metropolitan Hospital, one of only two inpatient drug treatment programs in New York City.[70] Fink described his work in the detox and treatment ward at Metropolitan Hospital:

I had one other experience in that era. That was the work with heroin. I think it's important to tell people that there was a time in New York, in the United States, when it was legal, possible for physicians, researchers to get heroin and to administer it to patients. We administered heroin and measured the EEG effects. We did blockade studies with cyclazocine, which was a narcotic antagonist. When naloxone was first described by a company in Long Island [Endo], one of the first samples was given to us at New York Medical College because we were known for doing this. We administered naloxone to our volunteers and then to each of us, finding that it, by itself, was a non-active drug except as a direct opioid antagonist.[71]

Fink and colleagues found that the "antagonistic actions of naloxone were particularly dramatic" in male heroin addicts who had been abstinent for at least a week.[72] The team administered naloxone shortly after heroin, which afforded them an opportunity to study EEG changes consequent to both agonist and antagonist. In addition to demonstrating naloxone's antagonistic actions, Fink and colleagues also tested it as a heroin blockade, their subjects complaining of a "precipitous withdrawal of euphoria, relaxation, and their 'high.'"[73] Naloxone, it became clear, was a way to take the high out of heroin.

But that was not the basis for its approval in 1971. Concluding that naloxone was an "effective narcotic antagonist that does not cause respiratory depression and therefore is an improvement over other currently available drugs," the FDA Division of Neuropharmacological Drugs recommended approving it for reversal of overdose. Discussion of narcotic antagonists as "emergency therapy" for overdose appeared first in the medical press and in the nascent profession of emergency medicine. News of naloxone's approval traveled slowly. One exceptional place where naloxone rapidly benefited "the addict in the street"[74] was at the Haight-Ashbury Free Medical Clinic, where physicians worked closely with members of the community experiencing overdose. Writing on the basis of their experience in San Francisco, George ("Skip") Gay, clinical director and founder of the heroin and drug detox clinics, and Darryl S. Inaba, director of pharmaceutical services, turned to the literature on emergency resuscitation, and decades-old studies of nalorphine by Eckenhoff and colleagues discussed in the previous chapter.[75] They opened with a typical case of an "unconscious youth" who had just shot up some "smack" being carried into the Free Clinic: "Every day we see at least one case of acute toxicity like this young man's, and we've yet to lose such an overdose patient—*who had a heartbeat when he arrived*."[76] Their patient was revived with an artificial airway, mouth-to-mouth resuscitation, and intravenous administration of 2 cc Narcan/naloxone. "The patient's color returned, spontaneous respirations resumed, and he regained full consciousness—all within about twenty seconds. Suddenly he spit out the airway, complaining only of a 'terrific headache.'"[77]

Detailing their emergency therapy, Gay and Inaba urged responders to begin breathing for the patient rather than taking the time to "look or call for endotracheal equipment—you are dealing with a true emergency and hesitation can spell disaster. Endotracheal intubation should be attempted *only* by a trained anesthesiologist or surgeon—and *only* when the patient has been previously properly oxygenated."[78] Their advice was not aimed entirely at medical professionals, but also at people who might perform first aid. After breathing was stable, they suggested administering "the Narcan test." This "test" was both diagnostic and therapeutic, enabling physicians to distinguish between opioid toxicity, and that of other classes of drugs. Gay and Inaba strongly differentiated Narcan (naloxone) from other narcotic antagonists, most particularly from the well-known negative effects of nalorphine/Nalline.[79] Within the social context of the Haight-Ashbury neighborhood, there was frequent extramural and experimental mixing of various drugs. Naloxone provided a clean assay for rapidly determining what was actually happening in a given case. If results were consistent with acute opiate toxicity, continuous monitoring was recommended. They recommended staying with a recently revived patient to ensure that they did not "lapse back into coma" should the opioid causing the initial condition outlast the naloxone that reversed it.

More remarkable than this precocious overdose reversal protocol was the authors' prediction that their colleagues would soon be treating great numbers of "typical middle-class junkie[s]." Although the Haight-Ashbury detox clinic treated up to 35 percent women, the typical junkie was characterized as "in his early 20s, white, male, ... and an Establishment-alienated dropout from suburbia."[80] As Gay and Inaba put it, acute drug reactions were "seen with increasing frequency in ERs servicing middle-class white suburbia."[81] They believed that the stuff of their daily clinical encounters would soon be experienced by physicians whose patient base was drawn from suburban, middle-class whites.

Drug "epidemics" are local, taking shape in relation to the cultural geography of cities like New York and San Francisco, which differed not only in terms of the populations most affected by overdose deaths but in terms of the kinds of drug treatment, illicit drug markets, and community organizations on offer. Naloxone's uneven availability when it was confined to hospital emergency rooms and emergency responders was no surprise. What was surprising was how long it took for naloxone to displace nalorphine. Naloxone had no abrupt "advent," but is more accurately portrayed as gradually trickling into the lives of those who needed it most.

5

UNNATURAL ACCIDENTS: The Science and Politics of "Reanimatology"

Historical tales of resuscitation often begin by relating the practices used by Amsterdam Rescue or the Royal Humane Society of London to reverse deaths from drowning—in defiance of religious proscriptions against interfering with God's will. Accidents increased with modern transportation systems such as railways and highways, accelerating development of trauma surgery and emergency medicine. State powers were invoked to mitigate the considerable variety of risks consequent to the automobile, as when local governments took on obligations to clear roads of snow and ice.[1] As emergency services professionalized to respond to traffic accidents, keeping people alive became a public service, an obligation, an expectation, and even a social norm in modern states.[2] Historians of accident have shown that the role of the driver became gradually perceived as one affected *by* risk rather than productive *of* risk.

Perceptions that drivers were acted upon by forces beyond their control prepared the public to accept high levels of death and injury as the price of modern living—and driving—today. But many people seem to think differently about the risks posed by opioid drugs, with drug users perceived as more to blame than drivers for the risks their behavior produces. Social justice concerns are embedded in societal handling of the risks and costs of traffic accidents.[3] Arguably, they are similarly embedded in societal responses to the risks and costs of using legal and illegal drugs.

Emergency response provides a site for seeing how questions of social responsibility, resource allocation, and thus social justice are worked out. Patterns of burden and benefit have shifted from individual drivers to state institutions that shouldered the obligation to create safer conditions within which traffic accidents could be prevented or reduced in severity. Automobile manufacturers, too, reluctantly at first and now with zeal, designed cars to mitigate the risks of driving.

This historical analysis was invaluable for my thinking about how social costs pertaining to risky behaviors move between individual and collective response. Health consequences have been the most central "cost." Drug users are playing new social roles in avoiding disease or managing risk, where once they were perceived as breaking the social contract that implicitly structured modern lives; as personally responsible for their "bad choices" to engage in risky behavior, including impaired driving. The Robinson case addressed in chapter 2 took place in the context of discussions about the state of California's handling of narcotics-impaired driving as well as alcohol-impaired driving. What social costs—or, rather, whose social costs—will be paid? For whom will social costs not be paid, shifted to pharmaceutical companies, or returned to consumers? "Bad choices" are compounded by aggravating conditions such as homelessness, "living rough" as it is called in Anglophone countries other than the United States, or poverty. Payment for social costs generated by people living in such conditions has often been extracted from them—fifty days in jail, or just fifty bucks—rather than amortized across society.[4]

The assumption that modern governments should prevent or reduce deaths from traffic accidents and other risks posed by automotive technologies involves a high level of historical amnesia. Traffic accidents occupy a "normal" category of everyday risks[5] by contrast to spectacular accidents such as aircraft crashes, terrorist strikes, or the sinking of the *Titanic*.[6] The assumption that traffic fatalities would continue to be the major cause of death for people ages 16 to 54 was enshrined in public discourse. Often when overdose death rates begin to increase in a given locale, the point is made that they exceed the number of deaths from traffic accidents. Since the early 2000s, stories announcing the margin of overdose deaths in excess of deaths from traffic accidents have appeared in state after state in the United States, as well as the United Kingdom.[7] Sample headlines read:

Drug Overdoses Killed 50,000 in U.S., More than Car Crashes (Associated Press, NBC News, December 9, 2016)

America, It's Time for an Intervention: Drug Overdoses Are Killing More People than Cars, Guns (Nick Wing, *Huffington Post*, August 2013)

Drug Overdoses Kill More People than Auto Accidents in 29 States (Reid Wilson, *Washington Post*, October 8, 2013)

Drug Deaths Exceed Traffic Deaths (Katie Moisse, ABC News, September 20, 2011)

Other headlines extended the analogy between lives lost to the war on drugs and those lost in other wars. Body counts, once familiar features of television news during the Vietnam War, began to reappear with increasingly gruesome regularity:

Opioids: Deadlier than Vietnam (David Brooks, April 4, 2017)

Since 2013, the US Has Lost More Lives to Drug Overdoses than WWI and WWII Combined (Kelsey Weekman, AOL.com, April 19, 2017)

In 2016, Drug Overdoses Likely Killed More Americans than the Entire Wars in Vietnam and Iraq (German Lopez, Vox.com, July 7, 2017)

Such headlines are condensed narratives designed to enable rapid grasp of the gravity and mounting death toll of ordinary as compared to celebrity overdose.[8] The YouTube channel is full of overdose videos made by bystanders or even police in the course of overdose calls. The seemingly nameless overdoses that comprise the vast majority of deaths are sometimes presented as a mass death, a missing generation. In 2013 the Centers for Disease Control and Prevention began reporting that deaths from pharmaceutical opioid overdoses had begun to exceed those from illicit opioids. Deaths from illicit opioids were the naturalized backdrop against which a seemingly new overdose drama was enacted—one that might strike unknowing and even innocent patients.

Modern societies have come to depend upon more predictable availability of expert rescue and resuscitation.[9] Road safety proceeded in as contingent a fashion as other harm reduction measures. Many presume that automobiles and roadways have become progressively safer because governments absorbed responsibility for road conditions and helped induce manufacturers to become committed to safety-related technological innovation. While systemic features safeguard against traffic fatalities, emergency medicine and the sciences of resuscitation have also reduced the numbers of deaths and thus the social impacts of traffic accidents.

THE HISTORICAL ABCS OF CARDIOPULMONARY RESUSCITATION

Those who develop sciences and technologies to reduce mortality do not work in a vacuum. Not until the mid-twentieth-century polio epidemic did a modern, biomedical "science of resuscitation" arise. Building upon what Russians called "reanimatology,"[10] modern scientists placed the capacity to resuscitate on the side of godliness rather than premodern deviltry. Instead of seeing resuscitation as interfering with the naturalized workings of a moral universe, it came to be viewed as a moral imperative that could even be performed by bystanders. Physician Peter Elam, working in Minneapolis in 1949, is said to have "rediscovered" mouth-to-mouth ventilation.[11] Elam's storied meeting with anesthesiologist Peter Safar, who had performed large-scale experiments on voluntarily paralyzed health care professionals, showing how important it was to first clear the airway before performing any resuscitative procedure.[12] Elam and Safar converged on the idea that their techniques could benefit both professionals and lay persons. Their alliance around mouth-to-mouth resuscitation and airway management shifted resuscitation away from the technological fix of installing prefabricated artificial airways touted by commercial firms.[13] From there advocates from the research and clinical communities moved toward a more socially effective model of "CPR for all" that leveraged a lay network dispersed in both space and time.[14] The universalistic nature of CPR training, chest compressions, and mouth-to-mouth resuscitation brought many a "bystander" into the social situation of the accident, and helped to solidify the moral imperative of rescue as lying on the side of resuscitation rather than noninterference.

Safar's *The ABC of Resuscitation* (1957) contained an unforgettable ABC protocol used to teach bystanders his technique.[15] Joining together with a Johns Hopkins University team working on chest compression for cases of cardiac arrest, Safar and Elam conducted efficacy studies and promulgated CPR to major organizations such as the American Medical Association, the American Heart Association, and the US military. But this process was slow; CPR guidelines were unpublished until 1966; and the Red Cross, the organization that became associated with the technique, waited until 1973 to adopt them. Inclusion of laypersons moved at a snail's pace.[16] CPR training was not in place in the United States until the mid-1970s; even then its availability was geographically specific. Over 100,000 laypersons, including 911 dispatchers, had been coached to provide bystander basic life support in Seattle, Washington—but training was less available elsewhere.

Although the story of CPR is typically told in the overdose rescue domain as a triumph of lay knowledge democratized with few barriers, Stefan Timmermans's *Sudden Death and the Myth of CPR* (1999) indicated that physicians remained proprietary over resuscitation techniques. Advocates sought support from voluntary organizations to further lay training, standard setting, and development of educational materials. CPR had to be recategorized as an emergency procedure, rather than a medical procedure, in order to widen the number and kinds of people allowed to legally perform it on others. The democratization of CPR to laypersons provides a useful analogue for thinking about naloxone. But CPR is "drug-free"; its proponents did not encounter the same barriers presented by naloxone, a drug that must typically be injected.[17] Social factors promoting democratization of CPR included its emergence from high-status medical professionals (anesthesiologists and cardiothoracic surgeons); production of an evidence base from its earliest days; and a series of paraprofessional groups poised to adopt it during the consolidation of emergency medicine. Finally, both out-of-hospital and in-hospital emergency medicine was portrayed on American television in the context of the Vietnam War and on the streets of major metropolitan areas.[18]

The social infrastructure for educating both emergency medical personnel and potential bystanders in the science and art of resuscitation is relatively new. Technologies for democratizing reanimatology have been developed and promulgated only since the 1970s. Another important step for out-of-hospital resuscitation was instruction of paramedics in advanced cardiac life support and changes to the legal framework permitting them to carry and administer drugs such as epinephrine, atropine, or naloxone. This step created a graduated form of expertise and defined roles in resuscitation: some paramedics were licensed to carry and administer drugs, but those on "basic" services were not. Credentialing distinctions led to differently equipped trucks labeled "advanced life support" and "basic life support." That distinction bode ill for naloxone, which was initially available only as an injectable solution packaged in small glass vials. Naloxone was confined to medical enclaves in part because of concerns about the skills necessary to inject it and the fussiness of the preparation itself.

NARCAN: THAT '70S THING

Naloxone in the form of Narcan became an accepted part of emergency medicine in the 1970s. One of the first manuals for paramedics, Nancy L. Caroline's *Emergency Care in the Streets*, stated that the protocol for treating narcotic overdose should follow the "principles of treatment for any comatose patient."[19] Noting that heroin overdoses tended to occur in "small epidemics" due to supply modulations, Caroline wrote, "When the paramedic team encounters one patient with heroin overdose, it is likely they will encounter others before the day is over."[20] The protocol was to maintain the airway, ventilate, start an intravenous line, and only "if the patient fails to respond to improved ventilation and glucose injection, and if there is reason to suspect narcotic overdose, give Narcan."[21] Narcan was to be injected "VERY SLOWLY" because it was undesirable for the patient to "come out of their coma in a combative mood and angry that the paramedic has just spoiled their $50 'fix.'"[22] The most important information came last: "So TITRATE the Narcan to the patient's respirations."[23] Emergency responders were to adapt to the patient's situation in the opioid overdose protocol, which was far simpler than the protocol for barbiturate overdose, which included insertion of an endotracheal tube. Caroline's textbook assumed that paramedics had naloxone and were willing to use it. It also indicated that clusters of overdoses were typically encountered in Pittsburgh, where Caroline directed Freedom House Ambulance Service when she wrote the book.[24]

Freedom House Ambulance Service served a largely African American neighborhood called the Hill District, and was perhaps the first ambulance service to be equipped with naloxone. The African American ambulance company had been founded by Peter Safar in 1967 to demonstrate that "so-called unemployable blacks can learn basic and advanced life support and thereby help people in their community."[25] Prevention Point Pittsburgh worker Ron Johnson remembered being revived by black EMTs after several overdoses in the 1970s:

I remember goin' out and then they brought me back. I can remember myself fadin' away ... the paramedics came and they revived me.... The EMTs were on The Hill, and they had this Narcan. I think they were the first EMTs in the country to have Narcan. They were just a group of, none of them had a lot of experience, some of them were users, some of them were nurses, busboys, every day people, but they trained 'em. There were overdoses then, massive fatalities, but they were in one community of just this specific color, so [overdose] wasn't really talked about. It wasn't in the newspaper. We more or less took care of our own. People looked after our own.[26]

Johnson recalled that overdose was regularly discussed among his close associates as a positive experience that resulted from "good product":

People were drawn to that product. Say if I said, "I got Red Devil," and you overdosed. People would say, "Go see Ron, he's got 'Red Devil.'" ... People were overdosing in galleries then. There weren't none of them cell phones then. Paramedics would come, 'specially on The Hill, or the police would come. They'd take you to the hospital because they didn't have that [Narcan] on all vans. Not like now. It's a great thing, but it seemed like they waited till the barn door was wide open before they really tried to do a lot. Because people been dying, I don't know how old you are, I'm 70 and they been dying 70 years from overdose. Even with Narcan.[27]

Thousands of articles covered overdose in the 1970s, but very few of them mentioned naloxone or Narcan. The drug was confined to advanced life support trucks and rapidly professionalizing paramedics.

Naloxone or Narcan was initially heralded not for its capacity to reverse overdose, but as an "off-label" method to "take the high out of heroin."[28] Naloxone was administered through pilot programs in Baltimore, Maryland, and New Haven, Connecticut. In February 1970, the New Haven, Connecticut, Mental Health Center Drug Dependence Unit directed by Herbert Kleber doled out daily naloxone to 150 patients, introduced through a pilot program that used naloxone to prevent them from getting high. Naloxone's short duration of action was improved by a longer-acting naloxone developed and tested by Blumberg, Fishman, and others in New York City.[29] Billed as a "drug that quenches an addict's yearning for three days at a time and may eventually end entirely his need for narcotics," this form of naloxone was described to the US Senate Health Subcommittee by eminent psychiatrist Alfred M. Freedman of New York University. Having tested it on thirteen confirmed heroin addicts,[30] Freedman found naloxone a useful form of relapse prevention because it created a window during which subjects could be convinced to avoid heroin. He speculated that after "several years on [naloxone]," the "hunger for narcotics" would likely be "extinguish[ed]" and lifelong maintenance treatment with methadone would be unnecessary.

Narcotic antagonists were held up as cures that might make long-term maintenance on agonists such as methadone unnecessary. Methadone was represented as itself an "addictive drug," and some addiction researchers recalled that heroin was prescribed as a cure for alcoholism and morphine addiction when it first appeared in 1898. Scientists modeled themselves guardians of the public health warding off another mistake like heroin; news of their quest appeared in *Science News* beneath the headline: "Another Heroin Cure Gets a Shot in the Arm."[31] Two million dollars' worth of contract research on narcotic antagonists was supported by the White House Special Action Office for Drug Abuse Prevention (SAODAP) and the National Institute of Mental Health (NIMH). Based on the pharmacologic principle that narcotic antagonists "shielded" ex-addicts from heroin's reinforcing effects, the scientific effort was to "make them less likely to get back into heroin."[32] The ambition was clear: "If they got no kick they would probably not continue with the drug. If and when an effective antagonist is produced a whole school or neighborhood could possibly be protected from heroin's effects this way."[33] But this was not feasible; cyclazocine's effects were unpleasant and naloxone's too short. "Even when perfected, however, there will be another drawback. A narcotic antagonist may be nothing more than a drug to replace a drug to replace a drug to replace a drug."[34]

Yet the dream of a technological fix to extinguish desire for opiates was widely shared and not entirely unrealistic. Addiction researchers recognized that the immediacy of craving played a role in relapse, as did the more prolonged cravings and preoccupations expressed by their patient/

subjects.[35] Many noticed that social and environmental cues—people, places, and things—were involved, but researchers differed in attributions of cause and therefore selected different targets at which to aim interventions. "As interested as I am in these treatment programs," said Freedman, "what we really ought to get at is prevention. We didn't wipe out typhoid until we did something about water supplies and sewage. Although this was much simpler than correcting the social conditions that breed addiction, the principle is the same.... Just substitute poverty and the disillusionment of youth [for water supplies and sewage]."[36] Bristol-Myers developed a methadone pill with naloxone in it; others experimented with mixtures of agonists and antagonists; buprenorphine was the favored candidate at the Addiction Research Center (ARC) because it had a ceiling effect on respiratory depression and so no one overdosed. Medications for addiction were one of many goals supported by the formation of the National Institute on Drug Abuse (NIDA), which rose like a phoenix from the ashes of SAODAP in 1973 and was placed under the direction of psychiatrist Robert DuPont.

Overdose was not uncommon in American life during the 1970s, and its impact was felt at the local level. In North Carolina, Regina Whittington, a *High Point Enterprise* reporter, typed the beginning lines of her 1973 story about overdose as a fairy tale:

Once upon a time, a very long time ago, when little boys wanted to be presidents when they grew up and little girls aspired to be no more than mommies, High Point did not have a drug problem.

Everyone knew, of course, that Somewhere there existed certain persons who were crazed, insane fugitives from justice who were addicts. But no one knew them. And they certainly weren't in High Point.

But it is no longer once upon a times and the pages of storybooks and fiction can offer nothing to explain the reality ... of High Point's drug society today.[37]

The previous year, High Point Memorial Hospital had reported 197 drug overdoses to the police in the cooperative spirit of a community attuned to the possibilities of rehabilitation. The hospital began the unique practice of directly admitting those who had overdosed into hospital-based treatment. An emergency room nurse observed, "You don't build or help people by destroying them. An overdose is an indication of trouble, mental trouble." Overdose was depicted as a medical problem that was relatively simple to treat with Narcan, but also as a symptom of trouble within an individual and community that required longer-term rehabilitation.

For the High Point emergency room to be dealing with drug overdoses every other day was understood—in 1973—as a serious problem requiring cooperation between authorities. Four decades later, High Point's overdoses and traffic accidents involving drivers under the influence of opiates again made news. In 2014 High Point experienced another "rash of overdoses—more than 100, including fourteen deaths," attributed to its status as a "traditional supply point" on a heroin pipeline.[38] The fact that the overdose rate was twice as high in 1973 as it was in 2014 seemed to have been forgotten. This amnesia illustrates how consistently overdose deaths are spoken of as harbingers of new troubles—even in contexts where recession, deindustrialization, and erosion of public spirit could not have been spelled out more clearly. Instead of telling fairy tales, the 2014 reporter served as Dante guiding readers through the circles of hell. There was a bright spot this time—the success of the North Carolina chapter of the Harm Reduction Coalition in distributing Narcan, tracking overdose reversals, and passing Good Samaritan legislation. This time around naloxone appeared in the form of rescuer, suggesting possibilities of redemption, revitalization, and resuscitation in the midst of an ongoing hell.

APPROXIMATING REALITY: OVERDOSE, DATA, AND NATION IN THE 1970S

National overdose death statistics were not kept in the United States until SAODAP contracted with researchers at the University of California at Irvine to create a method of national data collection on drug-associated deaths. Overdose concerns were widely reported both by wire services and local press, but these concerns were typically quite local. The city of Los Angeles had begun to keep overdose death statistics by neighborhood, recognizing that most went unreported. As recounted above, headlines warned "Drugs—Overdoses Kill More than Traffic Accidents."[39] Chelmsford, a small town near Lowell, Massachusetts, reported its first known heroin overdose case in 1973—on the very same page as a story about residents' campaign to keep drug treatment out of town due to fears of "large congregations of addicts" and "hippie rallies."[40] Readers were presented with graphic descriptions of the "unpleasant" experience of being treated "vigorously" for overdose with Narcan.[41] Much emphasis fell upon juvenile heroin use and novel drugs such as PCP, but pharmaceutical opioid overdose claimed some attention. The consumer rights organization Public Citizen requested that the Food and Drug Administration (FDA) remove from the market an analgesic implicated in overdose deaths, propoxyphene (marketed as Darvon and Darvocet).[42] Considerable attention was drawn to methadone poisoning or overdose, as discussed below. However, journalists did not present an orchestrated national chorus, but a cacophony of claims about populations that fit the stereotypical figure of the American junkie.

Overdose was represented as a naturalized consequence of heroin use and thus confined in effect to the heroin-using population. As a Columbia University Center for Social Research epidemiologist[43] noted, the ubiquitous knowledge construct of the "inevitability" and "irreversibility" of addiction rested upon how, where, and with whom knowledge about the use of heroin and other opiates was produced:

The early systematic studies of users of heroin and other opiates were primarily shaped by a small, prolific, and imaginative group of researchers associated either directly or, at least for research purposes, with the federal treatment programs at Lexington and Fort Worth.... It was a natural place for studying addiction because it was, at the time, the only place where large numbers of opiate users—in the later years, mainly heroin addicts—could be carefully studied and followed for various tenures of posttreatment adaptation. Connected with a government program, the authors had access to official records often difficult for outsiders to obtain, and with persistence and ingenuity they were able to locate discharged patients over long periods of time after their release.[44]

Given these advantages, researchers at the ARC held a virtually worldwide monopoly over human clinical research on addiction until the late 1960s; they framed addiction as a form of neurophysiological dependence and tolerance as a protective adaptation by the brain. For them drug dependence was a chronic problem of the central nervous system—a "brain disease" in which relapse was typically triggered by social cues.

However, new biomedical research on heroin addiction as a metabolic disorder was emerging from Vincent Dole and Marie Nyswander's pioneering methadone maintenance program at the Rockefeller Institute. Overdose and heroin use had grown a public face in New York City.

Historian Samuel K. Roberts has shown that knowledge emerging from community programs in New York City was not biomedical as much as it was framed in the social terms of community organizations seeking greater access to drug treatment for members of the black community.[45] Indeed, neighborhood-based treatment advocates mobilized against the form of treatment espoused by biomedical researchers in New York City, methadone maintenance. Community groups favored more socially contextualized and culturally appropriate forms of care.[46] For instance, the Narcotics Committee of the East Harlem Protestant Parish (EHPP) established a peer recovery support group that worked with Riverside Hospital's adolescent treatment program to limit the damage of heroin addiction within the social ecology of Harlem and the Bronx. Active since the 1950s, that group renewed efforts in an early 1960s campaign to promote holistic community-based care based on the Riverside model. According to Roberts, the EHPP lost out to the biomedical approach backed by the Health Research Council, a funding body developed to support biomedical research on etiology and treatment of heroin addiction in New York City. The more socially oriented Riverside program contrasted sharply with the biomedical approaches of Metropolitan Hospital, where the clinical trials of naloxone described above were performed by Max Fink in support of Endo Labs's FDA application for naloxone.

Methadone maintenance researchers championed treatment but administered an agonist drug on which individuals could overdose. By contrast the sociological research community viewed treatment itself as social control and dismissed treatment as often unnecessary.[47] By the early 1970s, these investigators were challenging the picture painted in Lexington, suggesting that a sizable contingent of heroin users did not cross the threshold to become socially problematic and therefore never sought treatment.[48] Such insights were validated by the results of the Vietnam Drug User Study, a landmark research project undertaken in the charged atmosphere of national urgency about heroin use in the military. Sociologist Lee N. Robins's studies of returnees from Vietnam showed that most did not relapse to heroin use once removed from the social ecologies in which they had initiated it. Due to assumptions about the chronic, relapsing nature of addiction undergirded by ARC studies, these results were unanticipated. Challenging long-held assumptions about the inevitability of relapse, Robins's defiant conclusion was cited as evidence that treatment may be unnecessary for many who leave off heroin use. As Robins put it, Vietnam was a natural experiment in exposing masses of young men to narcotic drugs. Although her findings did not fuel the fires of fear-mongering fanned by the Nixon administration, they also did not extinguish them.[49]

During this time, the drug treatment infrastructure was undergoing change. The centralized drug treatment system concretized in the federal narcotic farms from the 1930s to the 1960s was in the process of giving way to multimodality treatment programs. Decentralization of drug treatment was fueled by the Narcotic Addict Rehabilitation Act (1966), a federal civil commitment statute that shifted treatment away from federal responsibility and expanded community-based treatment and state oversight. That shift created the conditions for the Nixon administration's paradoxical emphasis on methadone maintenance as a response to growing public concerns about heroin abuse.

Only half a dozen people were enrolled in Dole and Nyswander's studies of methadone maintenance in 1965. But the numbers expanded dramatically in the 1970s when new regulations were crafted by the Drug Enforcement Administration (DEA), SAODAP, and the FDA. Concerns about the illegal diversion of methadone from clinics to the street led the DEA to take a strong stance towards this "new killer"; some feared methadone would become the "illegal narcotic of the century."[50] A syndicated news story from the summer of 1974, "Illegal Methadone Kills Hundreds of Americans," contrasted "good" methadone patients, who got up every morning, went to clinics, daintily drank methadone dissolved in Tang (that quintessentially American product), and headed off to jobs as dentists or secretaries, against "bad" methadone users, who neither worked nor regularly attended the clinic, and purchased the drug on the street to shoot into their arms.[51]

Twice as many deaths from methadone as heroin overdoses were recorded in the northeastern United States in the first quarter of 1974. The NIDA, the FDA, and the DEA set regulatory standards by which methadone maintenance treatment was governed in their own guidances and in an amendment to the Controlled Substances Act of 1970 called the Narcotic Addict Treatment Act of 1974 (Public Law 93-281). The act required that physicians engaged in "maintenance treatment" register with the DEA. It also established the NIDA as an independent institute, which soon found itself beneath a new bureaucratic umbrella in the Department of Health and Human Services called the Alcohol, Drug Abuse, and Mental Health Administration. Among the first ten research issues explored by the new NIDA was *Drugs and Death: The Nonmedical Use of Drugs Related to All Modes of Death.*[52] Overdose was a priority item because the NIDA wanted a baseline against which to measure its effectiveness as the first federal agency to integrate research, treatment, prevention, and education on drugs. *Drugs and Death* was little more than an annotated bibliography that summarized empirical research using case records from coroners and medical examiners, and characterized the variety of classification and reporting systems then in use throughout the country.

Methadone overdose statistics were often couched in ambivalent terms that could be used to justify *or* condemn the method of maintenance therapy. Borne in New York City, one of the first places to implement programs in criminal justice settings, large hospitals, and stand-alone clinics, the idea and practice of methadone maintenance has been contentious. An agonist like heroin or morphine, methadone can form a "poison cocktail" with other respiratory depressants such as alcohol and benzodiazepines: "When used alone, many of these substances are relatively moderate respiratory depressants; however, when combined with methadone, their additive or synergistic effects can be lethal."[53] A less public contention was that stable attributions as to the role of any given drug could not be assigned by toxicology screens at autopsy because "postmortem blood concentrations of methadone do not appear to reliably distinguish between individuals who have died from methadone toxicity and those in whom the presence of methadone is purely coincidental."[54] Such open questions did little to improve the public image of methadone, which developed a stigma all its own.

"Methadone overdose" and "heroin overdose" were unstable signifiers. Trying to differentiate between these drugs' varied contributions to overdose deaths was difficult and often left undetermined even as late as the 1990s.[55] Heroin was so short-acting that only morphine and other metabolites could be detected in postmortem human and animal tissue. More finely tuned techniques were needed for determining precise relationships between alcohol and opiates in deaths categorized either as deaths from "narcotic intoxication" or "narcotic and ethanol intoxication." Postmortem tissue samples could be examined to determine disposition of ethanol and opiates, but there were no techniques for determining ethanol's exact role in these deaths. Deaths from "heroin intoxication" could not be reliably distinguished from deaths from methadone or other opioid "intoxications" in the 1970s; relative contributions of each component of a "poison cocktail" could not be assigned with validity.

Gas-liquid chromatography became viable for use in postmortem examinations around 1970.[56] The method was adopted early by the City and County of San Francisco Medical Examiner-Coroner's Office. In San Francisco there had been only twenty-five deaths from heroin overdose between 1962 and 1965, most (80 percent) in young men (48 percent white, 40 percent black, and 12 percent "other"). By contrast, far more blacks than whites died from overdose in New York City, Detroit, Chicago, and other East Coast or Midwest cities. But since the 1967 Summer of Love, rates of opioid overdose had steadily risen in San Francisco, sometimes rivaling or exceeding those of New York City, which joined other East Coast cities in a heroin drought when the French Connection was disrupted in 1972 and 1973. Between 1970 and 1973, the San Francisco Medical Examiner identified 217 heroin overdose deaths; the gender ratio was the same, but 61 percent of young men dying were white, 35 percent black, and only 4 percent "other." Nearly half tested positive for ethanol.

Surprisingly, the blood morphine concentrations of heroin users whose deaths were categorized as "heroin overdoses" showed no difference from heroin users who died from unrelated causes.[57] According to the authors, "The results of the blood level study indicate that morphine blood levels per se are meaningless in attempting to assign a cause of death in a medical examiner's case, since morphine levels found in narcotics users dying of causes other than overdose averaged slightly higher than those of the overdose victims."[58] Although positive identification of morphine in blood indicated that opiates had been used within hours of death, there was no way to ascertain accurately what had been used and when it had been used: "Attempts at relating the actual concentrations to severity of narcotic effect have usually proven fruitless."[59] Basic physiological tests on tissue and bodily fluids—whether of blood, urine, or bile—were unyielding of confirmation.

Overdose death statistics were regarded as imprecise even by those producing them. A 1975 NIDA conference on state-of-the-art epidemiology captured epistemic problems involved in producing this data. University of California at Irvine researcher Louis A. Gottschalk argued for a uniform reporting system for "psychoactive drug-involved deaths," finding current systems in twenty US cities "rather haphazard and unsystematic"[60] after surveying medical examiners, coroners, and the laboratories on which they relied for toxicology testing:

We cannot aspire to an accurate epidemiological understanding of drug abuse patterns until we begin to acquire valid criteria for the human toxicological effects of these drugs and the human body concentrations of the drugs that are absolutely necessary to produce a fatal outcome. The assumption is unwarranted that we know, with certainty, what drug levels are fatal or, in combination with other drugs or pre-existing medical conditions, can lead to death.[61]

Concluding that "social scientists cannot safely assume that all biochemical measurements are automatically reliable and accurate,"[62] the team nevertheless produced detailed statistics and distribution by race and sex for eight cities. Skeptical of the value of information so produced, Gottschalk concluded that "epidemiological data, using the expert opinion of forensic pathologists and medical diagnoses of the cause of death, are not equally accurate."[63] In all but Philadelphia and Washington, DC, white overdose deaths far outstripped those of African Americans (although if New York City had been included, available data indicates that deaths in the black community would have similarly exceeded those of whites). Even these discrepancies convinced no one that local overdose statistics were far more accurate than those produced at a national scale.

Overdose deaths were "indicators" that obscured the racialized geography of public health reporting.[64] Hepatitis B reports to the US Public Health Service were not random as they came only from urban areas where infectious disease surveillance was vigilant for particular conditions.[65] In 1972 the Office of Science and Technology asked Lee Minichiello of the Institute for Data Analysis (IDA) to develop measures for the NIDA and the DEA using data from "ghetto users in Eastern cities."[66] The IDA also sampled "both the suburban and white young person," and in 1972 recommended NIDA implement what became the Drug Abuse Warning Network (DAWN) and a federal treatment database. The IDA used hepatitis B reports, DEA seizures, and "drug buys" to track the geography of heroin availability. The DEA kept precise chemical data on "source signatures" of drug seizures; so-called Mexican brown heroin had unique chemical characteristics by contrast to the white powders from Southeast Asia trafficked through the French Connection.[67] Differing types of heroin and routes of administration had different health consequences. Largely available in Chicago and the West Coast, Mexican brown was administered differently than the finer powders found in East Coast cities. Thus there had been an eight-fold increase in hepatitis between 1966 and 1972, and then a marked decline in East Coast cities when the French Connection was interrupted. What realities, then, did an increase or decrease in hepatitis or overdose deaths conceal or reveal? The 1970s turned out to mark a period of intensive investment in national infrastructure for disease surveillance and drug control—but the knowledge problems of those involved in creating such systems evaporated as their systems congealed and became taken for granted.

AN EARLY DAWN IN OVERDOSE EPIDEMIOLOGY

The fledgling DAWN was a national database of emergency room visits. Did these reflect "real" shifts in the drug-using population? Some interpreted the data as indicating that "underlying patterns of drug abuse show[ed] a real increase in the number of white male and female users in the United States since 1970."[68] But these were "probable" patterns taken to reflect real national trends, rather than being seen as probabilistic pieces of a puzzle that did not necessarily fit together. Premature and disjunct closure was common: researchers discussed the open and malleable aspects of their information systems with each other, but then blithely made claims based on the outputs of those very systems. The effort reinforced the sense that a national drug epidemic was unfolding.

The epidemiologist's job was "approximating reality," according to Mark H. Greene, who had worked with DuPont particularly on the need to control methadone by preventing diversion in Washington, DC. The usefulness of epidemiological approaches for producing and interpreting indicator data, he noted, did not "enable[] us to describe reality with absolute perfection."[69] Strongly arguing that local data was the only way trends could be discerned with real validity, he disagreed with DuPont over the validity of extrapolation from local to national levels. "There are enormous hazards involved in trying to make broad generalizations about trends at the national level based on the kinds of data sources that we have available right now."[70] His example was the overgeneralization of a regional phenomenon that was highly relevant for overdose deaths: the disruption of East Coast heroin supply chains by the disassembly of the French Connection by drug law enforcement and diplomacy. While data on drug availability could give communities a handle on the specific risks they faced, Greene was reluctant to rely on ER data because stigma might lead to underestimates of opioid abuse. In Pensacola, Florida, where opioid abuse was documented "at a fairly extensive level," hospital workers claimed to see no problems.[71] ER staff there claimed they had never seen an overdose and had never used naloxone—yet data indicated that hundreds of people were in treatment for heroin use, and there had been hundreds of deaths from overdose in that area. The disconnect was the result of drug users "go[ing] to any lengths they can to avoid showing up" at the hospital, and giving victims of overdose "milk and other good things to cure them at home—so they end up in the coroner's office."[72]

To perceive trends more accurately, Greene created an early warning system that relied on multiple indicators. It combined data from drug treatment programs, emergency rooms, coroner's reports of overdose deaths, hepatitis reports, and information from law enforcement. Yet "each of these sources of information has very serious limitations.... You are on tentative ground all the way."[73] Historical data on cohorts by year of first use was "cleaner" because the "first time someone uses heroin is a fairly memorable occasion," and could be used to analyze what was actually happening at the micro level. Greene favored "slicing" his data to trace the spread of particular practices through specific populations; he remained focused on drug-using behavior at the community level. Overdose was simply one indicator to Greene, whose patience for claims about increasing methadone overdose deaths issuing from the New York Office of the Medical Examiner was thin because they lacked technology capable of accurate verification, performed insufficient numbers of autopsies, and designated them on "presumptive and circumstantial evidence."[74]

These deficiencies were widespread and contributed to uncertainty about all overdose death certifications; this was a contentious field for epidemiologists and medical examiners. They disagreed over what counted as data and which indicators were valid, and the two groups waffled over whether it mattered how opioids killed and how the confounding co-occurrence of alcohol and other drugs should be handled. It was perhaps unsurprising that preventive public health responses to drug overdose were slow in coming when so many of the basics were up for grabs. It took two decades longer before anyone thought about asking drug users themselves how many overdoses they had witnessed or had.

WITNESSING THE RAISING OF THE DEAD

Naloxone was becoming standard equipment in emergency medicine only as the 1970s gave way to the 1980s. Anecdotally, Dr. Sidney Schnoll, then a medical resident, described setting up a drug testing facility at the outdoor Philadelphia Folk Festival in the mid-1970s.

It was right around the time that naloxone became available, not experimental but actually out. I brought some to the Folk Festival. They brought in this young woman who had overdosed on heroin. She was blue; she was dead. We pumped her full of naloxone, and she woke up. I mean, it was amazing—and a lot of the staff around had never seen that. Anyway, I didn't want to keep her there so I sent her out to a local hospital and I sent Narcan [naloxone] with her because I didn't even know if they had it at the hospital. About two or three years later I was asked to be on some television show in the Philadelphia area, talking about drug abuse issues. After I was interviewed this young woman comes up to me, and she said, "Dr. Schnoll, I'm sure you don't know me, but remember the person who overdosed on heroin at the Philadelphia Folk Festival? You saved my life, and I never had the opportunity to thank you. I want to thank you now."[75]

This anecdote highlights the uneven presence of naloxone in a city with longtime experience with a visible heroin-using population. Just because a drug is approved for a specific indication does not mean that it gets where it needs to go.

Efforts to make naloxone available were socially disorganized and restricted to organized medicine until the 1990s, when the harm reduction social movement discussed in the next chapters gradually got underway. Unraveling naloxone's role within various imaginaries of medical examination, emergency response, addiction therapeutics, and relapse prevention, this chapter leaves us with the question: Why did it take tens of thousands of opioid-related overdose deaths annually before a movement arose to put this technology into the hands of those who needed it most? Why didn't the steady trickle of overdose deaths that began in several US cities in the 1960s—and the attention to overdose paid by the early NIDA—seem like a clamor for resuscitation after naloxone's FDA approval for overdose reversal? If the opiates by then in regular circulation—heroin, morphine, and methadone—had been taken as technologies that generated risks and harms from which users should be shielded, the story would have been different. Instead, drug users were thrown to the "just deserts" of their own devices to deal with these intended and unintended consequences. Urban emergency services grappled with overdose, evolving local responses that only gradually phased naloxone into emergency rooms and prehospital care.

6

ADOPTING HARM REDUCTION:
Early Democratizations of Naloxone

Overdose was not on most political agendas in the early 2000s.[1] Where, when, and for whom did overdose deaths become something that public policy and ordinary people could aim to prevent? This chapter looks at several protohistories from the 1990s that developed in partial conjunction with HIV/AIDS movements, the push for needle exchange, and the approach to homelessness known as "housing first." Bundled together, these efforts reshaped the social organization of activism, advocacy, research, and clinical practice, shifting these collective responses toward meeting basic needs through harm reduction.

Early adopters of overdose prevention navigated expertise and authority within an embodied health social movement.[2] They helped create small-scale "gift" economies among drug users, harm reduction activists, and treatment providers, setting the stage for regular distribution channels that were eventually institutionalized. All this occurred well before naloxone became an object of research and action in an evidence-based era. Some of these protohistories are linked to organized efforts to ensure naloxone's availability for overdose prevention; others fell by the wayside or went more circuitous routes. By the early twenty-first century, knowing how to prevent overdose death with naloxone was becoming essential knowledge in some places. Who first cared about the affective and political stakes involved in distributing naloxone as a technology of harm reduction?

The story told in previous chapters showed that before there was a social movement to reduce drug-related harm or an "undone science" of opioid overdose, it was left largely to forensic examiners and emergency responders. When early adopters of harm reduction turned their attention to overdose, they found little that could be called science. Slowly, the patterned nature of overdose began to be revealed, leading to a more nuanced understanding and response than was possible when its folk status was that of an abrupt, singular, lethal event. Without the contributions from drug users and their harm reduction advocates, overdose possibly would have remained consigned to the dustbin of undone science. What were the material conditions and social relations that seeded change?

A PROTOHISTORY OF HARM REDUCTION: "DONE BUT NOT MENTIONED" IN ITALY

Activist-oriented doctors and medical students began enabling street outreach groups to distribute naloxone in 1991. Their activities spurred the Italian Ministry of Health to declare naloxone an over-the-counter drug in 1996, which opened the door to pharmacy-based distribution for the first time anywhere in the world.[3] Yet the Italian government issued no national policy, nor did it invest in the scantily financed regional efforts. Detailed only in retrospect by the nongovernmental organization Forum Droghe,[4] which summarized the Italian experience in a 2016 report, this unusual endeavor was largely delinked from the international harm reduction movement. Tantalizing references peppered early English-language papers encouraging naloxone distribution,[5] but the Italian health care workers who were the "first observers and experts of this intervention" were reportedly too busy to monitor or evaluate their own efforts. Reading their accounts with hindsight, one can see that frontline workers often were summoned abruptly to overdose events, and therefore worked under emergency conditions.

Although Forum Droghe reported harm reduction "done but not mentioned" in Italy, advocates were passionate about using naloxone to prevent overdose deaths by keeping on hand the low-cost, life-saving medication that naloxone was in this context. Because the relevant national ministries were almost completely passive, the effort worked through individuals who introduced naloxone during a wider process to bring about low-threshold services called SerDs (Servizio Dipendenze/"public addiction services") through increased collaboration between public and private sectors. The decentralized efforts brought about increased consensus with respect to drug treatment—including opiate substitution treatment and other practices that come under the harm reduction approach—which had until then been the object of political and ideological resistance.[6]

Teams involving peer leaders, health care professionals, and social workers undertook direct service in absence of emergency medical services:

In [] that time heroin was used a lot and therefore they asked for Narcan ... the young guys would ask for Narcan (....) the people in the needle exchange unit were mixed, there was a health worker, a nurse, a social worker, a psychologist, and then there were one or two outreach workers. We dealt with the overdoses on the number ten platform [in] an abandoned train carriage and it was there that we went to treat the overdoses ... an ambulance would never arrive ... it was those same guys ... or they would call out "run! run! there's an overdose!!" or else one of them was already trained to treat an overdose.[7]

Less a stylized "model" or "protocol" than a form of socially situated knowledge that arose in response to specific events, the so-called Italian model for naloxone distribution relied on the principle that

no matter the context and the conditions where psychoactive drugs are used, on the street or not, people have the ability to control their own lives and are open to new learning experiences ... notwithstanding the processes of marginalization and stigmatization ... work that includes accompanying people, installing a relationship with weak ties and an attitude of acknowledging the other person represents a constituent basis for every social and socio-sanitary professional capable of giving value to these processes.[8]

Beyond saving lives at immediate risk, Forum Droghe informants argued that a convergence between naloxone availability and interruptions of heroin supply led to reduced overdose deaths. Ensuring naloxone availability across social sectors helped put problematic opioid use into remission, reduced overall consumption, and infused safer practices into drug-using cultures. Naloxone alone did not achieve this. Investing a technology with too much agency would be a form of technological determinism. Yet naloxone is a nonhuman actant that "formats"[9] both those who give and receive it. As persons implicated in moral practices, those who give naloxone insert a twist into a story that shifts the plot from death to prevention.

Local and individual successes did not translate into official recognition of Italy's overdose problem. The decentralized distribution model suffered institutional neglect and tepid government and academic response. Naloxone went unmentioned in major Italian studies of overdose deaths from the 1980s to the early 2000s.[10] "Further, there is a certain weakness in the otherwise active movement of advocacy for HR [harm reduction], which in Italy has a strong alliance between health workers, associations, and PWUDs [people who use drugs], but that still today have little impact on orientating policy decisions."[11] The actual shape and extent of Italian efforts remained almost invisible both within and beyond Italy, an absent presence indicating how crucial it has become for social movements to amplify grassroots activities by producing knowledge, going "on record," and documenting their practices, models, and protocols—especially in peer-reviewed venues. Italian efforts exemplified the problems of undone science in an era when lay experts shape "a new regime of relations between science and society,"[12] advocacy groups advance "mutual learning," and PWUDs become "both subjects and objects of research."[13]

Writing about very different efforts of parents and patients to become involved in producing scientific research and "intermediary discourse" on muscular dystrophy, Rabeharisoa and Callon argued that "reflexive" forms of social organization are enacted by a social movement that "constantly questions the procedures and tools both enabling it to learn, i.e. to accumulate competence and knowledge produced collectively, and to evaluate this competence and knowledge so as to decide on future actions to undertake."[14] Yet movement actors who undertake the work of evaluation and documentation might feel themselves engaged in acts of reification more than acts of creative reflexivity. Stopping to collect, archive, interpret, structure research agendas, attract funding, cooperate with "real" researchers, and think about how to frame collective action as research is draining and requires institutional resources. It is bound to feel less like saving lives and more like gathering evidence or even performing for a distant bureaucracy. Such reflexivity might even become "confused with the action itself."[15]

The contentious politics of collective action may be seen "not only as struggles between institutionalized forces of power and knowledge, but— more important—as agonistic fields where demarcation between right and wrong, legitimate or illegitimate, are constituted or transgressed through the mundane but ideologically shaped practices of both science and poli- tics."[16] Evidence has become a weapon in public controversy. Producing it is a high-stakes game that moves away from direct action and the affec- tive politics of connection crucial to social movements. Making harm reduction compatible with the politics of evidence-based medicine leads quickly to questions like, "Whose evidence counts? Whose research mat- ters?" Feminist philosopher Sandra Harding asks, "Whose science, whose knowledge?"[17] Harm reduction is full of examples like the Italian nalox- one efforts: small, local, and reliant on lay expertise without thinking it necessary, desirable, or worth the transaction costs to translate activities into "evidence," "protocols," or "standards" demanded by policymakers.

ANOTHER PROTOHISTORY: RURAL NEW MEXICO

The state of New Mexico would become the first US state to sanction and support peer distribution of naloxone. This 2001 official action built upon a history of rural heroin usage that had become entrenched by the 1990s.[18] Before public health responded to that ongoing history by insti- tuting harm reduction, there was an example of an isolated version of naloxone distribution self-organized by Philip Fiuty, now a harm reduc- tion outreach worker at the Santa Fe Mountain Center and once a drug user living in a remote mountain community.[19] As the 1990s unfolded, members of Fiuty's community obtained naloxone to revive one another, a practice that emerged organically in this rural, isolated, hard-drug-using setting: "I found myself in what I didn't realize at the time was a really unique situation. For example, one of the folks from whom I bought her- oin would not sell me heroin if I had been drinking. I just assumed that those things were common knowledge or common practice. When I got involved in this work professionally, I found out that was really not the case."[20] This indigenous harm reduction practice of keeping alcohol and heroin use separate may be seen as just people taking care of each other.

Few sources of information about hepatitis, safe injection techniques, or wound care were available in rural New Mexico, so Fiuty relied on old medical textbooks. He found a Navy Seal medical manual from the 1950s that explained "everything about subcutaneous and intravenous injection, where was good to shoot up and where was not."[21] But he learned about naloxone from those who had it—the paramedics who regularly visited the remote reaches of the Ensenada Valley. Paramedics also shared knowledge and naloxone (see figure 6.1) in other social contexts.

Trying to purchase naloxone from domestic drugstores and those across the border in Mexico, Fiuty and friends were turned away by puzzled pharmacists. Approaching their problem as drug users might approach obtaining clean needles or drug supply, Fiuty recalled:

After our friend got diagnosed with AIDS, these things became a reality out here in the mountains and it was very important for us to figure out how to get clean needles into our community. We ... [got] ... our names added to the approved list of diabetes patients at a bunch of pharmacies in Santa Fe. When we were doing our dope hustle, we would collect extra money from people and go into one of these pharmacies where our name was on the list and buy syringes.... We liked the idea of doing exchange, and we definitely got it together to keep clean needles in the neighborhood. It was getting rid of them that was the problem. I remember at one point, we had a collective 55-gallon trash can that was just full of used needles. We didn't know what to do with it. It was just sitting there. I think it eventually just went to the dump. That would be my guess. We didn't think we were part of anything. We were just trying to respond to our situation in our neighborhood.[22]

EMTs were among the very few professionals who delivered clinical care in this remote reach of the US health care system, well-known for uneven patches of care in rural settings:

When the EMTs would show up for an OD, they had this drug that they would give people. We knew they used it in the emergency room. So word was out there. We thought, wouldn't it be great if we could get some of that out here, so we wouldn't have to do all of the guerilla stuff when people OD'd out here? We are thirty to forty minutes from the nearest hospital, and at the time a call to 911 meant about a half an hour wait. We had no idea what it was or how it worked. We just knew that it existed.[23]

Turning to their own resources, they realized that their community was well known for marijuana cultivation:

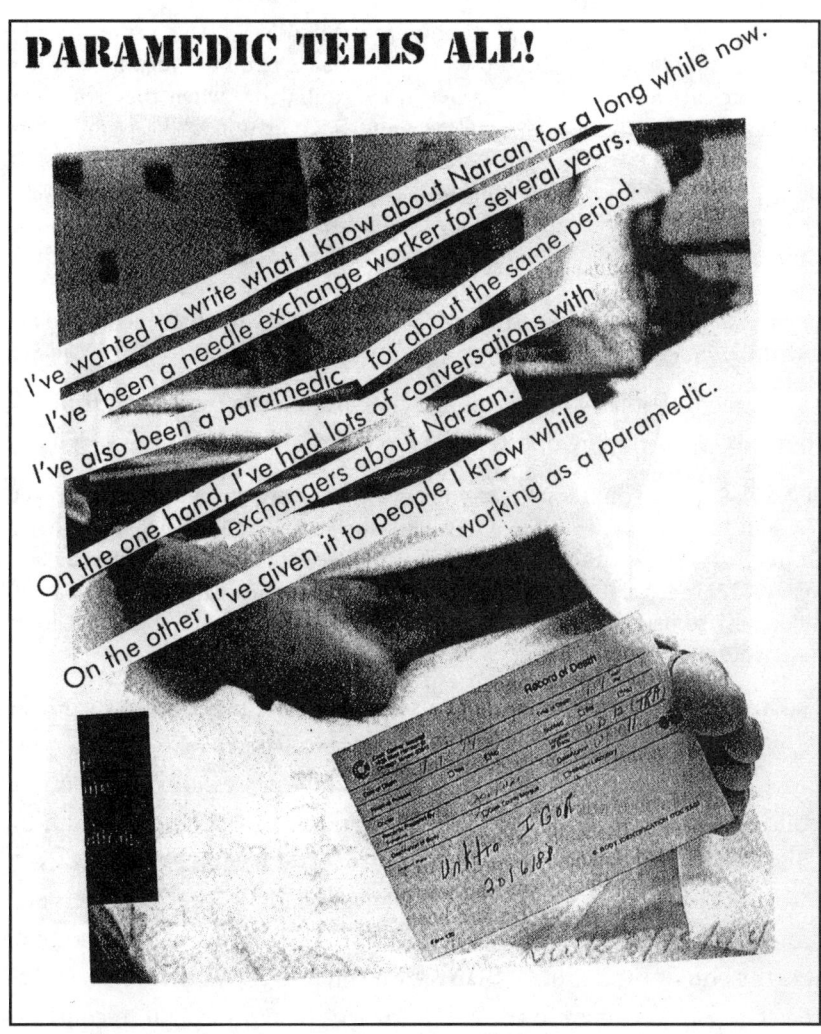

6.1 "Paramedic tells all!" Collage appeared in *junkphood: The Book of Death*, vol. 1 (Santa Cruz, CA: Santa Cruz Needle Exchange, 1997), 12.
Used with permission of Heather Edney and Brooke Lober.

Back then, it was pretty much a contemporary Wild West.... There were a couple of enterprising paramedics that used to come out here on a regular basis when they were off work. They'd come out in the ambulance when they got done with work, and they were always offering to trade us things that were within their realm.... They wanted pot. Usually because we couldn't think of anything better, we'd trade for a tank of nitrous oxide. They'd come back later and we'd give 'em the empty tank and we'd give 'em some more pot. Well, the next time they came out, we asked them if they had Narcan on the ambulance and they said, "Oh, yeah, you guys want some of that?" And they threw us three or four of the small cases of Narcan. They told us basically how to use it and we took it and disseminated it amongst our group of people.[24]

Fiuty's recollection was that this early 1990s practice resulted in three or four "saves," including one person who OD'd on the street:

Several of us were on the street, so one person ran to somebody's house to get the Narcan, came back with it, and we hit him with it and he woke up right in front of us. That was the first time we used it. He actually just passed away last year [2012] in his eighties. Old age finally got him, which was amazing to all of us because so many of our folks have died basically from alcohol and hepatitis and overdose.[25]

This prosocial account of drug users' collective response to overdose (see figure 6.2) was not Fiuty's only experience with naloxone. When he drove out of New Mexico with all his worldly possessions in his car, a policeman found an expired box of Narcan in his trunk, and asked him what it was: "I said, 'You know, it's Narcan.' And he said, 'What's it used for?' And I said, 'It's the antidote for an opiate overdose.' And he said, 'You're not supposed to have this, are you?' They confiscated it and that was the end of that. It didn't surface on the charge. All that surfaced on the charge was the heroin because it carried the most weight in court."[26] Once reaching his hometown, Fiuty learned more about harm reduction and legal needle exchange. He was relieved to encounter nonjudgmental or low-threshold services, finding it "amazing that people could actually legally do what we were sneaking around and working so hard at doing for ourselves."[27]

Both needle exchange and naloxone are often spoken of with the intertwined urgency of rescue and resuscitation, a do-it-yourself spirit, an agentic pragmatism attractive to new converts. Protagonists' harm reduction stories are emotionally moving accounts in which ignorance is

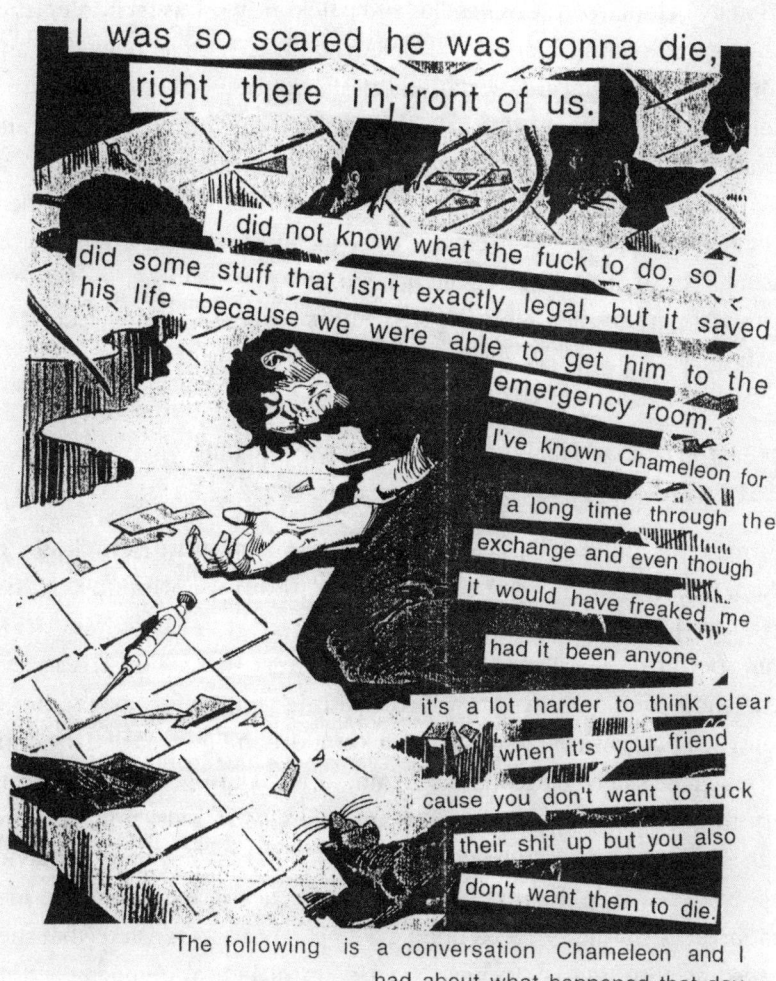

I was so scared he was gonna die, right there in front of us.

I did not know what the fuck to do, so I did some stuff that isn't exactly legal, but it saved his life because we were able to get him to the emergency room. I've known Chameleon for a long time through the exchange and even though it would have freaked me had it been anyone, it's a lot harder to think clear when it's your friend cause you don't want to fuck their shit up but you also don't want them to die.

The following is a conversation Chameleon and I had about what happened that day.

6.2 "I was so scared he was gonna die, right there in front of us." Interview with Chameleon by Heather Edney. Collage appeared in *junkphood: The Book of Death*, vol. 1 (Santa Cruz, CA: Santa Cruz Needle Exchange, 1997), 17.
Used with permission of Heather Edney and Brooke Lober.

displaced by knowledge, shame and stigma by purpose and pride, and isolation by belonging. Clean needles and naloxone work as technologies of solidarity, imparting agency to people who otherwise feel powerless and vulnerable in the face of social, political, and economic forces beyond their control. Naloxone acted as a material signifier of moral worth, and it was taken up differently in different places. However, as with the Italian and New Mexico examples, homegrown experiments muster no evidentiary muscle and have little chance of being taken seriously enough to be included in historical narratives. This forgetting then leads to distorted historical narratives in which it seemed that naloxone for overdose prevention evolved entirely within the urban Anglophone world.

"MENTIONED BUT NOT DONE": REVOLUTION, RESPECTABILITY, AND HARM REDUCTION

Overdose mortality rates vary across continents, countries, classes of drugs, and drug users because there is significant variation in drug use practices, public policies, legal codes, and illicit drug economies. However, specific conditions must be detached from the circumstances that generate them if evidence for a particular intervention is to be produced. Doing science in an evidence-based era produces more science to be done on an increasingly narrower set of questions. During the quarter century that naloxone has been distributed as a harm reduction measure, both clinical practice and public health policy have become "evidence based." Social movements pressuring public health embraced the idea that practice should be based upon evidence—but also resisted what they viewed as authoritarian guidelines in favor of radical democratization. The dialectic between revolution and respectability often manifested in profoundly personal ways. On one hand, activists attracted public health authorities who could institutionalize harm reduction practices directed toward reducing social stigma and health disparities. On the other hand, social forces undermining the cultural authority of medicine had been underway since the 1960s among patients' and consumer rights groups, culminating with the HIV/AIDS movements of the 1980s and 1990s. Such movements converged with harm reduction in terms of ideas, tactics, and people, but sometimes diverged in terms of normative visions of how public health policy, drug policy, and human rights should look and feel to drug users.

Hierarchies of evidence were naturalized in medicine and, increasingly, in public policy. The "gold standard"—randomized clinical trials (RCTs)—occupied the pinnacle.[28] Shaking naloxone loose from medicalized domains required new forms of activism and advocacy in expert communities consisting of professionals, drug users (also called "service users" and "experts by experience" in UK contexts), and movement actors who either themselves produced evidence to undergird their claims or courted researchers who could be convinced to do so. Ironically, taking naloxone out of medical confines required research designed to produce evidence that the intervention worked. The paradox of "evidence-based activism" is that evidence does not speak for itself but must be framed by and attached to interpretations and symbols mobilized in affective domains.[29] For naloxone to accomplish the cultural work it now does, it had to be wrenched out of its revolutionary social movement frame. Clarifying how these complex social relations and political commitments shaped the social organization of evidentiary science, medicine, policy, and activism is one of the overall goals of this book.

One of the earliest mentions of naloxone as a way of doing harm reduction occurred in an address titled "From Faith to Science" delivered by British addiction psychiatrist John Strang at the Third International Harm Reduction Conference in Melbourne, Australia, in March 1992. He meant the brief mention to be a "throwaway line," merely one low-hanging fruit that the harm reduction movement might grasp. At the time, Strang worked closely with Lifeline, a harm reduction drug outreach and prevention charity.[30] Framing himself as an agent provocateur, Strang chided the movement's vanguard for concentrating on oppositional politics, foolishly skipping "easy gains" because they were not revolutionary enough.

Casting his lot with quieter revolutions that "bring about fundamental alterations in key aspects of our lives, without necessarily evoking impassioned opposition from sections of the population,"[31] Strang urged distributing naloxone on grounds that "its potential for abuse is nil, the risks are probably minimal, and considerable benefit may accrue if drug users could give emergency doses of antagonists to fellow injectors who inadvertently overdose."[32] He suggested that the movement make itself credible in the eyes of potential allies in policy and clinical practice.[33] Soon thereafter Strang argued for establishing a "sound scientific basis" for naloxone in a 1992 editorial on the prestigious pages of the *British Medical Journal*; this would, he hoped, usher in a new phase in which the "critical researcher is not outlawed as a heretic but welcomed as a scientist."[34] At stake was a politics of respectability in which naloxone became a means to establish scientific credibility for harm reduction.[35] Over the next years, Strang and colleagues published a series of letters, articles, editorials, and talks about naloxone in which they built upon their own arguments and those of each other. For instance, the first paragraph of a 1993 book chapter by John Strang quoted his own earlier address:

Faith must be replaced by science, so it was argued in the opening address at the First International Conference on the Reduction of Drug-Related Harm in 1990: The time has come for the debate around harm reduction to move from something akin to religious dogma to a sounder scientific basis. The religious passion may have been important at one stage but a move to objective and scientific study of the policy and practice dimensions of harm reduction is now required.[36]

Arguing that harm reduction had the potential to be incorporated into personal practice and a "dominant perspective on drug use," Strang placed himself in the sometimes awkward role of a pragmatist using a harm reduction yardstick to determine which measures to support. He said,

If your heart is just not into such obvious but uncontroversial harm reduction measures, then why not give some thought to the idea of distribution of supplies of naloxone, the opiate antagonist, to opiate users who may at some later date be able to give a life-saving injection of the drug to a fellow drug user who has inadvertently overdosed.[37]

In portraying naloxone as a clinically obvious tool of harm reduction, Strang was aware that this seemingly mundane technology had revolutionary potential.[38] Orthodox clinicians were troubled by the suggestion; he recalled his feelings at the time:

Many of us are running treatment programs where half to a third of the people are going to die from hep B virus and we can do something about it. Maybe we don't think about it as harm reduction, but what an organizational change that would make. I went on to say that if that's not radical enough, what about this—here's an idea, we have naloxone, the antidote for heroin overdoses, why don't we give people a supply of naloxone so that when their friend overdoses, they can save the friend's life? That was early 1992.... As far as I'm aware, that was the first time the idea had been seriously mooted. I was looking for a throw-away example of what the existing orthodox system could do in the name of an extension of treatment.... This was my equivalent of my pushing the envelope and saying that it doesn't have to be revolutionary, you don't have to destroy the system or create an alternative system.[39]

Naloxone had not been taken up in the United Kingdom or anywhere but Italy by that time. But the potential seemed obvious to Strang, who compared it to other first-aid devices such as EpiPens or snakebite kits.[40]

Four years later, Strang and research associate Michael Farrell (see figure 6.3) coauthored another *British Medical Journal* editorial with Shane Darke, Wayne Hall, and Robert Ali, Australian colleagues who had visited London in 1995. Titled "Heroin Overdose: The Case for Take-Home Naloxone"[41] had a whiff of a manifesto. Subtitled "Home-Based Naloxone Could Save Lives,"[42] the BBC covered the story in 1997, but the message went largely unheard. However, the Australian National Drug and Alcohol Research Centre at the University of New South Wales in Sydney hosted an international symposium on opioid overdose in 1997.[43] The organizers fastened on variability in mortality rates among drug injectors according to place:

Heroin injectors who regularly consume large amounts of different drugs, face a risk of death which may be as much as twenty or even thirty times higher than non-drug users in the same age range (EMCDDA, 1996). In Glasgow, Scotland, drug injectors are twenty-two times more likely to die than their peers and mortality rates for drug injectors have been increasing since the early 1980s (Frischer et al., 1997). Mortality rates among opiate addicts in Catalonia, Spain, increased from 13.8 to 34.8 per 100,000 between 1985 and 1991 (Orti et al., 1996). In Amsterdam, the Netherlands, a mortality rate of 32.3 was reported for drug injectors recruited through low-threshold methadone clinics and a sexually transmitted disease clinic (van Haastrecht et al., 1996). In Milan, Italy, an overall mortality rate of 25.2 per 100,000 was reported for injectors attending treatment centres. These rates remained under 16 from 1981 to 1986 then increased rapidly to 63.8 in the first half of 1991, when the follow-up period was closed (Galli and Mussicco, 1994).[44]

That the sociology of overdose might be undone is not surprising, but the underdeveloped state of expert knowledge extended to the physiological

6.3   John Strang and Michael Farrell in 1996.
Used with permission of John Strang and Michael Farrell.

mechanisms involved in overdose—and in breathing itself.[45] Pharmacologist Jason M. White reported on the mysteries of normal respiration.[46] Forensic pathologist Johan Duflou gave a compelling demonstration of inconsistent death determinations from a survey of ten out of the thirty forensic pathologists then practicing in Australia. He asked each respondent to categorize a death based on the following scenario: "The deceased was a known intravenous drug user, found dead in a hotel room in Kings Cross with drug-injecting paraphernalia on the bedside table.

There were recent and old needle puncture marks in the antecubital fossae, and apart from chronic hepatitis C virus infection, the autopsy was negative." Each pathologist was asked to identify cause of death for each of three toxicological scenarios—(1) high opioid/no alcohol; (2) low opioid/elevated alcohol; and (3) low opioid/no alcohol. For none of the scenarios did all agree, and every pathologist made at least one finding with which the majority of his peers disagreed. The field lacked consensus, partly because much remained unknown:

Forensic pathologists accept that heroin injection kills acutely, usually as a result of the pharmacological effects of the drug. The exact mechanism of death ... is far from clear, and this has resulted in forensic pathologists using a variety of terms to circumvent the obvious fact that an overdose in the classical sense of the word did not take place. These terms can include, but are not limited to, narcotism and acute narcotism (Helpern, 1966), heroin or opioid intoxication, heroin or opioid toxicity, acute narcotic overdose, and narcotic/heroin/opioid poisoning.[47]

Taxonomic difficulties were not mere semantics; they connected directly to forensic practice. The exact mechanisms by which heroin kills remained unclear in laboratory and morgue, subject to each pathologist's inference or preference. Although this may seem astounding, it follows fairly directly from the fact that there is no bright, solid line between the "therapeutic, recreational, toxic, and fatal concentrations of opioids in blood [which] overlap, and in many cases are the same."[48]

Conference participants agreed on the goal of enrolling drug users in reducing heroin-related mortality by distributing naloxone. A small Sydney program that trained HIV outreach nurses to administer naloxone reported already having reversed eighty-six overdose events.[49] However, momentum on provision of naloxone for peer administration slowed,[50] and further pilots were not destined to be enacted in Australia. Some elements of the international research community appeared poised to undertake naloxone provision under the guise of study. However, such research was legally blocked in Australia, stymied by an unresolvable debate about the necessity of randomized clinical trials and how to conduct them.

The Sydney meeting presaged a major definitional problem that was to haunt the naloxone access and overdose prevention movement. Deaths were attributed to heroin overdose even when other respiratory depressants such as alcohol or benzodiazepines were present, [51] classifications that have social implications well beyond the forensic profession.[52] "Killer heroin" signals *both* vernacular uses of the term "killer," meaning "great" as well as "deadly." Overshadowing other depressants, which were typically present but also typically "disappeared," heroin has often been singled out where "multiple drug toxicity" would be a more accurate term. Such inaccurate death determinations provide a shaky foundation for epidemiological analyses, and heroin overdose invites media portrayals that impede reality-based responses to overdose. Researchers illuminated these knowledge problems early on, but also sometimes conflated what could be done to save lives with what they needed in order to produce evidence that their interventions worked. This remains a perennial problem in knowledge-based social movements—deepening social antagonisms between "doers" and those who study those who do.

UNMENTIONABLE: HARM REDUCTION AND NEEDLE EXCHANGE IN THE UNITED STATES

The pragmatic practices constituting harm reduction were politically "unmentionable" in the United States.[53] Unmentionable actions cannot be standardized. Designated a "dirty word" from the beginning, the concept spawned euphemisms and workarounds (such as using community outreach workers to quietly circulate harm reduction practices within drug-using communities). C. Robert Schuster, National Institute on Drug Abuse (NIDA) director during President Reagan's second term, mused that "Many people were concerned that if we made it safer for people to use drugs, more people would use drugs. If you used the term 'harm minimization' in the federal government at that time and even now to some extent, you would have to wash your mouth out with soap if you didn't get fired for it."[54] Nevertheless, the federal government, including the Food and Drug Administration (FDA), became a flashpoint for social movement response to HIV/AIDS. This took dramatic form beginning in the mid-1980s,[55] but even AIDS activists did not connect the dots on overdose until the mid-1990s or later.

This was partly because monitoring the number of opioid "addicts" in the United States has long been an epidemiological nightmare.[56] As discussed in chapter 5, overdose deaths were used as "indicators" signaling prevalence of heroin use during the 1970s. But by the 1980s there was a much graver problem. The New York State Department of Substance Abuse Services Street Research Unit, an ethnographic surveillance unit composed largely of ex-addicts, reported emaciated heroin users dying of "walking pneumonia" beginning in the late 1970s. Drug deaths spiked in the early 1980s, seeming to indicate a rapid increase in the number of heroin users. Epidemiologist Don DesJarlais related:

We really didn't understand it. We just couldn't see how the number of addicts would be increasing that fast. We had no explanation for why they would be dying at much higher rates. So we talked about it, and we worried about it, and we tried to see what we could do to understand it. But at that time there was really a total disconnect between drug abuse research and infectious disease research. We would get the death certificates and notice pneumonia on there, but we didn't have any infectious disease capability to go out and try to find out why these people were dying. We would get the death certificates when they were already deceased.[57]

Researchers struggled to put the puzzle together, but did not yet have all the pieces.

Surprisingly, the handful of overdose studies from that early period barely mentioned what became cruelly labeled "gay-related immune deficiency," cases of Kaposi's sarcoma, "walking pneumonia," or other sequelae of what would ultimately be identified as HIV/AIDS as early as 1981. In that era, it seemed like only a couple dozen drug users might have died of AIDS. Then a NIDA-funded risk factor study in New York City, the longest-running study of its kind in the world, enabled a capped, gowned, and gloved DesJarlais to interview patients dying of AIDS:

You tried to maintain as much human-to-human contact while you were dressed as if you were going to outer space. I have to say that the people I interviewed were really heroic. They knew they were dying. They were often in severe pain.... You could see them mentally going in and out as the pain would ... take over, and then they would fight their way back to consciousness and talk with me. And that really convinced me that, you know, these people are altruistic. They're taking the last few hours of their life often to try to prevent other people from getting this disease. They're doing it at great personal effort to talk to me, go through this two-hour interview despite the fact that they're in pain and ... many of them had dementia, and they were also on painkilling drugs. But they would fight their way to go through that interview. And I didn't have any one of them turn me down for the interview.[58]

As the 1980s wore on, connections to intravenous drug use and other conditions of so-called hidden populations began to become evident.[59] Building on knowledge of needle-borne hepatitis, DesJarlais realized that drug users who were not yet hospitalized also had the disease. Once HIV antibody tests could be used to confirm infection, it turned out that over half the intravenous drug users (IDUs) in New York City were infected. Unlike other lethal viruses, AIDS killed all who contracted it and was an international phenomenon that forced alliances between infectious disease epidemiologists, public health departments, and those studying substance abuse.[60]

Not until the mid-1990s did overdose begin to take on wider social significance. More definitive diagnostics and, most importantly, life-prolonging antiretrovirals such as AZT (zidovudine) made it possible for people in the global North to live longer with AIDS. Drug users began noticing that they and their friends were living with HIV but dying of hepatitis and overdose. Needle exchanges organized in response to HIV actually exchanged far more than needles. Education, information, health promotion, and prevention were intertwined with tacit messages of self-efficacy, moral worth, and a renewed sense of agency.[61] As generative sites of underground activism, many needle and syringe programs had an explicitly revolutionary aura in the United States. Yet they were also places characterized by exhaustion: operating a regular, ongoing, and well-stocked community program was very different from organizing marches, protests, die-ins, or zaps for which the AIDS movement was most vividly known.

Needle and syringe programs helped slow transmission and prolong lives among IDUs. Politically contested in the charged cultural politics of the 1990s, needle exchange demanded an evidence base; to develop it, researchers had to innovate methods to overcome views that drug users could or would not change their behavior, as well as overcome legal obstacles and lack of federal funding.[62] Some of these obstacles were attitudinal; for example, ACT UP activist Richard Elovich criticized the dehumanization of drug users as "noncompliant populations" implicit in the words of immunologist Anthony Fauci,[63] a label used to justify exclusion from clinical studies:

I just became enraged, and, hopefully in a more measured tone said, "There's no such thing as a noncompliant population. I'm part of that population." ... [T]hat was the first time I came out as an IDU. And the Treatment and Data and Policy committee were the same committee; they hadn't broken up yet into two

different committees. And I heard Tim Sweeney come and talk about the bud-
get, and he was talking about treatment slots. And it, suddenly this really spoke
to me, it spoke to me in a way that maybe the gay stuff didn't.[64]

ACT UP accomplished emblematic identity work; nearly everyone who
has written about it has recognized the ambivalent social bonds documented in a rich oral history archive held at the New York Public Library,
as well as in the signature graphic "Silence = Death."

ACT UP members thrived on putting the disobedience back in "civil
disobedience." Their involvement with needle exchange arose when
individuals intentionally got themselves arrested in states where it was
illegal to possess syringes without a prescription. Elovich recalled that
AIDS activist Jon Parker would come down to New York City from New
Haven, Connecticut:

[He] was like Johnny Appleseed. He'd go to different places, and he'd usually
buy … a whole chicken [a kilo of needles], and open it up, and people in the
SRO would come, and I'd get needles from him. And this happened for me—and
I know I made it happen for lots of other people …—if you see needle exchange
happening, it's no longer an argument. You totally get it when you see it. And
when I saw what Jon was doing, I completely got it. And I think within a week,
… we went to the floor of ACT UP, and described this, and said we had to do it.[65]

Soon thereafter, the ACT UP volunteers who met weekly to assemble
kits containing needles and bleach became the Lower East Side Needle
Exchange Program (LESNEP). Organizer Rod Sorge gained support for
LESNEP from an internationally known proponent of harm reduction,
Ernest Drucker, then director of methadone programs at Montefiore Hospital and the Montefiore Symposium on AIDS at the New York Academy
of Medicine (1986–1992).[66] Many feared ACT UP lacked the infrastructure to support an ongoing exchange. Alternatives were bleak: what other
organization would have been willing to risk violating drug paraphernalia laws in a New York State known for the draconian Rockefeller Laws
that formed the cornerstone of its "war on drug users"?

Needle exchange became ACT UP NYC's longest-running act of civil
disobedience.[67] Harm reduction efforts leveraged the historical presence
of urban street outreach necessary for researching drug use, part of the
city's fabric since the 1950s.[68] An evidence base was assembled; DesJarlais represented harm reduction as a "framework for incorporating science into drug policy."[69] Speaking to the US AIDS Commission, DesJarlais

testified to the effectiveness of the Port Defiance AIDS Project, one of the earliest needle exchanges in the United States; it had been started by Dave Purchase in Tacoma, Washington, in 1988.[70] Purchase organized the North American Syringe Exchange Network (NASEN), a cooperative buyer's club that purchased in bulk and served as a national hub.[71] NASEN extended start-up packages and technical assistance to "young" programs and conducted a national survey of need with Beth Israel Medical Center in New York City.[72] Lacking federal funding until the Obama administration, needle and syringe programs across North America depended on NASEN.

Drug control laws, including paraphernalia laws, posed barriers to harm reduction in the United States. Sterile needles were illegal in many states unless prescribed for legitimate medical use. Skirting these laws often meant operating underground, as did Prevention Point San Francisco, which imaginatively distributed clean needles from a baby stroller. In 1992 the San Francisco City Council declared AIDS a public health emergency and funded Prevention Point.[73] Formal laws were far from the only barriers; harm reduction was a stretch even in ACT UP. Prominent African American community organizations portrayed needle exchange as even more a tool of genocide than methadone.[74] Notoriously, New York City mayor Ed Koch dismissed needle exchange as "an idea whose time has not yet come."[75]

To overcome legal and cultural barriers and mount a long-running action, ACT UP combined forces with the Lower East Side AIDS Strategy Group, a consortium of HIV/AIDS service providers founded by Raquel Algarin.[76] After two years, a group of ACT UP activists known as the Needle Eight challenged New York City in court,[77] culminating in a favorable state ruling that stated clean needles were a medical necessity for curbing AIDS.[78] The case established the legality of needle exchange in the city of New York—and allowed activists to gain perspective and confidence that stood them in good stead during naloxone's later cycle from illegality to legitimacy. Elovich, who went on to build harm reduction programs abroad through Open Society Foundations, "saw something, especially through Rod [Sorge], watching it, which was a great lesson for me that I have been able to use in lots of other countries. Which is that needle exchange needn't be a minimalist thing. It's right as a minimalist thing. Meaning, just giving the needle out is enough.... But, you can build all these things around it."[79] The framing of needle and syringe exchange as

HIV prevention made needles tools that were connected to life, not death. AIDS activists had lived through many deaths, making loud and clear the full range of emotions responsive to loss, rage, and grief.[80] Railing against the sentimentalism and sensationalism with which AIDS was treated in mainstream media, Sorge reflected on his own approaching death: "My overdose from heroin will not be tragic, shocking, sad, pathetic, or any of the other words *The New York Times* uses when it reports on such unseemly matters. It will be the act of an empowered, purposeful human being committed to gaining ultimate control over his life and death"[81] (see figure 6.4).

Layered into needle exchange, naloxone was portrayed by the movement as a tool to directly challenge or "own" death—even their own deaths (see figures 6.4 and 6.5). Asked why it took so long for people to see naloxone as a basic route to overdose prevention, Daniel Wolfe, director of International Harm Reduction Development in the Open Society Foundations (OSF) Public Health Program, answered:

The answer to that question lies at least in part with the question, "Why does advocacy for illicit opioid users take so long to do anything at all?" HIV was obviously such a key force for multiple reasons. One is that it brings to the discussion people who have some experience engaging regulatory agencies and/or clinical trials ethics and findings critically, with an eye toward their own needs. One is that there was the whole idea of on- and off-label use, which was something again that had incredible importance for regulators and the FDA's self-image, but it was not something patients were always particularly focused on until it became important from AIDS onward. And lately as the story gains momentum, there has been a reinvigoration with prescription opioids and the "innocent" or "nice" people dying of overdose, as well as morally suspect people [who] are dying, and the blurring of those lines in ways that are politically important.[82]

An active participant in the movement, Wolfe had worked at Gay Men's Health Crisis before moving to OSF, and he had witnessed the transition from focusing on HIV alone to a more inclusive approach. For it became "clear that overdose was a much more feared and pressing health concern than HIV.... Overdose was a much more immediate and pressing health threat and in some cases was killing more people than HIV and at any rate was not a long latency condition. It was something that people had experience with—they had either had or had witnessed an overdose event."[83]

LESNEP did not work directly on overdose until obtaining a municipal grant thanks to pressure from the Injection Drug Users Health Alliance in 2004, when physician Sharon Stancliff was hired to work on naloxone.[84]

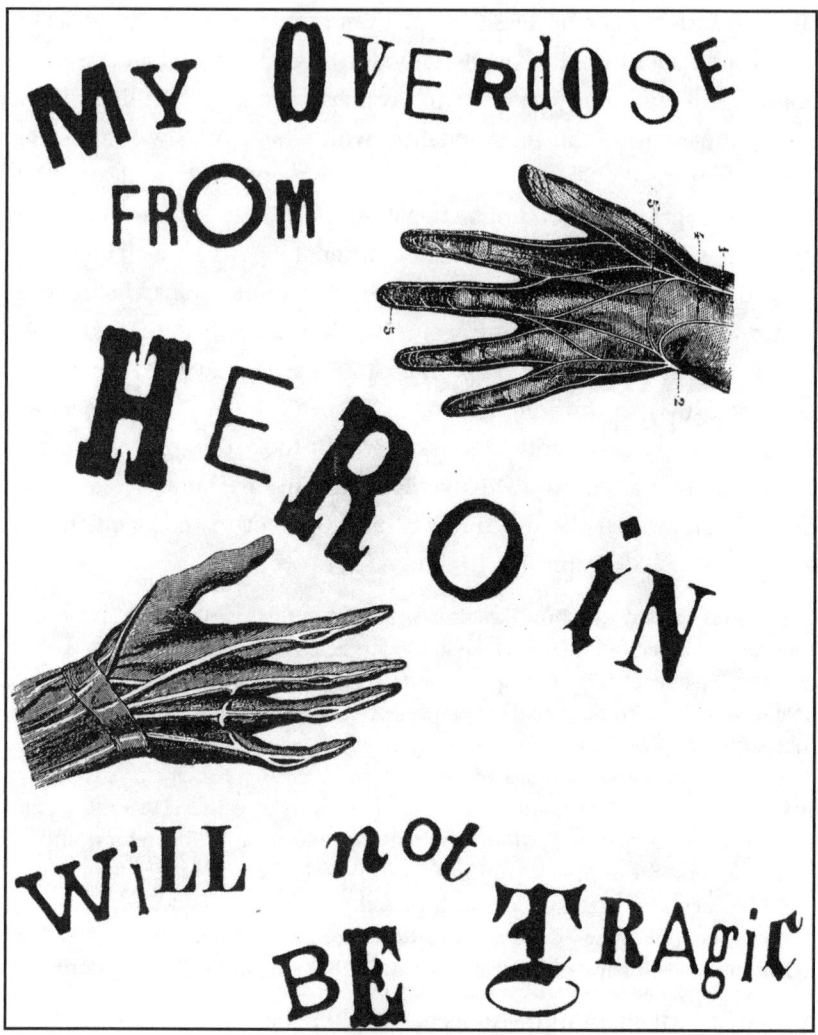

Figure 6.4 "My overdose from heroin will not be tragic." Sorge, *junkphood: The Book of Death*, vol. 1 (Santa Cruz, CA: Santa Cruz Needle Exchange, 1997), 8–9.
Used with permission of Heather Edney and Brooke Lober.

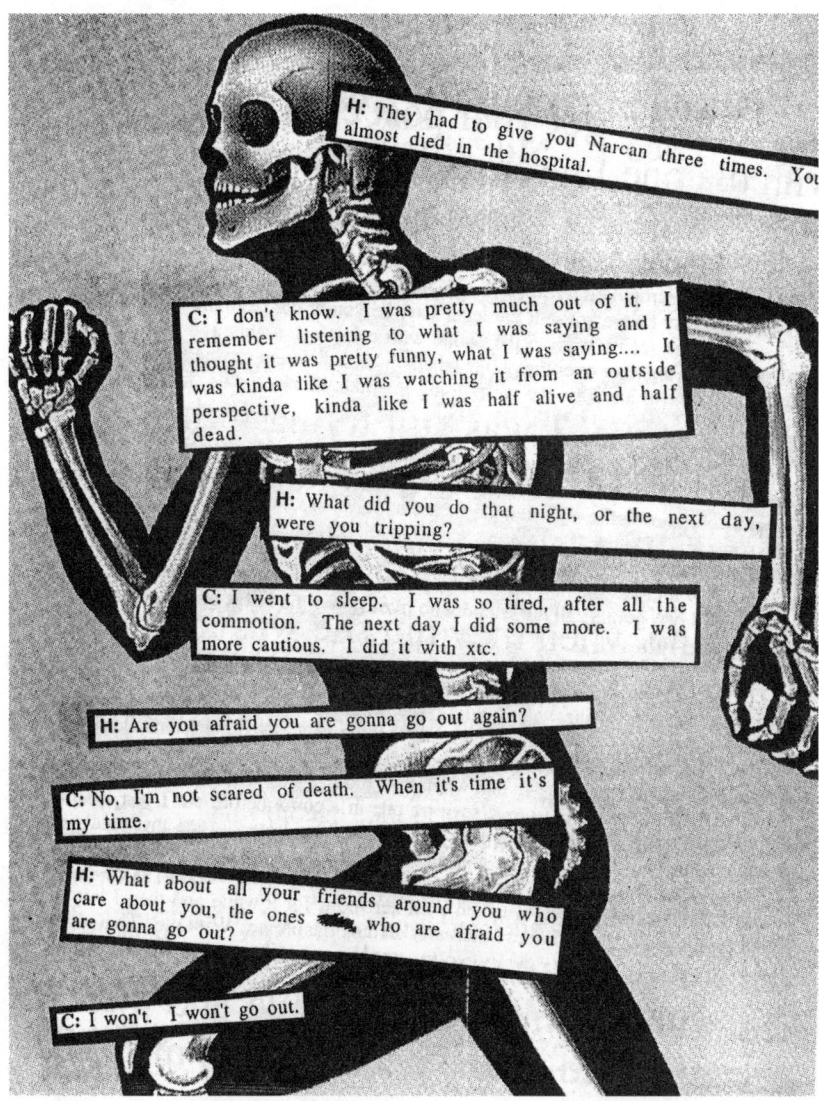

6.5 "No. I'm not scared of death. When it's time it's my time." Interview with Chameleon by Heather Edney. Collage appeared in *junkphood: The Book of Death*, vol. 1 (Santa Cruz, CA: Santa Cruz Needle Exchange, 1997), 19.
Used with permission of Heather Edney and Brooke Lober.

Active not only in the clinic but in municipal, state, federal, and international policy, Stancliff helped put naloxone on an "unstoppable path to being normalized"[85] by persuading public health-oriented professional associations, including local and state medical boards, to pass resolutions allowing naloxone to be prescribed via standing order instead of individual prescription. Those working on health care for the homeless and public housing also became concerned with preventing overdose deaths on their premises by ensuring availability of naloxone. The New York City AIDS Housing Network, founded in 1999 by Jennifer Flynn, Joe Bostic, and Jose Capestany, became VOCAL-NY (Voices of Community Activists and Leaders) in 2010. VOCAL-NY activists like Matt Curtis emphasized the concentration and intensification of HIV/AIDS in low-income communities of color; his efforts continued after his move to the West Coast. As the VOCAL-NY website put it, "Our members always recognized that HIV/AIDS was not an isolated health issue but rather a symptom of institutional injustices rooted in race, gender, and economic inequalities. VOCAL-NY was formed to shift attention from treatment to addressing the root causes of the epidemic, like homelessness and incarceration." Harm reduction drug policy was a major plank in VOCAL-NY's efforts, spearheaded by a User's Union working to ensure "universal access to harm reduction services, evidence-based drug treatment, and reducing and eliminating criminal justice involvement for people who use drugs."[86] Elizabeth Owens, a grassroots activist with VOCAL-NY, emphasized a need for educating law enforcement and reassuring people from communities of color that they would not be arrested for carrying a naloxone kit and "taking charge of saving someone's life."[87] From needle and syringe programs to housing and health care for the homeless, these organizations and individuals addressed two of the primary generators of poverty in the United States—health care and housing costs.

Overdose prevention called forth new forms of political action and cultural production that combined personal and political. As described in this chapter, localized and even singular efforts worked against pattern recognition, except in a few places where overdose came to matter because specific individuals worked to make it so. This form of labor is crucial to all social movements, but was particularly poignant and necessary in the case of harm reduction.

7

"ANY POSITIVE CHANGE": Naloxone as a Tool of Harm Reduction in the United States

Naloxone was initially controlled almost as if it were an opioid drug itself—rather than an "antidrug" or antagonist. Logistical and legal barriers to naloxone distribution abounded, causing law enforcement to treat it as paraphernalia or even proof of illegal drug use. To get around such laws, drug users sought out people who regularly dealt with naloxone—paramedics, hospital workers, pharmacists, and hospice workers—but found it difficult to build and sustain solidarity across professional, cultural, legal, and practical divides. Harm reduction activists gradually spun a web of trust and connection that became central to their identities. ACT UP activist Allan Clear, attracted to syringe exchange after hearing Richard Elovich speak about it,[1] spoke in clarion terms of harm reduction as a revolutionary practice. Naloxone education and distribution emerged roughly simultaneously in different places in the mid- to late 1990s, building on drug users' shared response to increases in overdose deaths in their communities.

SANTA CRUZ NEEDLE EXCHANGE AND THE VISUAL POLITICS OF *JUNKPHOOD*

Drug users had regular contact with paramedics due to overdose events. The Santa Cruz Needle Exchange (SCNE), founded in 1990 in Santa Cruz, California, by Heather Edney and Matthew Boman, built connections with paramedics in order to educate themselves and other heroin users about how to prevent or treat overdose.[2] Edney and Boman perceived overdose as a more immediate threat than HIV transmission, and set out to document user experience and offer harm reduction education by publishing *junkphood*, a "zine" that used "copy art" collage techniques to achieve a "hip junkie aesthetic."[3] The zine took the position that drug users were experts at what they were doing[4], and lovingly reinforced this expertise through regular infusions of practical education, fiction and poetry, interviews, and articles.

From its inception, *junkphood* discussed subjective experiences of overdose—and of revival via naloxone (see figure 7.1). By 1997, the SCNE crew had collected so many interviews and artworks on the theme of overdose that *The Book of Death* had to be published in two volumes. The first issue included a piece titled "Narcan Doesn't Have to Fuck You Up," written by an SCNE volunteer, a paramedic who had "fixed more ODs than I can count." The author recounted overhearing other volunteers at the needle exchange discouraging people from calling for help "because they give them Narcan, and it will kill them: it almost killed *me*." Admitting that less-professional paramedics were "trying to prove something" by administering naloxone at excessive doses, the author argued that it was unethical, uncaring, and punitive to use Narcan at a dosage level that precipitated withdrawal and instead advised a "gentler" dose: "Narcan does not kill people. It doesn't even make them sick. The way that it's given is what makes the difference between gently waking someone up and jolting them into a projectile vomiting, shaking, cold sweating hell." Adopting scientific lingo, the paramedic characterized heroin as a "vehicle for getting morphine molecules to your brain. When you fix, or more importantly, when you OD, the synapses in your brain are flooded with morphine molecules. Lots more of the morphine circulates around in your blood, waiting for its turn to get on the synapse." Too much morphine stopped the brain from "telling you to breathe," a signal to call for paramedics who could administer oxygen, support the airway, and give naloxone:

We can give it slowly, in little hits: First you will start breathing. Then, after a little more, you'll wake up. Or we can slam a huge amount, and do to you what was done to the woman at my meeting. We can just as easily keep you substantially stoned as wake you up like a fluorescent light bulb. Our job is to save your life. We are NOT there to teach you a lesson, or make you feel like catshit.

Text within collage image:

S: What I said, is okay fine, if you want to do all this, take as much as you can get. I think it's too much for you. If something happens and you die, it's on your hands. That's the thing with dope, you can just die, straight up, you are playing with death, I think that's part of the draw to it too.

B: Like what William Burrows said about heroin, 'nothing that feels this good isn't dying'

7.1 "That's the thing with dope, you can just die, straight up, you are playing with death, I think that's part of the draw to it too." Collage appeared in *junkphood: The Book of Death*, vol. 2 (Santa Cruz, CA: Santa Cruz Needle Exchange, 1997), 19. Used with permission of Heather Edney and Brooke Lober.

We can do our job with compassion, or we can do it with hate. Unfortunately, you don't get to interview us before you OD, and you can't pick your favorite paramedic when the time comes. The truth is that saving your life with Narcan is easy. Doing it compassionately takes a little sophistication.[5]

By contrast, "lame, unethical, and abusive behavior" discouraged people from calling for medical help.

Realizing the ease with which naloxone could be administered proved inspirational. "The Oxygen Project," as SCNE dubbed its overdose education program, "liberated" naloxone from local sources and appealed to the youthful drug users of Santa Cruz, a coastal community that took leisure pursuits seriously. The collages and zines contributed to a genre of feminist postmodernist confrontational art oscillating between refusal and embrace of popular culture.[6] Contributors played with the visual tropes of advertising and science fiction, rather than adopting the sterile tone of public health, and talked back, with an irreverent freshness, to the very safety-oriented messages they also paradoxically managed to deliver. The relative social isolation of Santa Cruz enabled a queer feminist culture to flourish in ways somewhat foreclosed in already-established spaces of harm reduction.

Visually distinctive and with unique authorial voice, SCNE materials appeared in *Harm Reduction Communication*, the Harm Reduction Coalition (HRC) newsletter. Edney wrote a column titled "User v. Addict/ Abuser," reflecting on labels that trapped drug users in disrespectful ways: "The language of 'addict' and 'abuser' leaves no room for everything else that I could be as a drug user. That language leaves me with no words to talk about my drug use.... I will never be able to learn from that type of language and I will never be able to understand that there are other ways of using drugs that aren't always about being addicted."[7] Thinking that "addict" subsumed every meaningful aspect of a person's identity was a common category mistake made by public health workers and college professors. Attention to reductive language was common in Santa Cruz, where Edney was a student—her undergraduate thesis was a video on young injection drug users. SCNE's was a brand of harm reduction that entangled feminist, queer, sci-fi, and, above all, youth-oriented language and imagery.

During the 1990s, the University of California at Santa Cruz (UCSC) buzzed with debate over identity politics, feminist and queer theory,[8] and the cultural politics of "new social movements."[9] AIDS protest was in full swing; activists traveled to the San Francisco Bay Area for demonstrations and pressured the city council at home. Overdose prevention flyers from Santa Cruz were circulated by the San Francisco Needle Exchange (SFNE), highlighting drug users' rights to self-determination and evolving a form of "oppositional capital," a do-it-yourself expertise.[10] Describing herself as once "bratty" toward researchers who got money and credit at users' expense, Edney began to see that research was "critically necessary" once realizing how common overdose had become by the late 1990s. The You or U-Find Out (UFO) study at the University of California at San Francisco (see figure 7.2) set out to study hepatitis B and C prevalence among young injection drug users in 1997, but along the way discovered that young injectors overdosed on average once every two years.[11] Fatal and

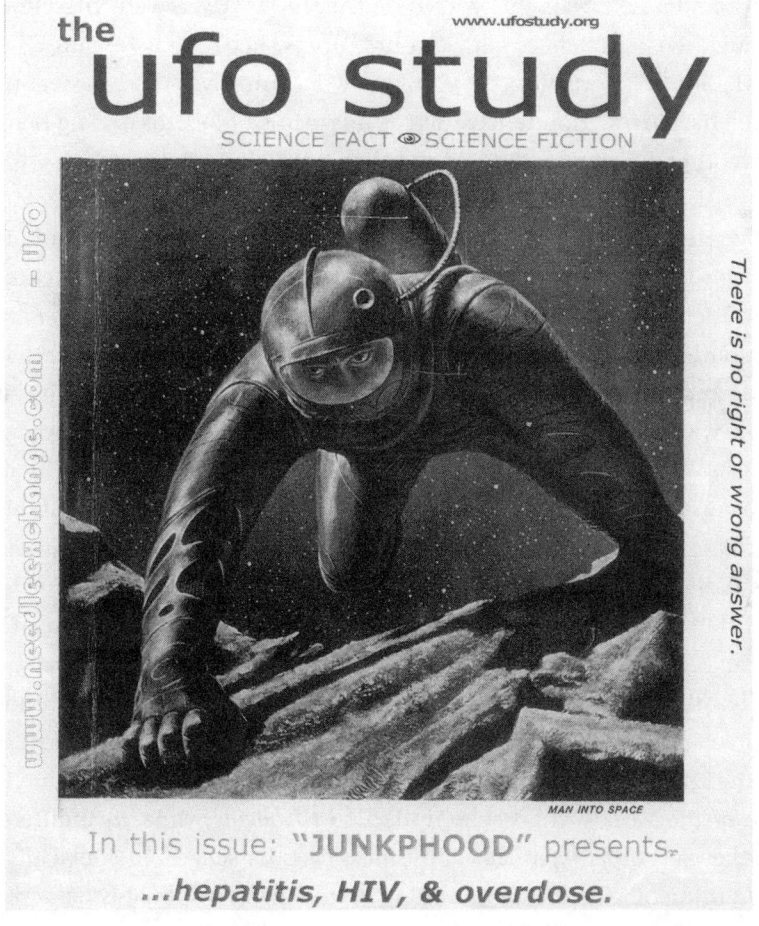

7.2 The back cover of the special issue of *junkphood* on the UFO study published in 2000.
Used with permission of Heather Edney and Brooke Lober.

nonfatal overdoses had increased to such an extent that 72 percent of the UFO sample had witnessed an overdose and 13 percent had seen that person die.[12]

The UFO study recruited outreach workers from the streets, who in turn recruited their peers. "Project staff also found that the young injectors were overdosing, their friends were overdosing, and they did not know what to do," wrote Rachel McLean, who observed HIV prevention funding shift toward overdose as a result of this new knowledge. She recalled knowing a dozen young injectors who had died during the three-and-a-half years she worked on the study.[13] Users were interviewed about their experiences with alcohol, drugs, hepatitis A, B, and C, sex work, and relationships; the findings went into flyers and posters that listed harm reduction tenets such as not using alone, not mixing heroin with other downers, and doing so-called tester shots. Activists recognized that risk of overdose was highest following periods of abstinence. Studies confirmed this claim, common sense to anyone managing a habit. *junkphood* offered a research agenda as well as a self-help approach and an outlet for expressing something of the experience of drug use. Edney attributed many of the tips on how to avoid dying from overdose to the Drug and Alcohol Services Council in South Australia. However, the zine itself was a homegrown vehicle, appealing to imagery that resonated with Californian LatinX culture, such as El Corazon (the heart). Brooke Lober gathered medical illustrations and engravings of the heart and the circulatory system, which frequently graced the pages of *junkphood* as the basis for her collages. The zine asserted a form of expertise that was heartfelt but no less a form of authoritative knowledge and expertise (see figure 7.3).

junkphood appealed to nostalgia via midcentury modern childhood characters such as Tinkerbell and an intense, even fiery, young girl sitting in the window seat of an airplane and holding a burning magazine the size of the weekly child subscription to *Highlights* while an indifferent mother and other well-heeled, business-class passengers look placidly at everything but the drama unfolding beside them. A special issue, replete with vintage science fiction imagery, rocket ships, robots, and astronauts "shooting up" or caring for the dead, and a plethora of syringes in all shapes and sizes, was devoted to the UFO study (see figure 7.4).

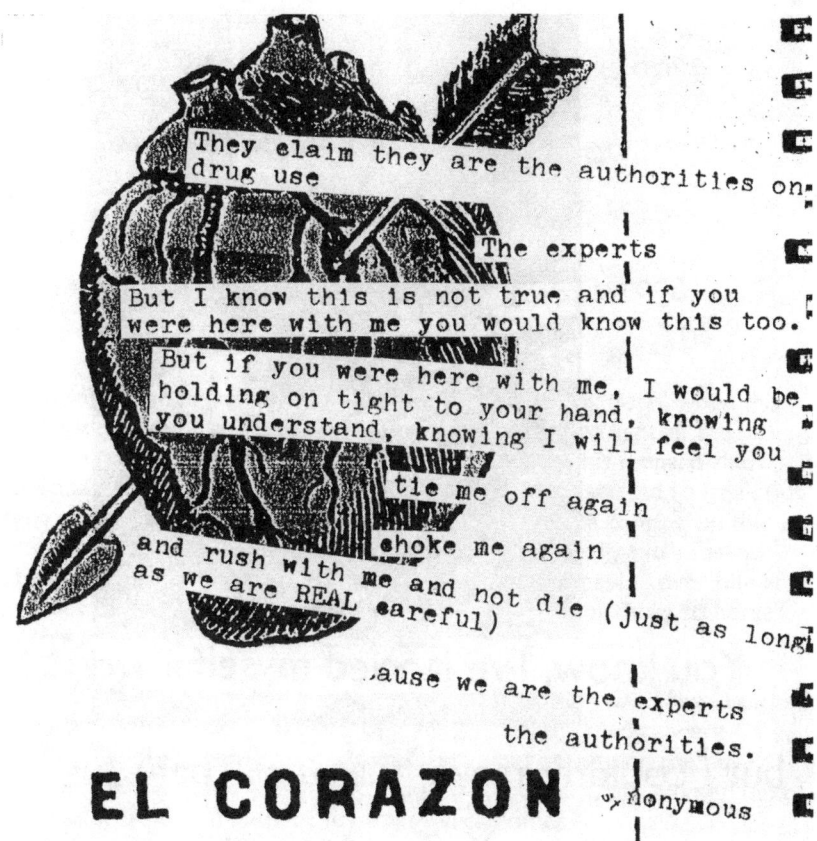

They claim they are the authorities on drug use

The experts

But I know this is not true and if you were here with me you would know this too.

But if you were here with me, I would be holding on tight to your hand, knowing you understand, knowing I will feel you

tie me off again

shoke me again

and rush with me and not die (just as long as we are REAL careful)

..ause we are the experts

the authorities.

EL CORAZON ~ Anonymous

7.3 "El Corazon." Collage appeared in *junkphood: The Book of Death*, vol. 2 (Santa Cruz, CA: Santa Cruz Needle Exchange, 1997), 51.
Used with permission of Heather Edney and Brooke Lober.

PARALLEL WORLDS: SAN FRANCISCO NALOXONE RESEARCH COMMUNITIES

Recognition that there was more to be understood about overdose developed gradually in the parallel universes of public health outreach and research. UFO OD data was among the first to be generated from an American study, according to principal investigator and epidemiologist Andrew Moss. The study began as a serological investigation with a vaccine program and was, according to Dante Brimer, self-described former heroin addict and alumnus of the Haight-Ashbury Youth Outreach Team, a "well-oiled testing, vaccinating, harm-reducing, data-collecting

So, it's kind of odd, it kind of sucks

because in my old group of friends,

people would think it was cool to OD.

And actually, a lot of my friends overdosed.

It's just, it's just not that cool ya know.

* Editors Note: Shooting cocaine is not an anecdote for a heroin overdose. Narcan is the only thing that will really work in this situation. See centerfold for more information about what to do in the case of an overdose.

7.4 "People would think it was cool to OD.... It's just, it's just not that cool ya know." Collage from *junkphood: The Book of Death*, vol. 1 (Santa Cruz, CA: Santa Cruz Needle Exchange, 1997), 40.
Used with permission of Heather Edney and Brooke Lober.

machine."[14] But it was not the only study working on overdose in San Francisco. Also working with the Haight-Ashbury Free Clinic, the Urban Health Study (UHS), directed by needle exchange pioneer John Watters until his 1995 death from overdose, not only produced an evidence base for needle exchange, but also directly connected with users.

Researchers who joined harm reduction activities were often motivated by overdose deaths in close associates. After Watters's death, UHS researcher Alex Kral, who got involved in needle exchange in 1993, joined SFNE cofounders Ro Guiliani and Maddy Luv as a "straight guy."[15] He participated in youth-oriented needle exchange in the Haight and in underground naloxone training programs,[16] an effort that evolved into the Homeless Youth Alliance under leadership of Mary Howe. Working with lawyer Bill Simpich, who represented needle exchange activists, Kral noticed overdose increasing among homeless youth and threw some of his credibility as a researcher behind setting up underground naloxone distribution in Golden Gate Park in an effort to reduce overdose deaths. Directed by Brian Edlin after Watters's death, UHS began studying naloxone's feasibility as an overdose intervention:

We just added some questions to our questionnaires to see if people were interested in [naloxone], if they would use it, if they would want to be trained in it, and so forth. That was the second thing we did. We started with the prevalence study, we found it was large, and that there was a community solution to this in terms of naloxone. Then we found it was quite feasible, quite acceptable among injection drug users. So then we went about trying to do a pilot study to see whether it's possible to do it.[17]

By spring 1999, SFNE had incorporated "naloxone administration training" into CPR trainings and was teaching young injectors to create "overdose plans" to "protect themselves, educate each other, and reduce drug-related harm within the community."[18]

Surprisingly few Bay Area drug users had heard about naloxone or knew what it was. "Nobody had ever had it in their hands. What I remember was that most people hadn't heard about it, and that some said, 'Oh, yeah, I heard about that from somebody in Los Angeles or New York' or 'Yeah, an ambulance did that to me.'"[19] Although drug user health was becoming of more interest to official public health agencies, activist researchers had difficulty funding their research on overdose and overdose prevention far into the twenty-first century. Kral remembered submitting a grant in the early 2000s that fared poorly in peer review:

One of the [peer] reviewers said, "I don't really understand how naltrexone is going to help overdoses." At which point, I did a word search on naloxone and I think I had the word "naloxone" 138 times in my proposal, but they read "naltrexone." So it did not get funded. People did not know what naloxone was. That played a huge role. I was the only one who had submitted anything on naloxone, and I felt really passionate about it.[20]

Eventually, Kral realized that reviewers were asking him to produce a randomized clinical trial (RCT), to which he was opposed on ethical grounds due to the necessary inclusion of a placebo control group in such a design. Believing it would be unethical to withhold a known life-saving intervention from participants assigned to the control group, Kral recalled:

Each time it was the issue of, I'm not willing to do an RCT and I would explain that we didn't have any data here, so it's better to have some data than none. People just wanted the RCT and I went to NIDA [National Institute on Drug Abuse] and said, "Hey, guys, you can do this. You need to fund this," and essentially the message was, "We're not touching that, it's too controversial." ... The proposals were too innovative, too ahead of their time.... People still didn't know what naloxone was at that time.[21]

Meanwhile, across the world and about the same time, UK researchers were gearing up to conduct an RCT, a story told in later chapters. Yet Kral recounted a conversation with a high-ranking NIDA official as late as 2009:

"I really don't understand the overdose and naloxone stuff. Why do you feel like you need to study this?" I said, "Tons of people are dying." And he said, "Well, it seems to me, we already know all that we need to know." I literally swallowed my tongue. I'm thinking, "So you think we don't need a scientific study that shows that we know this?" ... At that point, I was pretty sure nothing was going to get going with NIDA. At this point, nothing is passing through peer review because I'm not willing to do an RCT. And NIDA is clearly not interested in funding any of these things.[22]

The research community worked to counter pervasive ignorance. Not until 2010 would federal agencies begin to fund direct provision of naloxone, together with evaluation studies. Activists, advocates, and researchers finally broke through the orthodoxy that had settled in since naloxone's initial approval and its availability solely via emergency medicine and surgery.

Research and activism went hand in hand at the local level. For instance, in San Francisco, the UHS prevalence study and SFNE pilot had helped convince the City Department of Public Health to scale up an "aboveground" program alongside Rachel McLean's Drug Overdose Prevention and Education (DOPE) Project.[23] Seal's 2001 paper was the "only published paper on naloxone and we could use it to convince others around the country that this was a positive thing to do and something they could do to save some lives ... but the sample size was small and there was no control group. In epidemiological circles, and in health departments around the country, it was still very easy to dismiss because it was easy for people to say, 'The science is not that strong.'"[24] What was compelling was how clear it was that the "majority of all the overdose deaths in San Francisco were in a very small geographical area.... [T]he map was amazing. The map caught all our imaginations [because] ... it turned out the OD problem in San Francisco was quite localized to the single-room occupancy hotels in the Tenderloin."[25] In early-2000s San Francisco, there was a "loose-knit, mostly cooperative assemblage of researchers, community activists, public health officials, and community-based organizations all working on ways of responding to high rates of overdose deaths."[26] This was hyperlocal knowledge produced in specific sites by specific people, and in the early days of overdose prevention and naloxone access, there were few attempts to constitute overdose as a society-wide or national-level issue.

Protagonists—the activists and advocates who devoted themselves to democratizing naloxone access as a means to reduce preventable deaths—adopted different tactics depending on their own different knowledge bases. Some worked hand in hand with the pharmaceutical industry on new formulations or delivery systems; others developed underground distribution channels through needle exchanges and NGOs; still others worked with regional, municipal, local, state, or federal governments. But some expressed or reframed experiences. *The Book of Death* (1997; see figure 7.5) centered on loss and defiance, documenting moments of humor and despair. Filled with artwork that often bled off the photocopied pages of the zine, *junkphood* presented so many interviews with drug users who had overdosed or witnessed deaths from overdose that two volumes were produced, rather than the one originally planned. Many of the interviews were monologues on responsible actions to take in response to overdose, or things that people wished they had done, as well as offering detailed descriptions of failed attempts to revive. This was the self-presentation of an uncensored expertise. As Edney later wrote in reflection on the role of *The Book of Death* in creating a conversation about overdose:

The response to *junkphood* was overwhelming. While the first three were whimsical, the last four were about life and death, and for the first time drug users were reading their stories in black and white. The *Village Voice* did a feature article on *junkphood*. CNN did a story on *junkphood*. College professors were asking if they could include our zines in their curriculum on ethnography and social media campaigns. We felt that the community at large was finally beginning to grasp the deadly consequences that stigma and shame produce and the importance for an outlet that could potentially save lives.[27]

7.5 Cover of *junkphood: The Book of Death*, vol. 1 (Santa Cruz, CA: Santa Cruz Needle Exchange, 1997).
Used with permission of Heather Edney and Brooke Lober.

Expertise on nonmedical opioid use was typically claimed by those in high status positions of cultural authority such as law, policy, and medicine—and not by those with practical experience. Drug users were often reduced to the instrumentalism of their needs, their identities disregarded and their intersecting memberships in other social categories subsumed by their drug use. They were very rarely allowed to express the multiple dimensions of their encounters with life, death, morality, or promoted to partnership in producing useable, much less authoritative, knowledge. As *junkphood* illustrated so effectively, drug users formed self-help or mutual aid groups, minoritarian collectivities that understood drugs in ways that

exceed consumption-based, medical, or psychosocial definitions. That is why they were able to bring together the most heterogeneous of protagonists: none could claim to possess privileged access to the truth about drug use, a truth able to situate all other versions. One must learn about the uses of drugs, about what it is that uses are articulated around, because drugs are something to which attention must be paid if they are not to become destructive powers.[28]

Harm reduction protagonists formed heterogeneous alliances that were by no means monocultures in terms of race, ethnicity, gender, ability, sexuality, or the myriad identities configured around diagnostic catego-ries or consumption patterns. They did not share political beliefs, for theirs was a postidentity politics, configured by remaking the social rela-tions of the War on Drugs into a source of harm while trying to extend humanity to people who use drugs.

CHICAGO RECOVERY ALLIANCE AND THE POLITICS OF RECOVERY AS "ANY POSITIVE CHANGE"

The Chicago Recovery Alliance (CRA) was cofounded by Dan Bigg and Mark Parts in the early 1990s. Committed to harm reduction after Bigg heard Edith Springer talk about a trip to Liverpool, the organization relied on an advisory structure composed of active drug users involved in a process of participatory design:

The first thing that came up to us was syringe exchange, as far as an activity that would embody [both AIDS and substance use].... [W]e just started listening to the people we were seeing as to how to do that, where to do that, what it would look like, where to start, all aspects, the materials we were going to use. We just follow that same premise to this day. Whenever we can, we ask people repeatedly, what do we do that you like, and what could we improve? Then we get feedback.[29]

A few years into CRA's organizational life, antiretroviral therapy began to enable HIV-positive individuals to live long enough to die of something else—and that something else was often overdose. CRA stalwart John Szyler, who had redefined recovery as "any positive change," died alone of an overdose in May 1996.[30] His death was a tipping point for Bigg (see figure 7.6), who described himself as "practicing harm reduction in earnest" in wanting to go beyond mourning his friend.[31] CRA had taken a research-oriented approach in order to learn what drug users knew about safer injection as a form of harm reduction. After reading Strang et al.'s early 1996 take-home naloxone (THN) manifesto (discussed in chapter 6 on page 137), Bigg began searching for physicians willing to prescribe naloxone, and he began carrying it with him wherever he went. That fall, CRA designed a T-shirt to sell at the first National Harm Reduction Conference, which was held in Oakland, California, on September 17–21, 1996.

7.6 Dan Bigg. Photograph by Nigel Brunsdon.
Used with permission of the photographer.

The front of the T-shirt proclaimed harm reduction as "Any Positive Change"; the back listed everything from "Start Brushing Your Teeth" to "Keep Narcan Around." When Alan Marlatt, author of *Harm Reduction*, arrived at the conference without his luggage, he bought a shirt and turned the experience into a book chapter, "Lessons from a T-Shirt."

Bigg's commitment to supplying naloxone and a track record of successful "saves" propelled naloxone to the top of CRA's priorities. Bigg found Sarz Maxwell, a physician who was comfortable with the role of maverick. Maxwell described having been drawn to "this man who kept coming by the [methadone] clinic and telling us we weren't doing enough, weren't doing enough for people."[32] After Bigg returned from a harm reduction conference in Europe and told her that doctors at Fixpunkt in Frankfurt, Germany, had begun prescribing naloxone and letting people have it at home for overdose, she said, "'What a great idea!' And he said, 'Do you want to do it?' And I said, 'Absolutely!'"[33]

Safety concerns often dampen such enthusiasm or are translated into technical barriers that impede novel or experimental forms of treatment—notoriously so in the case of AIDS medications and, in the drug world, methadone maintenance. Drug control concerns about diversion and overdose dominated the regulatory agenda when it came to methadone, which was stigmatized and marginalized to such an extent that it was effectively withheld from a large fraction of those who stood to benefit from it. Because of naloxone's documented inactivity when narcotics are absent, Bigg and Maxwell presented it as having no harmful effects and not needing special legal status for licensed physicians to prescribe it: "I always say that if you want to hurt yourself with naloxone, the only way to do it is to break the vial and slit your wrists with the shards. You can pump a pint of it into a newborn baby and no problems. It's a hundred times safer than aspirin. So I sat down and talked with Dan to figure out how to do this."[34] Meanwhile, Bigg was giving naloxone "out to people on a one-to-one basis, people I'd known over the years. We'd ask, what do you think about this, take some and let us know what experience you have.... We've never had any legal problems, to me this has always been good medical practice and we've needed no special laws for it."[35] Known for showing up at gatherings with bags of naloxone, Bigg hoped to interest people in starting distribution programs of their own. Embodying an ethic of care and a radical commitment to inclusion, CRA distilled, codified, and transmitted that wisdom via an unforgettable protocol:

Stimulation (sternal rub, etc.)
Call for help
Airway
Rescue breathing
Evaluate

Muscular injection
Evaluate and support

As CRA gradually expanded to mobile methadone, vaccination, and other harm reduction services, it formed the hub of a vast naloxone distribution network. The organization had to tinker with how to supply naloxone legally and train people to administer it properly. Maxwell determined that in order for her to prescribe legally,

> someone needs to be a patient and in order for someone to be a patient, I had to open a chart. So I got a one-page history form and I would fill that out. What I would do was talk to a person and get the basic history of their addiction and other medications. I'd give them the bottle of naloxone with the needles and write a prescription for it at the same time. Of course, I can't write a prescription and them take it to Walgreens because Walgreens won't carry it. So what I did was write the prescription so they could keep it with the naloxone and they could see that it was prescribed.[36]

Early in the program, participants were hassled by police, for whom the presence of needles typically signaled heroin. Even when drug injectors had legitimately prescribed naloxone, they were sometimes reluctant or afraid to use it. Negative experiences circulated: "Most people had some stories, firsthand or secondhand, about naloxone in emergency rooms. In emergency rooms, the standard treatment was to give 2 milligrams, 10 cc's, intravenously, and people become screaming, puking, shitting, breaking out of full leather restraints, very suddenly, in seconds."[37]

Narcan's reputation thus was hardly stellar when administered intravenously in high doses by emergency responders. An anonymous overdose survivor had heard such horrible things about Narcan that the very name made her nauseous:

Getting hit with Narcan is fucked up. One minute, you're unconscious, and the next minute you are completely straight. The main thing I remember about it is that I really wanted to get high again, and I couldn't stop shaking or get my teeth to stop chattering. I was also super agitated, a feeling I spend a lot of time, energy, and drugs trying to avoid. (I've been with people who have been brought back from an OD with Narcan and just hearing the word "Narcan" makes them stand up and bolt for the door.)[38]

Recognizing on the basis of such stories that naloxone might be rejected by the very people who would benefit from it, CRA tried lowering the dose and turning to intramuscular rather than intravenous injections. This slowed naloxone's action but still restored breathing. Titration of the dose, as initially recommended in *Emergency Care in the Streets* (discussed in chapter 5), became part of the CRA ethos. So was reporting and record-keeping of "saves" or "rescues."

Yet Maxwell found the process required to widen naloxone distribution in Chicago disturbingly slow. In January 2000 her trainings included basic opioid neurophysiology; the pharmacodynamics of opioids and opioid antagonists; overdose risk factors, prevention, and symptoms; aspiration prevention; rescue breathing; naloxone administration; and dosing guidelines.[39] She wrote a standing order to enable CRA staff to distribute a 10-cc bottle of naloxone plus four syringes and a checklist documenting that a client had received training. Meanwhile, Bigg negotiated a bulk purchasing deal with the major supplier of naloxone. Blanket standing orders modelled upon or utilizing Maxwell's were put into place in dozens of jurisdictions across the country. Bigg's dedication and Maxwell's moxie made for heady days as both spread word of their "OD juice" with an almost religious zealotry. Maxwell recounted her feeling when she realized what a transformative technology naloxone was:

Often, with a young patient who would come with a parent, we often gave it to the parent and parents would just dissolve [saying], "This was my nightmare, to find them, alone and cold." It changes the relationship because a lot of the anger was about fear and … about being totally disempowered…. Naloxone saves lives, yeah, that's nice, but it does more than that. It changes lives. Particularly for the people who administer it. The person who overdoses, it's not much of an experience. They shoot up, they're high, and the next moment, they're sick. We are pretty sure from research that overdose does not keep people from using. It doesn't convince people not to use. But the person who administers it, the person who does the save, has an incredible experience. Shame is one of the most lethal components of the disease of addiction. It hides the disease and keeps the person from asking for help. Everyone in a dope addict's life is shaming them, telling them that they're worthless, and they begin to believe that. Because dope addicts are so disempowered, this is the real thing that makes the naloxone prescriptions work.[40]

The CRA experience offers an opportunity to reflect on the history of how social relations between drug users and sellers change when market conditions do. Maxwell contrasted opiate users to users of other drugs, describing empowerment she witnessed as participants or "patients" became willing to administer naloxone to others. "Dope addicts have a sense of community, of caring for each other that you don't see.... [N]o one else wants to take care of them and they know they have to take care of each other. That's what makes naloxone work, that willingness to take care of each other.... When they're given something that works, dope addicts are thrilled to take responsibility and help."[41]

This was the premise—one that many regarded as a leap of faith—on which CRA proceeded to spread word of naloxone's efficacy. Although US physicians could legally prescribe naloxone, they rarely did so even when they knew a patient was an active opioid user. Bigg wanted to break what he saw as a "collusion with silence" by giving people naloxone and teaching them how to use it: "We started small and we were very paranoid, and above all I promised the physicians that I would protect their identity.... Those medical people are very glad they helped us out back then as they see how much it has grown, confirmed their own views, and saved so many lives."[42]

NALOXONE AT THE TURN OF THE TWENTY-FIRST CENTURY

Bringing naloxone vials along to a hepatitis C training in New York City, Bigg showed them to Ethan Nadelmann, Jean-Paul Grund, and Glen Backes of the Lindesmith Center, then in the process of becoming the Drug Policy Alliance. Backed by Open Society Foundations (OSF), they joined Marlatt's University of Washington Alcohol and Drug Abuse Institute to cohost the first international conference on naloxone access shortly after the so-called Battle of Seattle, the protests targeting the

World Trade Organization meetings in late November and early December 1999. In early January 2000, the world's naloxone protagonists found themselves—and each other—discussing pragmatic approaches to "Preventing Heroin Overdose" and reading *junkphood*.[43] By then Bigg had spoken publicly about naloxone at the 8th International Conference for the Reduction of Drug-Related Harm in Paris in 1997, where Pat O'Hare, a founder of the International Harm Reduction Association,[44] had asked him what he had to say to physicians about overdose, and his response had been, "Do your fuckin' job!"[45]

In Seattle, after complimenting Strang for "paving the way" with his papers suggesting THN, Bigg challenged Strang to "stop writing about it, and do something about it." Both described their first meeting as a friendly rivalry. Strang had just published a "prelaunch" study on the possibility of THN.[46] He returned from Seattle thinking that he should pilot naloxone in London, an idea that he directly attributed to "Dan's intervention."[47] As heir-apparent to the academic directorship of the National Addiction Centre at Kings College London, Strang was in a position to act upon his convictions. He set up the first mainland UK naloxone distribution program in South London at the Maudsley, where he was then clinical services director.[48] And the National Health Service addiction treatment provider there began supplying naloxone to those exiting detox starting in 2001.[49]

Naloxone distribution was tightly intertwined with Strang's ambitions to set THN on an evidence base. "Waxing lyrical in some lecture at the time," Strang recalled, "I wrote up a paper about the necessary symbiosis, that what you wanted was to have a clinical practice with research endeavor where each one was asking questions of the other side in a two-way dialogue."[50] The naloxone research conversation became far more than symbiosis; pricing, production, and distribution posed multifaceted research questions as well as practical problems. Answering them was like a global game of Whack-a-Mole: each time researchers produced new evidence, the questions they were expected to answer changed. As the implications of naloxone distribution clarified, a broader and more global social movement evolved than could have been anticipated by the protagonists who gathered in Seattle. Especially in the United States, War on Drugs policies criminalized paraphernalia and rendered needle exchange and naloxone illegal, public figures promoted abstinence-based recovery (who can forget Nancy Reagan's "Just Say No" campaign?), and harm reduction was framed as "condoning" illegal acts. Hence activists had to go to extremes in rhetoric, often embracing the idea of revolution, and in the everyday risks they took to provide naloxone when its quasi-legal status remained untested. Grounded in the ethical stance of meeting people where they are, harm reduction does not intervene to bring about massive behavioral change to abstinent lifestyles and identities. Practitioners start with actual, concrete experiences of harm and work pragmatically toward practices that would reduce harm in specific sites and social contexts. While deploying the more abstract language of risk reduction, harm reductionists dwell not in the realm of potential but in the everyday—often irreverent toward conventional public health outreach on grounds that it disrespects people's own choices about ways of living.

As the center of the world's first large-scale naloxone distribution program, Bigg ensured that CRA documented distribution and "saves."[51] He and Maxwell freely dispensed advice on how to start naloxone training and access programs.[52] Despite lack of controlled studies, no adverse events or repercussions appeared.[53] CRA Research Director Greg Scott, a drug-using filmmaker and "embedded sociologist" on the faculty of nearby DePaul University, caught an overdose on film. While CRA's gritty realism contrasted to the sci-fi and anime-inflected cultural productions emanating from SFNE and SCNE on the West Coast, there was value in watching Scott and the partner of Steve Kamenicky, who had overdosed on camera, revive him on camera.[54] The resulting film was titled "LIVE! Using Injectable Naloxone to Reverse Opiate Overdose," posted online in lightly edited form. It demonstrated key aspects of CRA's SCARE ME protocol, including recovery position and sternal rub.[55] An interview with Scott himself emphasized learning from experience, and subtitles were inserted so that viewers could see how the steps of the protocol would lead to rescue. Research inspired by CRA was premised on gaining knowledge from actual experience and using it to inform public health via harm reduction, rather than following research conventions. Suzanne Carlberg-Racich, who succeeded Scott as research director at CRA, had conducted interviews with thirty-one participants recruited from three CRA needle exchanges (80 percent of whom were male opiate injectors with a median of ten years' injecting experience) in 2004.[56] Systematic efforts to understand how people responded to administering naloxone rather than receiving it joined anecdotal evaluations recounted through the lens of each responder's dignity, pride, and sense of making a difference by saving a life.

CRA presented naloxone as an antidote that was as simple to administer as glucagon for insulin overdose or adrenaline/epinephrine for anaphylactic shock—as a simple first-aid measure that almost anyone—even a child—could administer. Bigg's own writings read with a moral forcefulness designed to persuade others that naloxone was a public good, an easy recipe for saving a life. CRA also developed its public health face in parallel to vaccination and viral testing programs:

Some participants, after being dispensed naloxone, have returned to be tested for HIV and HCV [hepatitis C virus], telling us that they are now feeling a greater sense of hope that they may live to see a long-term future. As one participant eloquently put it, "People who overdosed used to be past tense—I *knew* a guy who overdosed. Now we can talk about them in the present: I *know* a guy who overdosed and he's ok now." Our finding of improved personal health care is only anecdotal at this point, but one fact is irrefutable: *Dead addicts never recover.*[57]

The undeniable facticity of the statement "dead addicts never recover" contrasted to putative objections from just-say-no proponents. By contrast to just saying no as the only path to recovery, harm reductionists like Biggs saw lifesaving measures as obligatory. All manner of life-saving measures have typically been withheld on the basis of social status markers such as race or ethnicity, age, imputations of "moral" worth and character, and features (such as the smell of alcohol or track marks) understood to mark membership in a stigmatized population.[58]

Holding no patience for those who held drug users in contempt, Bigg treated users as experts in their own right who should be treated with respect. "Just as our participants are not stereotypical, pathological drug users, they are the experts with the answers that we need to know."[59] He saw participation in harm reduction as a step toward a "thoughtful life. It's an attraction, a mutual appreciation, a mutual reliance on expertise. That's the power of it." Naloxone empowered those who refused to say no to ward off death together. In spreading its power, Bigg cast himself as the Johnny Appleseed of naloxone.

Saving a life became the necessary ground upon which all other interventions, determinations, and documentations of effectiveness of naloxone were founded. The stark simplicity of the harm reduction stance stood in sharp contrast to the complexity of naloxone's medicolegal status.

US-based naloxone access advocates ran up against statutes governing everything from sales and use of paraphernalia to who was empowered to administer an injection[60] and to attitudes toward needle and syringe exchange among pharmacists and physicians.[61] The state of Illinois did not make it legal for nonmedical persons to administer naloxone in order to prevent an overdose from becoming fatal until January 1, 2010. After more than a decade of providing naloxone, CRA was no longer operating outside the law. By then, programs throughout the world had visited Chicago to observe, obtain naloxone through Maxwell's standing orders, or consult with Bigg on the design of local naloxone distribution channels. The organization became legendary. Two decades later, Bigg's own overdose death, surrounded by cases of naloxone, on August 22, 2018, was publicly grieved by a diverse movement otherwise riven by differences of identity and strategy.[62]

8

HARM REDUCTION RESEARCH AND SOCIAL JUSTICE: Working Naloxone into Public Health USA

Tagged as "an unidentified heroin user in the process of killing himself," the man in the photographs sat in the broad light of day on grassy Boston Commons. *Boston Herald* photographer John Wilcox caught this "Death in the Nation's Garden" in 2005. His photographs unleashed public reaction ranging from outrage and disgust to the staunch conviction that something had to be done by the City of Boston.[1] Naloxone access advocate Maya Doe-Simkins was pulled into the Boston Public Health Commission response team. Having grown up attending marches and rallies focused on social justice, Doe-Simkins had become a public health professional. Although she worked "aboveground," she recognized that the underground naloxone access movement "had tried again and again to bring it into the light of day. It was a unique situation in Massachusetts because the people who were doing the underground work, which, depending on your interpretation, was probably illegal, [and] those folks had an actual relationship with public health administrators. They were sometimes in the room together."[2]

"DEATH IN THE NATION'S GARDEN"

Clandestine naloxone distribution had occurred within Massachusetts harm reduction networks since the early 2000s via a channel between

Chicago Recovery Alliance (CRA) and Cambridge Cares. But activists had not been able to gain traction until the Wilcox photos appeared. Working with Andy Epstein, Peter Morse, Alexander Walley, and others, Doe-Simkins helped initiate distribution of nasal naloxone rescue kits from 2005 to 2007, as well as researching effectiveness of the program. Although fluent at implementing public health interventions and assessing outcomes, Doe-Simkins perceived herself a "reluctant researcher,"[3] partly because she saw researchers playing "catch up" just to document what activists already considered common knowledge in their circles. Doe-Simkins later came to appreciate what researchers could bring to the public health table, however, and her commonsense approach helped catalyze alliances between underground actors and city and state officials. These relationships helped the Commonwealth of Massachusetts develop a multipronged, state-level effort years before federal agencies were willing to supply naloxone or research funding.

Building on relationships with needle exchange staff at the Massachusetts AIDS Service Organization, naloxone protagonists gained a foothold in the city and state-organized public health structure. Despite initial skepticism toward naloxone as a means of overdose prevention, many in the official public health apparatus proved to be "incredible allies," according to Doe-Simkins:

Interestingly, weirdly, the tragic convenience of a front-page, very public fatality that came out in the newspaper was this galvanizing, perfect moment of throwing a fit and sending an anonymous email to the anonymous mayor's helpline saying, there was this big, huge public death in the Boston Commons that is a tourist eyesore. It was an anonymous tip to the mayor, and within 36 hours a meeting was convened.[4]

In their retrospective slide show titled "From Underground to State-Funded: The History of Overdose Prevention/Naloxone Distribution in Massachusetts," activists Adam Butler, Jon Zibell, Kathy Day, Monique Tula,[5] and Gary Langis[6] sketched the story of community-based underground naloxone distribution in the state. Peer educators and needle exchange staff began overdose prevention trainings and brought Massachusetts legislators to the table. But they did not gain traction with Boston's Department of Public Health until the glare of publicity in 2005. From there, state-funded programs began in Boston, Cambridge, New

Bedford/Fall River, Quincy, Lynn/Gloucester, Brockton, Provincetown, Hyannis, Springfield, and Northampton in late 2007. By 2010 naloxone distribution had been expanded to four additional communities experiencing high rates of overdose: Lowell, Lawrence, Holyoke, and Worcester.

The statewide question of which communities would have naloxone programs was asked carefully, with an eye toward research, community readiness, and capacity building, along with reliable data on overdose deaths. A second unusual aspect of the expansion was exclusive reliance on intranasal naloxone, the only formulation available on emergency medical services trucks in Massachusetts. Intranasal naloxone lacked Food and Drug Administration (FDA) approval and was controversial due to technical questions about bioavailability.[7] Intranasal naloxone gained public support partly because of the unique ecology of academic medicine and public health in the Cambridge/Boston metropolitan area, home not only to Massachusetts General Hospital and Harvard Medical School, but also to global health equity activist Paul Farmer's public health program (in which Doe-Simkins had been a student).

Physician Alex Walley was drawn into research on overdose education and naloxone distribution after attending meetings between the Boston Public Health Commission (BPHC), Boston EMS (supervised by the BPHC), and the Boston Medical Center. Although he recalled witnessing naloxone used during his emergency medicine rotation in the mid-1990s,[8] he said, "To these paramedics, it was no big deal. As a basic EMT, I was aware of naloxone, but I didn't see many ODs in Boston at the time. Even then, I didn't think, 'Oh, this guy's girlfriend should have had this.'"[9] Walley had "covered naloxone in pharmacology as a standard tool you have in the hospital. It is surprising I didn't make the connection that it could be used as a community-based intervention."[10] Central to undone science, this phenomenon of "seeing but not seeing" was described repeatedly by clinicians interviewed for this book, and is a good example of what Ludwik Fleck called "directed perception."[11]

Naloxone was an "old" drug. Those committed to its "new" uses often turned to tactics drawn from the repertoire of ACT UP described in previous chapters. For instance, the slideshow noted above played upon the ACT UP slogan "Silence = Death." In Massachusetts, two sisters, Mary Wheeler, program director of Northeast Behavioral Health's Healthy

Streets Outreach Program in Lynn and Gloucester, and Eliza Wheeler, who ran the Cambridge Needle Exchange in Cambridge until relocating to the San Francisco Bay Area in 2003, planned overdose vigils and remembrance walls, delivered naloxone trainings, and tracked lives saved. Programs moved from cultivating awareness to directly training people how to use naloxone. Heroin use had become endemic in Massachusetts and Rhode Island, drawing researchers eager to document the effectiveness of community-based naloxone distribution programs:

The intranasal was a political strategy. Boston was the first public health department to use intranasal naloxone. At that time, Chicago was community-based, New York City had the HRC [Harm Reduction Coalition], and San Francisco had nurse practitioners at the needle exchange. So I knew that some public health departments had gotten the nurses and doctors out of it.... We maintained a database of rescues. I persuaded the state to fund data management at the Boston School of Public Health. I wrote a grant to CDC [Centers for Disease Control and Prevention] to evaluate the state program. We got money to hire biostatistics people. Xuan joined and Al H came up with an analytic plan. The shift to statistical analysis made a big difference. I could not do that by myself. We had the idea for an interrupted time series analysis, for an ethical natural experiment. I was not enthusiastic about a controlled experiment. But naloxone will not be accepted as evidence-based until RCTs [randomized clinical trials] are done.[12]

This reference to the discomfiting ethics of naloxone trials will be taken up in chapter 11 on the N-ALIVE pilot study then being organized in the United Kingdom. While interviewing US activists and researchers, I encountered considerable concern about the ethics of research and particularly the idea of subjecting naloxone to randomized research design because it would require essentially withholding naloxone from some in order to confirm its efficacy for others.

In the United States there had been no calls for research proposals specifically mentioning naloxone or overdose prevention at the time of my 2013 interviews. Relevant federal agencies still did not fund harm reduction research; proposals that mentioned it continued to be referred to federal waste baskets.[13] Some potential investigators described frigid reception of proposals, recounting dismissive references made by science administrators and peer reviewers that rested on pharmacological confusion, in addition to political or ideological rejection of naloxone for purposes of reducing drug-related harm. Given the explosion of attention to naloxone that emerged just a bit later due to aggressive reporting of rising numbers of overdose deaths across the United States, this lack of prescience in the federal research apparatus suggests that priorities were misplaced due to scientific orthodoxy and deference to senior agency staff.

Government agencies had trouble establishing good working relationships with harm reductionists—to the extent that federal agencies had relationships with "users," "consumers," or "clients," they typically involved people in treatment or abstinence-based recovery. Indeed, most substance-abuse research has historically been conducted with treatment-seeking populations rather than with active users of the sort who found that the harm reduction movement spoke to their lives, goals, and values (see figure 8.1). In some ways the lack of research and funding opportunities paradoxically benefited harm reduction; shaping research and demonstration projects in order to attract federal or state funding, or to appeal to corporate sponsors, can co-opt grassroots activities.[14] Imposing clear-cut leadership or structured lines of authority can breed conflict within movements that had more lateral relations. When harm reduction organizations were brought in as contracted partners, this may be seen as supporting a grassroots social infrastructure—or as the state "using" social movement organizations to do work that it should do. Perhaps because of a balance between top-down and bottom-up planning in Massachusetts, mutual benefit was perceived. According to Walley, "Activists couldn't have accomplished what they did without the state, and the state certainly couldn't have accomplished what it did without the activists. Harm reduction activists are actually doing the work, actually saving people."[15]

"DOING THE WORK": OVERCOMING SOCIAL, STRUCTURAL, AND REGULATORY DIVIDES

"Doing the work"—and in the harm reduction movement, it is always referenced as an active form of work oriented toward social change—was not primarily about doing the research. It was about harm reduction as a form of affective labor and pragmatic practice, a kind of work that has become a way of life in some urban centers. Activist work also refers to efforts to address racism, sexism, ableism, homophobia, and

8.1 Created for the *junkphood* special issue on the UFO Study published in 1998, this graphic, which appeared on page 31, was created by modifying the cover of a science fiction novel. It also appeared on the front cover of *Harm Reduction Communication*, no. 9 (Fall 1999).
Used with permission of Heather Edney and Brooke Lober.

other differences within work environments.[16] Dealing directly with the array of differences encountered in contemporary social movements has brought changes in language and behavior, policies and protocols. However, overdose prevention relies on different modes of expertise needed to bring about legal and regulatory change. When naloxone was quasi-legal, trainings and supply were undertaken by those willing to operate in a legal gray area—but also by those who could afford to do so. The privileges of whiteness served activists well, whereas those from communities of color had to be more circumspect.

One event that galvanized regular interactions between harm reductionists and the CDC was a cluster of more than a thousand deaths involving nonpharmaceutical fentanyl (NPF) that occurred between April 2005 and March 2007.[17] Reported via the CDC Epidemic Information Exchange, the initial alert came from Poison Control in Camden, New Jersey, in spring 2006. That cluster elicited similar reports from elsewhere and enabled retrospective analysis of an earlier spate of mischaracterized "heroin overdoses" that had occurred in Chicago and Detroit. Fentanyl was identified not only by medical examiners and law enforcement, but also by drug users and needle exchange workers at CRA and Prevention Point Pittsburgh.[18] Although fentanyl later became ubiquitous in the US heroin supply, throwing a wrench into naloxone access advocacy because of speculation that higher naloxone dosages might be necessary for reversal, it was unclear in 2006 and 2007 which agencies should respond to fentanyl:

Overdose had always been seen as a drug treatment problem or a law enforcement problem, and there were people that were saying, "This should be a public health problem." So Ken Thompson, who worked for SAMHSA [Substance Abuse and Mental Health Services Administration] at the time, convened these phone calls that were people from different federal agencies—FDA, DEA [Drug Enforcement Administration], ONDCP [Office of National Drug Control Policy], different federal agencies that were involved in this issue—but then there were people from the harm reduction world who got invited to participate.... that, I think, was a unique opportunity where you had people from ONDCP and FDA and NIDA [National Institute on Drug Abuse], all these federal and state agencies talking weekly with me, or Dan [Bigg], or Mark [Kinzly], people who were running needle exchanges, that were talking with people in federal agencies. Where else did you have people from all these different programs that were able to hear what we were doing?[19]

Nontrivial social divisions bedeviled efforts to cooperate between activists and regulators, but some, such as Harm Reduction Coalition medical director Sharon Stancliff, could speak both languages. Activists received some unexpected federal support, also encountering the resistance they expected from others. For instance, ONDCP's Bertha Madras indicated that naloxone distribution was "sending the wrong message," a position she broadcast to national media throughout her tenure at the White House. However, as Alice Bell recalled, at the 2006 meeting, overdose prevention activist Mark Kinzly said in response, "This is not what I had intended to talk about today, but I have to say, after those remarks, if it wasn't for naloxone, and friends of mine with naloxone, I would not be here today."[20]

Building upon the raw emotion of Kinzly's disclosure, activists forged even stronger connections with one another. The extent to which those who ran needle exchanges or other harm reduction programs sought official recognition or considered themselves vanguards of a "revolution" varied, as Alice Bell from Prevention Point Pittsburgh explained:

I don't think I would have ever used the term "revolutionary" at that time to talk about this.... [W]e were like a little needle exchange program doing our thing barely above ground at that point. I never imagined that we were going to get any kind of official [recognition], the hope was more that we would be allowed to do what we were doing on a small scale and that we would be funded to do it, but never that we would be officially recognized as any kind of larger public health policy, but that was what we were fighting for. To me, my upbringing was in fighting for social justice, a social justice activist, and that's how I came into this work, seeing it as social justice work.... In the work that I had done as a social worker, in my personal life, I felt like people who used drugs were treated very badly, were treated as second-class citizens who didn't get the kind of services other people got. So that was the hook for me, the social justice angle.[21]

Official recognition first translated into the nurturing of ties between activists, the CDC and other relevant federal agencies through the weekly conference calls that began in the wake of the 2006 meeting. Collectively referred to as the NOPE calls—weekly conference calls with occasional face-to-face meetups[22]—they multiplied points of contact, served as a clearinghouse or repository for the assembly of an evidence base, and, perhaps most importantly, enabled the sharing of strategies for responding to mischaracterizations of naloxone, overdose, and even "addiction" and "addicts" in the mainstream media. These calls proved crucial for the expansion of naloxone access. In Pittsburgh, which had experienced endemic heroin use and overdose prior to the fentanyl deaths, Prevention Point had begun pulling people aside during its weekly needle exchange in the summer of 2005 to give them naloxone. That fall when reports of fentanyl deaths began coming in from New Jersey and Chicago, Bell recalled there were news reports of five deaths overnight in Pittsburgh:

I remember that we put together a flyer saying, "This is happening in New Jersey and this is happening in Chicago, so we think it's likely that we're going to see it here. We haven't seen it yet, but we think we're going to." We had those flyers ready to go, and Alex [Bennett] and I were in my living room with the TV on, and heard on Saturday—the needle exchange was on Sundays—and we heard this news report and there was this cluster of overdose deaths in Pittsburgh, and we said, "Okay, that's it, it's here." We realized we had this really slow start with people doing this hour-long training … so we had said, "We have to do everything we can to get this into people's hands." It was not working to do hour-long trainings. We decided to do it simultaneously with the needle exchange, so that people could come into the needle exchange and we could say, "Step over here, we'll do the training, and then we'll give you naloxone."[23]

As more people became interested in receiving naloxone supplies, trainings were shortened. Each step in the process was studied to see how short an effective naloxone training program could become, and how long information was retained. Activists, familiar with studies demonstrating the effectiveness of needle exchange for slowing HIV transmission, urged colleagues to collect data from the beginning. Although it was not always easy to publish these data (some programs were better connected than others to social scientists who knew how to assemble, analyze, and publish such studies), many parts of the naloxone access movement were aware of the need to produce evidence from the beginning. The commitment to social justice in a data-driven world, in which public health activists had to remodel themselves as lay epidemiologists, went hand in hand once the longer-term need for evidence became clear against the backdrop of the movement's day-to-day pragmatics.

Although non-pharmaceutical fentanyl-related deaths trailed off in early 2007,[24] key federal agencies had become aware of naloxone distribution efforts.[25] As need for more accurate and less assumption-driven epidemiology of overdose became obvious, the nonprofit sector stepped into the breach, funding a two-day conference in 2008 organized by Scott Burris at Temple University in Philadelphia, Pennsylvania. Most activists and advocates interviewed for this book pointed to this gathering as a significant meeting ground for the nascent national coordination that had begun in 2006. Because bureaucratization tends to deaden social movement interactions, the affective ties that hold social movements together are crucial to consolidate the sense that disparate activities occurring in different places are somehow linked. These ties were further stewarded by the harm reduction conferences and the international movement.

Nevertheless when the FDA held its first hearing on possible regulatory approval for naloxone as an over-the-counter (OTC) medication in 2012, activists had still not achieved enough visibility to be officially invited to the FDA table.[26] The people who knew the most about naloxone in the United States had to crash the hearing:

Actually, Nab [Dasgupta] had heard it was going to be announced and he said, "What we need to do is, as soon as they announce it, we all need to call to ask to give community comment." So a bunch of us did that.... At that point, there really weren't states that had passed laws around naloxone. If there were, not many. I remember what I talked about was, we've been giving out naloxone since 2005 and I get calls all the time from parents who say, "I just heard about this [naloxone]. Do you mean to tell me that if I had had this in my medicine cabinet, my kid might be alive today?" I said to the FDA, "I'd like to have something to tell them, besides, 'We actually can't give you naloxone because we're only allowed to prescribe naloxone to people who are themselves at risk.'"[27]

The nonsensical nature of naloxone supply drew parent groups into the conversation. Often from suburban or rural areas, such groups have a complex relationship to the urban and queer harm reduction infrastructure that evolved out of the HIV/AIDS movement. There is ongoing debate about whether parents whose children have died from overdose should be supplied naloxone, given that it has become a more expensive and limited resource with increasing demand since the FDA began to consider making naloxone OTC. Naloxone offers symbolic comfort at best to those who have already lost loved ones to overdose. Parents like Susan Gregory demanded, "Why didn't I know about this when my child was alive?" They were recognized as having legitimate moral claims to the medication.[28] Activists were later recognized as bringing knowledge to the table; they were invited to a second FDA hearing in 2015. Peter

Lurie co-organized a 2016 hackathon between the FDA, SAMHSA, and the NIDA to crowdsource "apps" to link naloxone to those who need it.[29] Although the goal of making naloxone OTC was left on the table, Lurie and colleagues documented a 1,170 percent increase in naloxone prescriptions filled at US retail pharmacies between 2013 and 2015.[30]

Regardless of their seeming failure to gain OTC status in 2012 and 2015, activists felt the once-chilly federal climate thaw as naloxone research, service, and supply began to be funded. Not only were protagonists' efforts translated into legitimate public health aims, they gained traction with first responders, including police and fire departments, county legislatures, sheriff's offices, and EMTs, which once rejected the notion of bystander naloxone. In 2013, SAMHSA released its first Opioid Overdose Prevention Toolkit and in 2017 was recognized as "a leader in efforts to reduce overdose deaths by increasing, through funding and technical assistance, the availability and use of naloxone to reverse overdose."[31]

WHY? THE RACE FOR PROOF IN AN ERA OF EVIDENCE-BASED ACTIVISM

The rich material culture of harm reduction includes editorials, letters to editors, blog posts, press accounts, training videos, websites, photo journalism, zines, and interviews. Voices in these materials are playful, biting, and yet deadly serious about mourning losses, recounting lives saved, and reducing lives lost. The hard-won knowledge of experience was marshaled in tips for preventing overdoses and for educating drug users about what *not* to do in the case of an overdose: don't put them in the bath or shower because they could drown; don't give them coffee, alcohol, or other liquids; don't shoot them up with speed, coke, salt water, or baking soda.[32] Hard-won knowledge was offered up to engaged researchers, also becoming a platform for cultural production (see figure 8.2).

Although opioid overdose was rarely addressed explicitly in this body of literature, the Center for Urban Epidemiologic Studies (CUES) at the New York Academy of Medicine did conduct one study on drug-related deaths based on files from the Office of the Chief Medical Examiner in New York City from 1990 to 1998.[33] Finding such a paucity of other studies on "changes in overdose patterns among different racial-ethnic groups

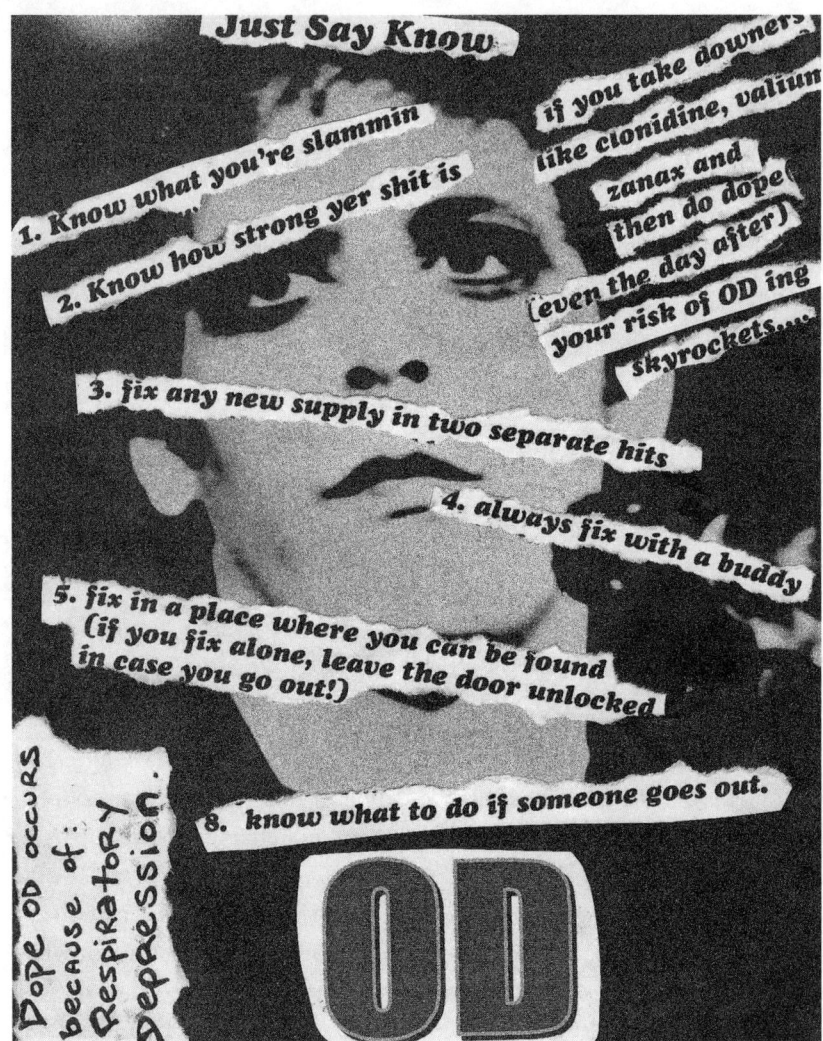

8.2 "Just Say Know." This flyer using the album cover from Lou Reed's second solo album, *Transformer* (1972), was produced by the *junkphood* crew as part of a series of flyers called "Superstar" from the early 2000s. Accessed at http://heatheredney.com/. Used with permission of Heather Edney and Eleanor Herasimchuk.

in an urban context," the CUES pitched its own findings as "hypothesis generating" and suggested "further avenues for research that might shed light on the complex etiology of some of these differences."[34]

Far from headline news in the year 2000, opioid-related deaths numbered ninth among the most common causes of death in New York City. CUES identified all deaths involving alcohol, cocaine, and opiates, then calculated a rate of overdose death using six gender/racial categories. The overdose death rate was highest among black and Latino men; cocaine-positive deaths were far more common for black women at the time than were opioid overdose deaths. "Rates of overdose death in New York City were high among all racial/ethnic groups and consistently higher among blacks and Latinos than among whites. In addition, cocaine was more frequently detected in black decedents, while opiates and alcohol were more common in Latino and white decedents throughout the period studied."[35] Men of all ethnicities died at a higher rate than did women of all ethnic groups; black women's death rates exceeded those of white women and Latinas. The real question was why. National surveys of drug consumption—which had been conducted in the United States since the 1970s—had not revealed consistent differences in drug use between racial/ethnic groups.

All racial/ethnic groups tend toward "simultaneous use of multiple drugs" rather than favoring any so-called drug of choice.[36] The CUES study left readers with far more questions than it answered. By focusing on individual deaths, distinct racial/ethnic categories, and single drugs, researchers could only conclude that contextual aspects might help understand these elusive patterns. Despite many uncertainties, the study showed that overdose mortality in New York City was rising among African American men and women in the 1990s. That had been obscured by CDC vital statistics, which are used to generate national pictures, including spatial visualizations of overdose "hot spots," that can be utilized to understand change over time, and the people and places most affected. However, the experiences of individuals, families, and communities are shaped by changes in legal and illegal drug market dynamics such as the regulatory structures and patterns of drug law enforcement. Such experiences cannot be detached from the use of incarceration as drug policy, or what we might call wars on drugs.

Naloxone protagonists were critical of drug policy—they questioned the continual role of law enforcement, sentencing disparities, and mass incarceration. As skeptics of criminalization, they problematized old political divides between those who saw drug addiction as a crime and those who saw it as a disease. Harm reductionists saw drug use as neither crime nor disease, approaching it nonjudgmentally and pragmatically as an identity formation and way of life. For them, undone science was not simply about understanding drug-using social worlds in order to control or police, but about governing drug users differently by prioritizing health and well-being. They remade naloxone as a "technology of solidarity" within a social movement that saw drug users as agents of change in their own lives and the lives of others. They sought to return agency, humanity, and room for negotiation to marginalized people who were often represented as acted upon by others bent upon controlling drug users either by enforcing law or treating disease.

In embracing the agency of drug users by casting overdose as a health problem that could be prevented and treated, harm reduction advocates and naloxone proponents distinguished themselves from health care professionals, law enforcement, and even first responders such as the EMTs for whom overdose cases had long been commonplace. Theirs was a form of evidence-based activism[37] that took overdose as a matter of social justice, producing knowledge about it for the sake of determining how best to go about structuring future interventions that would change the narrative. Their activism also changed research priorities in the relevant federal agencies, pressuring the monopoly on research agenda setting exerted by power dynamics within the agencies themselves. They revealed the limits of so-called evidence-based policy in an arena where much of the science was as yet undone.[38]

"Undone science is not only about absent knowledge; it is also about a structured absence that emerges from relations of inequality that are reflected in the priorities for what kinds of research should be funded. It is about the contours of what funders target as important and unimportant areas of research."[39] Science and Technology Studies (STS) scholars study aspects of scientific controversy; technological innovation or momentum; policy analysis of new technologies and their contentious politics; or movements for social change centered on embodiment, regulation, industry, and the state. Sociologist David J. Hess uses the framework of "undone science" to situate social movement actors within design processes that engage both professional and lay expertise. While he sees antagonisms between actors as based in "social structural conflicts of race, class, and gender," these are often compounded by social divisions between those who identify with discharging conventional obligations and social norms, and those who see themselves as challenging them.

Design conflicts over opioid overdose prevention with naloxone are structured by an absence of "why" questions. Willful as well as unintentional ignorance and amnesia play a substantial role in what is taken as known—supported by evidence—and what is taken as unknown. For those developing knowledge and practice around and about overdose and naloxone, "why" questions seemed obvious—people were dying from preventable deaths. Depending on where and how they were situated, researchers encountered different barriers to research and obstacles to the democratization of naloxone. They systematically designed studies to address barriers, determine why they persisted and what could be done to brook them; they sought funding for such studies; and at first they were typically denied it. Funders were not similarly situated, to say the least. Early grant proposals were rejected; industry sponsors balked at the lack of evidence; but the regulatory bodies such as the FDA and the CDC became open to "capture" (or at least "cajoling") by activists. A few foundations such as The Tides Foundation or the Comer Family Foundation funded early outreach, education, and research in the overdose arena in the United States.

When areas of undone science are identified, funders typically have to be dragged by activists and researchers into funding newly articulated knowledge problems. Overdose was remade as an interesting scientific problem, remodeled not as a simple matter of mortality, but as a common occurrence consisting of social processes that could be broken down into separate components for the sake of study. Overdose has had to be undone—or in my terms "unsettled"—for it to be remade as a problematic worth looking into. It has had to become a forensic matter, a metabolic matter, a scientific matter, a therapeutic matter, a matter of life not death. As a matter of life, overdose also had to be framed as sentimental in order for its stigmatization to be overcome. Increasingly a matter of public record, into which municipalities, states, and nation-states inquire, the object of overdose has become paradoxically elusive. It cannot be reproduced in a laboratory setting or studied *in situ* due not only to ethical limitations, but also to its very rarity. Endemic and yet rare, epidemic and yet unpredictable, overdose has been made tractable to "the work" of prevention and research. But the real "work" of harm reduction was voiced by one of the activist artists (see figure 8.3): "We made it possible for people to live."

Making it possible for people to live was also the goal of research communities that became involved with overdose throughout the world. Researchers sought to use their talents in service of reducing drug-related harms by comparing different policies, approaches, practices, and methods. But as a group, "researchers" are no more homogeneous than "users," "artists," or "activists." The movement engendered a variety of research modalities when it remade naloxone as a tool for harm reduction. The first researchers out of the blocks had obvious ties to the movement. Motivated by oppositional claims that there was no evidence that harm reduction approaches would prevent drug-related deaths, they raced to prove it. Because they were repeatedly interrogated in public venues about how they knew naloxone worked, activists either became epidemiologists or public health researchers, or allied closely with them as occurred in Boston, Chicago, San Francisco, Seattle, and other US cities, in addition to those in the United Kingdom discussed in the next three chapters.

This was no race for the cure, but a race for proof. US naloxone advocates became qualitative researchers or even epidemiologists out of the need to evaluate programs with which they were personally associated. Nabarun Dasgupta, who worked as an epidemiologist for Purdue Pharma from 2003 to 2005, returned to graduate school upon realizing that there were no solid national numbers on the actual prevalence of overdose even at the CDC.[40] He researched the relationship between licit pharmaceutical opioids and illicit opioids, becoming deeply involved in the institutionalization of harm reduction in the state of North Carolina through Project Lazarus and other organizations. Within the harm reduction penumbra, social justice activists such as Alice Bell, who modestly characterized herself as a "social worker who runs a needle exchange in Pittsburgh," despite acting as the energetic linchpin of a large national

8.3 "Remember: Two holes in your arm are better than one in the ground. Don't OD!!!" From the *junkphood* special issue on the UFO Study (1998): 39.
Used with permission of Heather Edney and Brooke Lober.

network and an inveterate and incisive research critic, spoke approvingly of "people who are in the harm reduction world who are researchers."[41] Although preserving the activist/researcher distinction was important, volunteering at needle exchanges or conducting naloxone trainings was a marker of commitment and credibility for movement-based researchers. Harm reduction practice remains important for generating the insights on which research questions are based.

Evidence for clinical judgments about nasal naloxone's role in overdose management was quite limited when Walley got his research underway. No federal money had been designated for research on overdose, overdose prevention, or naloxone; and he perceived that his proposals went into the "harm reduction wastebasket."[42] As discussed in the previous chapter, only two naloxone pilots had been published before 2005.[43] When in 2007, John Auerbach, Epstein, Michael Botticelli, Kevin Cranston, Peter Moyer, and Sarah Ruiz expanded the City of Boston's program statewide, Walley became chief medical director of the Massachusetts naloxone program. Although a physician, Walley sought to "get the nurses and doctors out of it."[44] He and his team also convinced Massachusetts to maintain a database of all overdose rescues; they set up a data collection infrastructure with support from the CDC. They published "Saved by the Nose," a field action report grounded in the work of AHOPE Needle Exchange that in 2009 was the first article on naloxone to appear in the *American Journal of Public Health*.[45] "The shift to statistical analysis made a big difference. I could not do that by myself, [but I] had the idea for an interrupted time series analysis, an ethical natural experiment. I was not enthusiastic about a controlled experiment. But naloxone will not be accepted as evidence based until RCTs are done," Walley predicted.[46]

Harm reduction-oriented researchers found themselves engaged in the infinite regress of an evidence-based era. They identified questions that the movement had to answer if its practices were to be justified and its interventions evaluated as effective. Opponents continually raised uncomfortable questions that motivated researchers to spend valuable time and resources figuring out how to answer antagonistic questions. They therefore had little time to conceptualize and assemble resources to address questions that were *not* posed by antagonists. Tailoring his research to address criticism of naloxone, for example, National Health Service analyst Andrew McAuley, who ran one of Scotland's first naloxone pilots (discussed in chapter 10), recounted response from Scottish general practitioners:

One of the drivers for me in terms of my research on naloxone has always been the "criticism papers," as I call them … saying it [take-home naloxone] was a bad idea. They were saying things like people would use more drugs, people would use it as a weapon, it would have unintended consequences, people would have adverse reactions to it, and one of the people also said that people would stop phoning ambulances so people would be at more risk if they had naloxone because people wouldn't get the medical attention they needed because people would start self-managing overdoses. I always had that in the back of my mind.… I thought it would be really good to answer that question or a lot of those questions.[47]

Doing undone science required systematically responding to charges that drug users were unreliable and disorganized.[48] Initial studies were designed to ask whether drug users themselves found naloxone socially acceptable, whether they retained knowledge for substantial time after learning how to use naloxone or whether possessing naloxone deterred people from calling for professional help. McAuley and his group found that having naloxone on hand and knowing how to use it did not discourage people from calling for the ambulance.[49]

"Why" questions are phenomenological, requiring qualitative approaches to the lived experiences of those who administer naloxone:

We found what we expected in terms of people's willingness to intervene to save their peers, but we also found a lot of surprises in terms of their tendency to administer the whole dose at once even though they're trained to titrate the dose to avoid withdrawal or "dope sickness," as you would call it in the United States. Everybody we interviewed who had used naloxone inevitably gave the whole amount. The sheer panic of the situation led them to give the whole amount. The result of that tendency to do that was that the response that they got from the individual was very rarely positive. In some cases, which tended to be the exceptions rather than the norm, they got some positive feedback and sometimes that was delayed. It wasn't until a couple days later that people realized their life had been saved and said "Thank you." The immediate response was usually very negative. It could often be verbally or physically aggressive. A lot of people have seen relationships break down on account of the administering naloxone to people they've known for quite some time.[50]

People who administered naloxone felt undermined and unappreciated, as if they "had been through a traumatic experience.… They were really

annoyed at the lack of gratitude, a lack of awareness and acknowledg-
ment about what they had done."[51] This unexpected finding could only
have become known through qualitative study. The perils of engaged
research on emergent matters also surfaced:

Things like that, I wasn't necessarily prepared for that. I was prepared for the
person who'd been saved to be negative, but I wasn't prepared for the person
who'd administered it to also relay negative experiences. So that was really
interesting in terms of that dynamic. Then there's another whole chapter I've
got which is about perception of role, role adequacy and role legitimacy. A lot
of people talked about how they felt their peers perceived them when they
found out they had accessed a naloxone kit. A lot of that was underpinned by
their existing credibility within a network so if they were a credible peer, they
were accepted as someone who could have naloxone and that was fine. Some of
the people we interviewed who were quite young or had only recently started
injecting did not have that credibility, and they were mocked or laughed at
when they tried to discuss naloxone with their peer group.[52]

In addition to reluctant researchers identified as activists, who are enthusiastic harm reduction-affiliated allies, naloxone research attracts bona
fide scientists and clinical researchers involved in producing "evidence"
within the parameters of evidence-based medicine. These professionals
move easily in the more rarified international echelons of public health.

PROTAGONISTS UNITE! PARTISAN PROFESSIONALS, ENGAGED RESEARCH, AND UNDONE SCIENCE

Relationships between public contention and good science are often cast
as if the public lacks understanding of "good science," while those producing "good science" oppose the public good. I have not seen much evidence
to support either claim. To the contrary, intense, ongoing, bidirectional
interactions between official publics and counter-publics changed political
opportunity structures, rendered some actions more or less meaningful,
and enabled harm reduction institutions, organizations, and coalitions to
be built inside or alongside public health, emergency response, and police
departments in places hardest hit by opioid overdose. Social movements
are knowledge projects, where "activists and advocates strive to convince
others to embrace their definitions of a situation and their prescriptions
for action, and their task of persuasion entails getting the information out

to them and making the information credible."[53] Like the environmental activists who embraced a "politics of scale,"[54] seeing how data aggregated at too high a level masks local effects made naloxone advocates acutely aware of the mismatch between the scale at which knowledge is produced and the scale at which regulatory policy is made. A politics of scale works in harm reduction, which focuses on microlevel social interactions, even as the macropolitics of regulatory policy and the War on Drugs unfold.

While pressuring the FDA to make naloxone over the counter, US harm reduction activists defiantly went on handing it out via needle exchange. In the United Kingdom, Scotland mobilized its smaller-scale population and commitment to accurate data collection and evaluation to maximize the impact of its National Naloxone Program. The politics of scale is problematic in places where the sheer scope of the overdose problem outstrips the infrastructure to respond, as has occurred in many US states. Variation between states in terms of access to health care, including the full range of drug treatment and reproductive health options, is such that the effectiveness of naloxone distribution does not always show up and efforts are represented as futile. Lacking national coordination, the United States depends on the motivational energies of harm reduction activists and advocates. This has, in turn, propelled researchers into advocacy positions; they have become in the terms of this book "protagonists."

The hierarchies of credibility at work in the naloxone access arena generate epistemic conflict deriving from multiple interpretations of harm, risk, and safety; different ways of organizing efforts and evaluation; and conflict over the moral worth of drug users and political implications of supporting their activities. Actual harms differ from abstract notions of potential risks. Movements sometimes prefer to take precautionary approaches to scientific uncertainties, limiting technological development or diffusion. Protagonists realized the necessity of preserving relations with pharmaceutical suppliers in order to ensure low-cost access to naloxone.[55] Positioned in a complex way relative both to the pharmaceutical industry and its regulators, protagonists serve as industry watchdogs. The FDA must be alternately enrolled, engaged, pressured, or protested. Activists deemed OTC naloxone in the public interest, appearing uninvited but carefully prepared during the public commentary period at the 2012 hearing, after which both the FDA and the CDC warmed toward them. The naloxone access movement operates in close kinship with patient advocacy groups seeking to widen therapeutic options and realistic solutions. Such organizations broadcast from narrow bandwidth to broad human rights activism. Focusing narrowly on naloxone access has enabled an intensified and globalized movement mobilization that closely follows the "mass" dimensions of problematic opioid use. As these have become apparent, the movement has shifted toward "mass" delivery of overdose prevention, naloxone training, media advocacy, and legislation.

Naloxone access is also tied to drug treatment access,[56] and thus it matters what counts as evidence-based treatment. In England, the naloxone access movement ran up against political critique from Conservative Iain Duncan-Smith and others promoting abstinent recovery as the only goal of public policy.[57] Access to medication-assisted treatment (including buprenorphine, methadone, and naltrexone) is restricted in the United States due to uneven coverage by health insurers. Although under the Obama administration some states expanded Medicaid coverage for evidence-based drug treatment, many later pulled back from this commitment. Similarly, access to naloxone depends on social location and readily maps to intersectional inequalities including racial-ethnic formation, sexuality, dis/ability, class, and urban, rural, or suburban setting. The only thing that seems to make a real difference in terms of access to naloxone is the presence of protagonists. Where there are people who understand naloxone's significance, there is naloxone. That some of them are scientists and clinicians means they must maintain credibility across changing social and governmental contexts.

Scientific protagonists develop strategies to maintain the perception of their neutrality, strategies that often foreclose alliances with social movements. While opting for this strategy may appear politically neutral, the commitment to "whither goes the evidence, there I will go" is not neutral but instead advances the idea of science as a "culture of no culture."[58] When governments get scientists to deploy technocratic risk assessment, efficacy assessment, and cost-benefit analysis, all seemingly politically neutral ways of analyzing technology's effects upon populations, the terms of engagement are closed to question. The harm reduction movement questions what counts as costs, benefits, and harms, who gets to define these terms, and who gets to define what count as harms. User groups and allies create different venues and processes for making technological innovations and assessments more participatory. An expanding literature documents tokenism, co-optation, and other nondemocratic forms of "participation" in government venues. In the case of naloxone distribution, as with other once-revolutionary movements that find themselves implementing service delivery, enduring questions about the effectiveness of naloxone or the opening up of "saves" to other interpretations arise.[59]

Protagonists who are "well-positioned insiders"[60] have to be very well positioned indeed to withstand scrutiny in the midst of crossfire from both dominant and subdominant groups. Audit cultures bent on accountability are also accusatory in ways that produce (dis)identification with social movements and put experts on the defensive. As evidentiary "holes" are plugged and once-undone science gets done, new questions arise about how much evidence is enough and if there is a reasonable end to the process of producing it. On high alert in relation to research, activists focused on overdose prevention with naloxone have become research-literate critics and contributors to the relevant sciences undertaking research. Institutional logics often change with political transitions, and the demand for naloxone research took quite a while to yield results, particularly for those reliant on federal funding. The NIDA was slow to consider funding naloxone research, seeing it as a technology that had reached closure decades ago, and thus no longer offered interesting avenues for research. Activists and scientists catalyzed not only industry response, but also governmental response to newly articulated uses of and needs for naloxone. Older social movement responses to the pharmaceutical industry—the women's health and consumer's movements in particular—cast it as greedy, corrupt, and exploitive. The AIDS movement shifted such framings in order to lead the charge for so-called fast-track regulatory approval. This process helped clear the way for more inclusive practices in the federal research apparatus[61] and for legitimating questions concerning applicability of findings across populations. Harm reduction-oriented naloxone access activism originated out of HIV/AIDS activism, but the movements differed due to the differing political climates that shaped them. AIDS activists cultivated expertise within their ranks in order to pressure the FDA and the federal research apparatus, creating a new form of public health capital from which many emerged with advanced degrees in public health and epidemiology.

Climate shifts are slow, and the project to cultivate scientific credibility and cultural legitimacy for overdose death prevention was no exception. Researchers who wanted to study peer naloxone distribution enjoyed little support. Theirs was a form of "undone science" attributable to institutional inertia, overinvestment and over-enthusiasm for neuroscience at the NIDA, and lack of support for study design that might reveal how people go about reducing the harm of what, after all, remained an illegal activity until recently. Activists appealed to private foundations and local health departments to fund direct distribution of naloxone. At the federal level a "service delivery" model based on a division of labor between research and service developed; at times, "research materials" that could be used for harm reduction such as bleach were approved on grants to be distributed by outreach workers and research assistants. This kind of circulation of "research materials" brought harm reductionists engaged in outreach and needle exchange together around the needs of drug users.[62] As a "boundary object," naloxone is contentious; negotiation over its meanings and cultures of use has often taken conflictual form as actors who have formed strong commitments to particular design trajectories have very different operating assumptions about how naloxone as a cultural artifact should be configured.

In addition to breathing "life itself" into an overdose situation, naloxone protagonists believe that both giving and receiving naloxone sends affirmative messages about moral worth and social value. An anthropological term, "gift" introduces a phenomenological resonance conveyed by embodied enactments occurring in the spaces of harm reduction research and practice. These enactments are configured around naloxone as a material technology of solidarity. "Every material is a relation with a milieu," says philosopher Isabelle Stengers,[63] who was trained as a chemist and thus holds expansive attitudes toward molecules, materials, and their inter- and intra-actions.[64] Materiality and meaning are both acquired in the middle of an experimental milieu, as the materialist enactments and pronouncements of protagonists in harm reduction movements exemplify. In an interview with Brian Massumi, Stengers stated:

In politics, I think that the question is less microperceptions than procedures, practices that are apt to produce new perceptions carrying new consequences, to produce collective assemblages of enunciation experimenting with ways of combining creation with an active, experimental taking into account of their milieu—a milieu that is by definition unhealthy—experimenting with how to "think through the middle," through the milieu, in the way that collectives for direct nonviolent action have done, knowing that the cops will provoke in any case.[65]

Almost by definition, immersion makes it tricky to discern how procedures, practices, and perceptions unfold. A very concrete "technology of solidarity," naloxone carries symbolic meanings that affirm the moral worth and social value of drug users. Immersive and experimental, social movements constitute their protagonists' realities by providing the material stuff for ongoing contestation. They do not sit still and congeal.

Health social movements vary in goals. Some seek access to health care, others aim to improve the quality offered. Constituency-based movements address health disparities through a vocabulary of social or environmental justice; "embodied health movements" respond to direct experiences of disability or illness.[66] Such collectivities organize self-help, mutual support, and alternative therapies as ways to live with chronic conditions. Unaligned with biomedicalized approaches, such movements enable those identified with them to overcome prevailing "epistemologies of ignorance."[67] Yet they, too, produce knowledge "gaps" or "epistemic holes." A new focus on embodiment and knowledge making has emerged in STS around the crafting of protocols and procedures from "discrepant histories of politicizing life and health" that respond to "already existent practices, infrastructures, commodities, epistemologies, and subject positions available in the landscape of its emergence."[68] While studying the epistemological assumptions that undergird existing knowledges and providing genealogies for how they came into being, STS focuses on forms of knowledge that are "forbidden," "hidden," or otherwise suppressed.

While it would be overly dramatic to claim overdose research was "forbidden knowledge,"[69] getting politically contentious research funded sometimes requires self-censorship and attempts to "cleanse" titles and texts of controversial words and ideas. As Kempner showed in a novel study of how politics affected National Institutes of Health researchers (including drug and alcohol researchers), scientists studying controversial topics perceive themselves in conflict with "intolerant" publics and a political culture that opposes research and encourages "intemperance."[70] Such perceptions fuel researchers' sense of victimization, a sense that was disclosed in many of my interviews with naloxone researchers. US-based researchers described how they made grant proposals acceptable to the NIDA by couching their research in terms unaligned with harm reduction. They created behavioral health "interventions" in order to get projects funded years after initial proposals obviously pitched from a harm reduction stance were turned down. For naloxone research to be publicly funded, proposals had to signal intention to interrupt the behavior being studied rather than allowing it to continue in ways that appeared to "condone" it.

Training the spotlight on health social movements as research aspirants reveals that they attempt to revalue the moral and social worth of those who identify with them, all the while producing evidence for knowledge claims. They confront questions about how "local" knowledge made on the basis of lived experiences can be generalized so as to be viewed as a contribution to systematic knowledge. Harm reduction movements do not fit the typical frame of actors arrayed on a spectrum running from radical/revolutionary to incremental/accommodationist, but are instead hybrid adaptations in these respects because they adapted to the imperatives of an "evidence-based" historical moment. Movement participants question the validity of "evidence" produced without reference to lived experience. Like feminists who urged women to "seize the means of reproduction," protagonists of harm reduction promote "counter-conduct" by crafting protocols adapted in complex relationships to social conditions.[71] Protocols order and evoke actions within a pragmatics of counter-conduct that constitutes participants as persons who are not subject to hegemonic rules of behavior. They resist or otherwise diverge from prescribed conduct—as even a cursory brush with the affective dimensions of harm reduction movements makes clear. Drug users engage in complex exchange relations of reciprocity and care. The capacity to contribute to the assembly of the evidence base required for activism and advocacy today takes a fair amount of cultural capital. For harm reduction movements to reproduce themselves, they must value "counter-conduct" in ways that become more difficult as they become more institutionalized and less "revolutionary." To be taken seriously in an evidence-based world, they must produce protocols and research that meet very high scientific and ethical standards, while also appealing to the affective grounds of counter-conduct. What it takes to produce acceptable scientific evidence, technology assessment, or program evaluation differs from what it takes to embody harm reduction.

Harm reduction movements are an interesting species of an embodied social health movement. They aim to change what people do without invoking the terms of abstinent-based recovery or behavioral health, both dominant discourses in the substance use realm. They encourage individuals to integrate "any positive change" without requiring commitment to rigorous new identities exacted by what they see as unrealistic proponents of a "drug-free lifestyle" or long-term recovery. Harm reduction therefore occupies a position of ongoing epistemic conflict relative to abstinence-oriented and recovery-based movements, which in the United States have targeted the federal apparatus to bring about a recovery-oriented treatment infrastructure. US harm reduction and naloxone access activists continually come into epistemic conflict—often misrecognized as social conflict, generational conflict, or identity-based conflict—with recovery-oriented groups intolerant of the unique brand of pragmatism and radical social acceptance of difference that is harm reduction.

The United Kingdom, to which the next three chapters turn, was structurally more hospitable to harm reduction in its statutory services sector; naloxone access advocates could afford to be less "revolutionary" in these places. But public health governance is rapidly changing in the United Kingdom as the Tory government devolves such matters to local authorities with uneven effects. In Brixton, where organized "service users" have gained voice, harm reduction and drug treatment remain on the agenda.[72] Ironically, in Liverpool, once a harm reductionist stronghold, naloxone access was not prioritized.[73] Brexit may change the meaning of harm reduction, which has typically allied with European and global organizations.[74] Harm reduction targets preventable conditions—HIV/AIDS, hepatitis C, opiate overdose—and works via self-organized, community-based organizations offering low-threshold services. Medical professionals working in such environments emphasize dignity, rights, and collective response. In the United States, where needle and syringe exchange was illegal, activists convinced local law enforcement to ignore their activities while they meanwhile worked to change state laws governing naloxone access. As they gained allies in state legislatures and medical licensing boards, they pressured successfully for changes to Good Samaritan laws and policies governing access to medicines, clean syringes, and naloxone. By challenging mainstream notions and insisting on drug use as a way of life, such movements came increasingly to collaborate with professionals such as public health officials; police, sheriffs, and other law enforcement; EMTs; and emergency department clinicians. They appealed to overarching commitments to human rights, social responsibility, and social justice in implementing programs that undertook concrete and practical steps to improve near-term experiences.

Preventing overdose deaths brought new energy to public health and moved it toward harm reduction. As activists institutionalized peer education, recovery support, or "postoverdose response," new forms of interactional expertise emerged to integrate "lived experience" into the evidence base.[75] New ways of relating the "evidence of experience" were embodied in the knowledge of how to save a life that grounds this form of interactional expertise. Where people who have experienced harm are included in the design process, remedies are more likely to be tailored to actual situations rather than invoking ideological concepts such as "deterrence." Yet this is getting ahead of our story, for many of the events recounted in the next chapters focus on events in the United Kingdom that sharpened both theory and practice, if bracketing the racial politics of overdose raised in the United States.

9

RESUSCITATING SOCIETY: Overdose in Post-Thatcherite Britain

Overdoses are typically called drug-related deaths (DRDs) in the United Kingdom. Knowledge about them was structured by the so-called British system, a national registry kept by the British Home Office of all patients on medically prescribed morphine or diacetylmorphine (heroin). As many on the list had become addicted to pain relief from war trauma, "therapeutic addicts" were considered able to live "a relatively useful and normal life."[1] Although the British system did not outlast the 1960s,[2] the policy of maintenance was invoked as a humane alternative to criminalization in the acrimonious US drug policy debates of the 1950s and 1960s. According to a British Ministry of Health Interdepartmental Committee on Drug Addiction, the total number of therapeutic addicts in Britain never officially exceeded 760 persons. This committee's 1965 report (named after its chair, eminent neurologist Walter Russell Brain) responded to an emerging generation of drug-taking youth.[3] The Brain Committee had noticed that the clubs of London's West End were enjoying a "vogue among young people who can find in them such diversions as modern music or all-night dancing," as well as "pep pills" (amphetamines), "reefers" (cannabis), heroin, and cocaine.[4] The 1965 report characterized addiction among this generation as a "socially infectious condition."[5] By contrast to the older, "relatively useful and normal" therapeutic addicts, members of the younger generation were cast as "totally different," "unstable," "psychopathic personalities," "poly-addicts," and "junkies."[6] They were to be controlled via special treatment centers called Drug Dependence Units.[7]

This new generation of "nontherapeutic addicts" was the focus of I. Peirce James's 1967 examination of "Suicide and Mortality amongst Heroin Addicts in Britain," which found a death rate twenty times higher for them than for people of similar age who did not inject heroin. Young addicts were dying at a rate substantially higher than that of the older, iatrogenic heroin users on the registry. In addition to overdose, diseases such as "syringe-transmitted hepatitis" and sepsis loomed large.[8] This chapter tells the story of how British protagonists—mainly professionals and researchers with occasional involvement of drug users themselves—built harm reduction into drug policy and moved toward peer naloxone provision in the United Kingdom.

The United Kingdom exhibited an exceptional profile. New recruits into heroin use increased in the late twentieth century. Nor were the countries homogenous: DRDs rose by 25 percent from 1997 to 2000 in England and Wales and remained much lower in Northern Ireland at that time—but deaths increased by 150 percent in Scotland.[9] The United Kingdom undertook a series of strategies to "tackle" the national problem, beginning with a 1985 Conservative government white paper, *Tackling Drug Misuse*, that was updated in 1995 as *Tackling Drugs Together*.[10] Tony Blair's New Labour government then set out a ten-year strategy in 1998 titled *Tackling Drugs Together to Build a Better Britain*.[11] Then-Prime Minister Blair believed that "change is happening and Britain is becoming a better place to live in. But it could be so much better if we could break once and for all the vicious cycle of drugs and crime which wrecks lives and threatens communities."[12]

Assiduous historians have pored over documents and regulations structuring the British drug situation; my purpose is not to duplicate their work but to sketch the background of what came to be called take-home naloxone (THN). Seeming continuities among the "Tackling Drugs" documents belied marked ideological differences. Labour's approach elevated medicalized strategies such as health and harm reduction, and centered the causative role of social exclusion.[13] Conservative approaches emphasized punitive strategies based on abstinence and criminalization. The drug treatment policy community, composed largely of liberal-leaning researchers, busily produced the empirical base for evidence-based policy, and THN was nothing if not evidence based. According to historian Sarah Mars, Conservative politicians acted divisively and opportunistically to "gain political capital from taking up drug issues."[14] Marked contrasts in the policies and practices of the parties have whip-sawed addicts and clinicians alike over the past quarter century.

PAVING THE WAY FOR TAKE-HOME NALOXONE: UP CLOSE AND ARM'S LENGTH

When rising overdose rates among injecting drug users began to be noticed in Britain, it was just before the advent of HIV/AIDS, which was often presented as changing the meaning of everything preceding it. Recognizing HIV as posing greater dangers than drug misuse, the UK government adopted harm reduction as national policy in the late 1980s.[15] The McClelland Report (1986) prioritized safer use over abstinence, justified better educating drug users in safe injecting practices, and became the basis for government-supported needle and syringe programs in Scotland and England.[16] Although met with moral outrage by the Conservatives who had commissioned it,[17] the Scottish Committee on AIDS and Drug Misuse, which McClelland chaired, was the first national group in the world that said:

We need to give out needles and syringes. That was 1986 and that was a game changer. That introduced needle exchange in a big way. Again, that was ministerially led. There was a task force set up by a conservative minister, a right-wing minister, but there was a crisis—and the crisis in 1986 was HIV. The crisis in 1987, 1988 was continued drug injection. The crisis now ... was Drug-Related Deaths, the figures were always going up. Crises are good if you want to change a policy.[18]

The United Kingdom mounted a swift and effective response to HIV/AIDS. Policy pragmatism prevailed over political and cultural differences, partly because matters requiring expert advice were insulated from public opinion.[19] Tight relationships between medical experts and civil servants marginalized the opposition.[20] Dominated by addiction experts and created by the Misuse of Drugs Act (1971), the Advisory Council on the

Misuse of Drugs (ACMD) was the responsible party and the obligatory passage point through which policy guidance for the UK government passed.

Unofficially, Scotland was a bastion of harm reduction. One of the first needle exchanges in the world began operating in 1986 on Bread Street in Edinburgh, where an intense concentration of HIV among intravenous drug users had been identified by a doctor whose practice was located near Muirhouse, the housing estate in north Edinburgh that gained a reputation as the epicenter of Scotland's twin epidemics. Roy Robertson diagnosed the first cases signaling the presence of the AIDS virus among intravenous drug users in Edinburgh.[21] He knew his patients well enough to trace HIV infection back to the early 1980s, and to identify intravenous drug use and widespread needle sharing as the culprit. As early as 1983 scoring heroin had become more difficult in Edinburgh, and so injectable amphetamines and buprenorphine in the form of Temgesic entered the city's drug scene. Increased law enforcement and harsher sentencing propelled these changes in local illicit drug markets, which had in turn fueled transmission of blood-borne viruses like HIV and hepatitis.

To understand how these forces affected drug user health, Robertson's practice, the Muirhouse Medical Group, followed patients who turned out to be HIV positive in retrospect:

We know that [HIV/AIDS] started in 1982, but the test didn't become available until September 1985 and that was the first time internationally the antibody test was available. Everywhere in the world used the antibody test and we found out that there were cases all over the world. We found out our cohort had been badly affected, but none of them were symptomatic. They were all asymptomatic and didn't know they had it until 1985, 1986 until we tested. We published in *AIDS Journal* and *BMJ* and suddenly we had people visiting from all over the world. It really was extraordinary.[22]

Handing out clean injecting equipment and carrying naloxone wherever he went, Robertson became a compassionate advocate for drug users and a perennial participant in policy making in Scotland, the United Kingdom, and Europe. As a media spokesperson for harm reduction, he often checked the sensationalism of stories in which the "explosive spread of infection" was over-dramatized. On a late 1986 visit to New York City, Robertson was the subject of a *New York Times* story in which he characterized attitudes toward AIDS in Edinburgh as a "grinding combination of public bewilderment and press sensationalism. Camera crews arrive looking for deathbed tableaux, a phenomenon that is perhaps months away at the rate of AIDS infection."[23]

There was something to the charge of sensationalism. Edinburgh had been dubbed the "AIDS capital of Europe" in a *Sunday Telegraph* article by C. Dawson on April 13, 1986. The "dubious accolade" stuck so stubbornly to "the Athens of the North," formerly represented as a place of "old-fashioned virtue," that two Scottish physicians felt compelled to defend their city by tracing media "misapprehensions" and recasting other European cities as those where AIDS first appeared.[24] On September 24, the short training film *The Edinburgh HIV Epidemic* by BBC television reporter Clive Ferguson echoed the phrase, showing as backdrop moody photographs of Council Housing on the Muirhouse estate as Gerry Stimson, founding director of the International Harm Reduction Association, noted that the "early intelligence about the spread of HIV in Edinburgh didn't come from a big surveillance study, but from a family practitioner in Edinburgh who had become concerned about rising incidence of hepatitis B among injecting drug users."[25]

Then the young Roy Robertson, sporting dark hair and a sport coat and striding confidently to the front door of his practice, began to speak about how drug treatment was once in the hands of social workers before the advent of HIV made it a medical issue with a role for physicians. Over the years Robertson had collected hundreds of blood samples from patients to test for hepatitis B, repeating them as many as ten times for each patient to ensure that infection had cleared for the entire cohort. Once HIV tests came out, Robertson realized that he could begin from an individual's most recent sample and move backward in time so as to pinpoint the exact moment when the person had become infected with HIV or hepatitis B. Tracking the Edinburgh HIV epidemic in this way, Robertson and colleagues found that fully half the cohort had become HIV positive between September 1982 and 1985. "Here we were describing untreatable, subclinical disease in a group of people who were young, who weren't gay men, who had sexual partners, children, a lot of them were women, so it was a different population, a population that really hadn't been described before with HIV."[26] The Edinburgh HIV epidemic sparked the UK government to act in uncharacteristic fashion due to the implication that many people might be HIV positive and not know it. The idea that heroin users were sharing injecting equipment and thereby creating an epidemic moved the whole country toward harm reduction.

Before antiretrovirals made HIV/AIDS survivable in the mid-1990s, the grim knowledge that everyone progressed to AIDS made prevention a prime directive.[27] Methadone maintenance was made available for injectors thanks to physician Raymond Brettle, an infectious disease consultant at Edinburgh City Hospital, but this was not enough to counter the shooting galleries that heroin dealers set up in order to avoid police detection—a tactic that promoted needle sharing. By the mid-1980s, most of Edinburgh's injection-related HIV infections had already occurred; by contrast, HIV prevalence in Glasgow injectors remained low because of different policing patterns and different social conditions. One Edinburgh study traced all "individuals thought to be drug injectors" between 1980 and 1985 for a two-year period from October 2005 to October 2007.[28] Injectors and controls (the latter also recruited from Robertson's practice) participated in interviews concerning early life experiences, substance use, health, and social histories. Cases and controls were linked to medical records, criminal justice records, death registrations, and hospital contact statistics. This approach provided for a far more socially situated form of knowledge than is typical of epidemiological studies. This cohort is still being followed.[29]

Enabling longitudinal studies in a hard-to-follow population, Robertson's "Edinburgh Addiction Cohort" played a substantial role in enabling overdose to be discerned despite the spotlight on HIV/AIDS. A 1990 study showed few deaths from AIDS among a group of known drug users but "increasing deaths from overdose, especially in young women."[30] As a primary care physician in a place with a high concentration of problem drug users, Robertson indicated that he kept naloxone ready to hand and administered it when necessary on the streets. Enfranchised to administer naloxone like any other prescription medication, physicians needed no special dispensation to carry or use an ordinary part of the physician's kit. The question was: Who else would be granted that opportunity? Could pharmacists dispense it? Staff of homeless hostels? Drug users themselves? Robertson's own views toward his patients exemplified a combination of pragmatism and humanism that pervades Scottish policy making: "Drug users, far from being exclusively manipulative, violent, and antisocial, provide a rewarding and fulfilling patient or client group who present with a challenging array of clinical and psychological problems."[31] In the face of this human element, Robertson felt that dithering over drug policy was inconsequential:

HIV swept all that away. It all became medical. It was our business. We are doctors, so get out of our way. All of a sudden, we were in charge and we had the support of our colleagues in infectious diseases and sexually transmitted diseases. Of course, the activist movement, the Terrence Higgins Trust, the Scottish AIDS Monitor, the gay community in Scotland, they were breathlessly, daringly challenging authority. Again, saying that we represent this marginalized, this oppressed community and we demand that something's done about it. The gay community led the charge in so many ways and at so many levels. They penetrated every level. They had people in politics, senior politicians.[32]

Robertson came to believe that drug users were an oppressed minority that lacked advocacy of the kind then cultivated by the gay community. "Drug users are crowded in by the law and criminal justice. They're oppressed by inequalities, they're marginalized by lack of education, and because of that huge stigma, no one is going to support them."[33] Realizing the probable fate of the Edinburgh Addiction Cohort,[34] Robertson called for prevention and intervention in Scotland's high DRD rate.[35] As recounted in chapter 10, he would find himself in a position to democratize naloxone far beyond his own professional practice as chair of the National Forum on Drug-Related Deaths (NFDRD).

One of the cyclic questions of drug policy history is, who will decide—will medical or public health expertise prevail, or will law enforcement determine the policy fate of drug users? Depending on how such questions are answered, research priorities and resource allocation follow. In the United Kingdom, a social welfare state grounded in policy-making by expertise, a "fevered atmosphere" prevailed after former prime minister Margaret Thatcher resigned, to the point that some working in the drugs field feared the welfare state itself might implode: "The fears that treatment services would be cut and that an American-style abstinence agenda would come to dominate were widespread in the drugs treatment field in the 1990s, along with fears that needle and syringe exchanges and other harm reduction services would be banned."[36] This concern became especially pointed when Conservatives tried to import an American-style, abstinence-based recovery agenda in the early years of the twenty-first century. *Tackling Drugs Together* announced the Conservative Party's goal of a drug-free nation—but also indicated that steps to reduce HIV and other harmful consequences of drug use would be based on evidence rather than rhetoric.[37] Public-private partnerships called Drug Action Teams were formed across the drug treatment and criminal justice fields, general practitioners were urged to become more involved in specialist treatment services,[38] and realistic approaches to evaluating treatment effectiveness were set out. Maintenance-based treatment—specifically opioid substitution therapy (OST) using methadone—was repackaged as harm reduction. The overall message of each of these measures inspired by *Tackling Drugs Together* was "treatment works."[39]

Then Britain's rising DRD rate began to become a concern. In *Reducing Drug Related Deaths* (2000) the ACMD quantified the heavy "burden of years of life lost"[40] to overdose and outlined gaps, uncertainties, and weaknesses in knowledge about DRDs, together with shortcomings in reporting them. Medicine, of course, had a remedy on offer—naloxone—which the AMCD recommended be added to the United Kingdom's exempt list of prescription-only medicines, meaning that anyone may use such drugs to save a life in an emergency. Although that change did not occur for another five years, the ACMD began to pull together the constituencies, "cohesion and common purpose,"[41] and knowledge base necessary to support the change. The ACMD also recommended that a change to the statistical basis for DRD reporting in the United Kingdom. Deaths related to drugs were not to be reported on the usual basis of occurrence within the general population (deaths per 100,000), which made them seem extremely rare. Instead the more salient value of deaths per 100 problem drug users or injectors was to be used.[42] That way regional variations in the prevalence of problematic drug use would not render DRDs invisible.

The Blair administration also created another agency important for THN in 2001. This was a UK Special Health Authority called the National Treatment Agency for Substance Misuse (known as the NTA). Rather than being subordinate to the Home Office or Health Ministry, this was a so-called arm's-length body, a nongovernmental organization granted statutory powers, upon which any government body could call for advice and guidance. In 2007 it was the NTA that recommended naloxone be used more consistently in emergency services.[43] Then in 2011 the NTA reported on sixteen pilot THN trainings set up to transmit "overdose and naloxone knowledge" in diverse community settings. Parents in particular reported experiencing feelings of empowerment, confidence, and a "measure of control over their son or daughter's drug use by being able to intervene in an overdose—whereas previously they had felt powerless."[44]

As Director of Quality at the National Treatment Agency for Substance Misuse (NTA), Annette Dale-Perrara oversaw England's expansion of drug treatment services even when the Conservative Party was attempting to impose time limits on maintenance therapy. Convinced that getting more drug users into treatment was key to reducing DRDs, the NTA doubled the number of UK heroin users in treatment:

The role and remit of the NTA was to oversee the new money that came in. [Our goal was] to double the number in drug treatment because we had a heroin epidemic.... From 2001 to 2007 we doubled the numbers of heroin users in treatment. I think the total numbers in treatment got up to just under 210,000 per annum, of which about 165,000 were heroin and crack users. So it was a massive expansion and, so a nongovernment department body, the NTA, was given the task of overseeing the money, keeping the performance steady, driving the performance, and overseeing the quality."[45]

Despite good results, the NTA was besieged for supporting OST and THN, both of which were contested by opposing politicians on the basis of the quality and quantity of evidence.

FROM THROWAWAY LINE TO DIRECT PROVISION: THN IN UK DRUGS POLICY

Research conducted on THN often originated from people with tight connections to the harm reduction movement, but activists' priorities did not match the cultural and professional imperatives of government agencies, clinicians, and researchers. For example, one influential set of researchers was located at the Kings' College London National Addiction Centre (NAC), an enterprise that produces evidence-based guidelines and professional outlooks, rather than outsider art or social protest. Directed by addiction psychiatrist John Strang, the NAC is a clinical research center.

Strang's first "real, proper research project" traced how heroin users fared when turned away from treatment; he found that many moderated or quit using drugs without the help of professionals:

[They'd] moved away from the problems they'd had. The problems had moderated. Some of them had quit, and for some of them it was just less bad than it had previously been. We were struggling for a language like harm reduction. We wrote a paper about "habit moderation changes amongst heroin users" and if we were writing it a few years later, we would have called it "harm reduction." We showed that people had moved from injecting their heroin to smoking it, or they'd reduced the dose, or sometimes it was the frequency per week which they'd eased back on.... Change was possible, even though continued use might be occurring. It opened up the intervention landscape.[46]

People engaged in processes of social change have various theories of how to avail themselves of levers and pressure points. Strang evolved a form of clinical research that simultaneously fit the culture of evidence-based medicine and that of harm reduction advocates. But the terminology, principles, and practices of both sides were nascent in the 1980s, and Strang still speaks of his surprise at the results of his first study:

It was quite a learning exercise for me because I think, in that sort of over-confident way that you do in the beginning of your career, I just presumed that anybody who was turned away from treatment would do very badly. But actually we discovered people did quite well and it's a good thing to find out that people find other ways of making changes, and therefore, maybe there's another way of working where you might look to facilitate those changes that they are themselves able to make. That's where the habit moderation paper went—"turned away from treatment and denied access to care, and yet they find a way of reducing the damage of their drug use."[47]

Believing that "incorporating harm reduction was what any competent holistic practitioner should do,"[48] Strang has investigated questions often perceived as unsettling to orthodoxies, whether they issued from the harm reduction movement, government bureaucrats, or pharmaceutical companies.

When he attended the First International Harm Reduction Conference held in Liverpool, England, in 1990, Strang argued harm reduction needed to shift toward "science" and away from "faith" or "dogma." He began promoting naloxone in 1992 at the Third International Harm Reduction Conference in Melbourne, Australia, where he painted naloxone as a "clear-cut" and "easy gain" for the movement. Taking aim at a vanguard that he saw embracing radical and provocative tactics without evidence, Strang gestured toward naloxone as one of several "less glamorous" interventions that sounded more like public health measures than revolutionary tactics. Along with other conference participants, he and Michael Farrell wrote the 1996 *British Medical Journal* editorial on naloxone discussed in chapter 6. In 1997 the idea reached a broader audience when BBC *Newsnight* aired a special investigation of THN on November 13, 1997, in which reporter Jeremy Paxman questioned whether THN might temper the then-current government's "just say no" policy. On that show naloxone was presented as a form of "protection" similar to condoms and clean needles and syringes—all of which were socially acceptable means for controlling HIV transmission in the United Kingdom. THN was designated controversial because it gave "drug addicts the means to take heroin without risking their lives," in the words of investigative journalist Jackie Long, who characterized DRD as an "occupational hazard," but one that emergency services already effectively and efficiently treated with naloxone.

Newsnight then cut to a brightly lit sequence that placed audience members in the virtual position of having overdosed, but being one of the "lucky ones" for whom an ambulance was called. Viewers watch blurrily through the distorting lens of an oxygen mask as a paramedic prepares to inject naloxone, pictured in its cardboard package and defined as the "antidote to heroin." The surreal footage with a techno-music backdrop evoked the subjective effects typically attributed to psychoactive drugs, contrasting starkly with the grim realism of brooding housing estates in Glasgow. Here overdoses were said to occur at the highest rates in Europe. In silhouette, "an addict named Steve" recounted the routine deaths of a dozen of his mates: "It was like, you know, as if you said, 'The kettle don't work.'" This moral minimalism—in which Steve's mates' lives were devalued by comparison to a small household appliance—framed overdose in a way that most viewers presumably had never considered. Without denying the emotional, moral, and philosophical implications of the heavy loss of life, Professor John Strang came on screen to propose "the simplest course of action": "just ... equip[ping] our drug users in treatment with an emergency supply at home." His disarming simplicity and sincerity invited viewers to step right over any qualms they might have about naloxone's lack of social acceptability. But the BBC was not going to appear one-sided. Sitting on a park bench, the reporter confronted Strang, noting that the Glaswegian drug users she had interviewed had pointed to fundamental difficulties such as the fear that carrying naloxone would mark them as heroin users to the police. Without missing a beat, Strang indicated that a "large proportion of the addicts we see wish to be trained [and] are enthusiastic about the ability to prevent the unnecessary deaths that occur."

Looking back at BBC *Newsnight*, perhaps the first piece of broadcast journalism to depict THN in practical terms, the stated objection was striking: the proposed intervention would make drug users into "part-time paramedics," thereby "trusting the life of others to people who seem barely able to look after their own." This realist stance embodied the voice of moral reason and the consensual view that only those who can govern themselves can govern others. Such assurances notwithstanding, early THN efforts were haunted by concerns about moral worth, self-regulation, and actions undertaken to preserve lives that were somehow not worth preserving. These perceptions did not give way easily. It took naloxone access advocates many years to gain a foothold, although by the time this book neared completion, some twenty thousand naloxone kits were being distributed in England each year.[49]

OFFICIAL NUMBERS: DRDS AND RESUSCITATIVE RESPONSE IN THE EARLY 2000S

Baldly noting a "strong positive relationship" between social deprivation and drug-related death, ACMD 2000 called upon cultural authorities to use their expertise to mitigate the effects of deprivation, which "can breed social conditions that encourage the more dangerous forms of drug misuse."[50] Police were urged to keep naloxone on hand and desist from regarding overdose situations as crime scenes.[51] Prisons were expected to make an organized response.[52] Drug treatment agencies, ambulance services, and even those "likely to witness opioid overdoses" were to keep naloxone supplies "at home" or "on hand."[53] Embedded in the ACMD's call for a collective "resuscitative response" was the assumption that DRDs would be reduced if naloxone got into the hands of the "right" people in the "right" circumstances via "proper arrangements."[54] These included the strategic addition of naloxone to the United Kingdom's list of exempt medicines described above. Although not quite offering a blueprint for a national naloxone strategy, the ACMD laid out facets of a society-wide "resuscitative response" to opioid overdose. Naloxone availability was put forth as a matter of principle that also clarified a set of assumptions that would structure the United Kingdom's response to DRDs for decades.

Overdose death prevention via naloxone played a role in wide-ranging efforts to reach drug users who were not yet in treatment by enrolling them in harm reduction efforts.

We knew that drug treatment was protective of overdose deaths, and therefore the call strand was, "Let's get treatment out there to as many heroin users as we can, let's train people in overdose prevention, let's train people in recovery position, [get them] calling ambulances, let's get agreement that if an ambulance is called, police won't be called. Let's get the word out there that people's lives can be saved."[55]

The NTA supported a public awareness campaign to teach would-be rescuers how to put an overdose victim into the recovery position. A stylish interpretive dancer is seen lying on his side against a geometrically patterned linoleum floor; the dancer repeatedly rearranges himself into the recovery position while house music played in the background. In a short film commissioned by the NTA called *Going Over* (2001), overdose stories were recounted by credible witnesses who were clearly present during the incident and able to share their observations. Rather than centering on THN, the film focused on how to recognize the signs of overdose and urged people to call the ambulance. These short films played in UK drug treatment services and were screened at the International Harm Reduction Conference in Barcelona in 2007. Making and sharing such materials was important for an increasingly global harm reduction, drug user health equity, and human rights movement that had begun to solidify collective identification outside of the United Kingdom in Australia, Europe, and Eurasia. Ironically given its historical importance in harm reduction, England began to shift away from harm reduction in the 2010s.

TIPPING POINTS: IMPORTING RECOVERY INTO BRITISH HARM REDUCTION

Numbers of UK drug users in treatment peaked in 2009, indicating that the NTA was meeting its stated goal of getting more drug users into treatment. Soon thereafter, however, the costs of arm's-length bodies such as the NTA were reviewed: readers of *Liberating the NHS* (2010) might be struck by the tone of rebuke with which it emphasized collaboration and cooperation, "efficiency" and "streamlining" as steps to reduce government bureaucracy and perceived intrusiveness. Under the guise of gaining greater autonomy, freedom, clarity, and transparency for the National Health Service, restructuring eliminated the NTA, which was absorbed into Public Health England in 2013. NTA commitments to expanding OST and THN were clearly thorns in the side of the Conservative, recovery-oriented agenda focused on abstinence as the sole path to recovery. Advocating harm reduction, drug user involvement in the services sector, and evidence-based treatment, the NTA fell victim to "rationalization" in a regulatory climate that abolished several agencies relevant to drug treatment. Remnants were subsumed into a new Public Health Service that gives considerable autonomy to regional local authorities. Since then the ACMD has continued to oppose restructuring out of fear it will translate into decreased treatment coverage and more overdose deaths:

We're telling the government that unless you protect this investment, you will have more deaths. We're saying you will have more deaths anyway because of the aging cohort, but if we get more heroin into the country [or] you lose coverage for OST, we will have more overdose deaths. We're saying to them, not just get naloxone out there, but get the coverage and treatment, keep it there, increase it, and increase open access. But it's very difficult because it's not now down to the government, it's down to local areas. So you can't tell people locally what to do. The NTA was observed to have dictates, almost. But now with Public Health England, you can't tell people locally what to do. We've got much less hold over the system than we had.[56]

British "audit culture" has been much studied in terms of "'new subjectivities' produced by neoliberalism, with a reduction of real-life complexities to 'metrics' and 'indicators.'"[57] Advocates took audit culture on its own terms and tried to create a more effective national treatment infrastructure that included key performance indicators (KPIs). But they now lament the loss of the "levers" by which they institutionalized harm reduction as a national endeavor, and look askance at recent devolution to local authority, seeing drug programs weakened when "you can't tell people what to do locally," as voiced above:

I don't think the prognosis is good. Drug users are very unpopular groups locally. They're competing for money with mandated services for children and old people.... But the only kind of saving grace is people like the head of the NHS are saying, unless you prioritize this group, they will place a greater burden on the NHS through Accident and Emergency attendances, etcetera. But nobody funds this group because, ethically, they should. Even in the times of Blair, drug treatment was funded because of the crime rates. Prior to that, it was funded because people were frightened they would give the general population HIV.... I think we're at a real crossroads [in 2017]. It's a tipping point, and I think it's a very frightening tipping point.[58]

Tipping points are recurrently politicized in British drug policy. When the NTA reported evidence showing that treatment retention was the single most beneficial tactic for reducing DRDs,[59] its efforts were undercut by a new national recovery agenda advanced by former Conservative MP Iain Duncan Smith.

The Centre for Social Justice (CSJ) was founded in 2004, ostensibly to make good on Smith's promise to a mother whose son had died of overdose shortly after prison release. Painting a picture of the "explosion in addiction" in Britain's "self-harm society,"[60] the CSJ presented numerous anecdotes concerning the inexorable rise in negative consequences from increased consumption of alcohol and drugs, including drug deaths, particularly among the poor: "Drug deaths have risen exponentially—a hundred fold since 1968 when there were just nine." The Conservative-leaning CSJ advocated abstinence and self-help as the blueprint for "revers[ing] social breakdown and poverty." This, they opined, would enable "such individuals, communities, and voluntary groups to help themselves." Sharply critical of the "government's harm reduction policies," the report presented ordinary Britons as "witnesses" instead of experts.[61] Witness testimony was marshalled into a sharp critique of the "progressive skewing of policy" since 1998; the harshest criticisms were directed toward the NTA's support of treatment expansion in the form of OST.[62] Repudiating existing experts and evidence, Duncan Smith's organization consistently took a punitive, abstinence-based policy approach in reports including *Breakthrough Britain II: Ambitious for Recovery: Tackling Drug and Alcohol Addiction in the UK*. Another actually urged inmates themselves be charged for stays in "drug-fueled prisons."

The CSJ reports illustrate how an abstinence-oriented recovery agenda based in the alcohol field converged with drugs policy in Britain.[63] To counter this, harm reductionists insisted that conflating opioid recovery with abstinence from alcohol made little sense. They charged that recovery was a loaded ideological agenda rather than a pragmatic response. As one harm reduction nurse educator noted, "It's not that we don't hope for abstinence for those who want it and those who can achieve it. Abstinence is a fine top for the hierarchy, but it's not the only end point in my opinion. With abstinence comes a greater risk of overdose because we see the maintenance program as being protective against death, against unintentional overdose due to lack of tolerance and misjudgment."[64] Harm reductionists think of abstinence as an heightening vulnerability to overdose and overdose death, which in their view produced greater need for naloxone and training to use it.

Recovery is a contested political space in Britain,[65] whereas in the United States, an inclusive and bipartisan coalition of recovery advocates, women's health advocates, and researchers created a consensual "recovery-oriented treatment system."[66] In the United Kingdom, recovery was hitched to the diminished state responsibility of neoliberalism. Despite DRDs in England and Wales declining between 2004 and 2007, the politically conservative recovery agenda wormed its way into British politics.[67] Advocates for abstinence-oriented recovery have contended against protagonists seeking harm reduction-oriented recovery, approximately since Blair left office in 2007. Even with a chorus raised for abstinence, and even with Conservatives governing again starting in 2010, harm reduction remained the actual policy in evidence-based Britain. Part of the explanation was that naloxone champion Strang chaired all four of the UK Clinical (Orange) Guidelines revision processes from 1991 to 2017. While not legally binding, clinical guidelines are extremely influential in the United Kingdom, articulating continuity and change in ideal clinical practice and standards of care. The 2007 Orange Guidelines were the first to mention naloxone, offering a cautiously positive view, saying something along the lines of "many clinicians have found it appropriate to consider giving naloxone," but they left open the opinion that many clinicians might not. However, in the new 2017 Guidelines, there's more of a presumption that [naloxone provision] should occur. That's part of how a field moves on, and in the UK the Orange Guidelines are part of that process. Way back in time, the original Orange Guidelines were a consensus document of clinicians, saying, "we are wise clinicians, trust us," whereas now they are much more an evidence-based synthesis.[68]

Naloxone was written into these influential clinical guidelines in part because of protagonists such as Dale-Perrara, who led the NTA contribution to the revision process. Also contributing was a 2005 revision to the UK Medicines Act.[69] According to some, the change meant little; whereas for Strang, the significance of the 2005 legislation broke a bottleneck:

It was driven by the ambulance services who wished to correct the absurdity that the crews of many of their ambulances, even though they carried naloxone, were not allowed to administer it as they were not "ambulance para-medics" but were at the lower grade of "ambulance technicians." … I was approached by London Ambulance Service to lend additional support to their case, specifically for families and peers to be able to administer [naloxone], and I gave my support.[70]

Some physicians seized on the revisions to widen uptake of THN. Birmingham general practitioner (GP) Judith Yates began prescribing naloxone kits to her many patients who were using opioids, and she has urged fellow practitioners to do so ever since. Yates recalled:

My patient had been in for [residential] detox, … and he came out to see me in my GP surgery after his detox, drug-free, and he reached into his pocket and took out a mini-jet, which was a Narcan, naloxone, kit, box, and he waved it at me and he said not a word. And I looked at that, and he looked at me, and I said he could have done with that three years ago because one of my other patients had died in this chap's flat three years before, a young man who was also my patient at the time, and he died ten days after leaving a rehab drug-free, I'm afraid. He relapsed when he got back to Birmingham when all kinds of the usual troubles hit him. He'd been in my second patient's flat, and my second patient had tried his hardest to resuscitate him and called an ambulance and everything. But by the time the ambulance got there, it was too late and the young man died. My patient and I both realized that if he'd had a naloxone kit at the time, he'd would have actually been able to resuscitate the chap.… Seeing that kit in his hand and him not even speaking but just waving it at me was part of what got me thinking, "This is a good idea. Why don't we get these kits out more widely?"[71]

Birmingham was one of the sixteen sites where Strang's "train the trainers" pilots had been conducted; once evaluations were published in 2008–2009,[72] Yates had assumed there would be a flood of naloxone prescriptions. Instead, "nothing happened in England."[73] In campaigning to get other GPs to help get THN to patients, Yates encountered reluctance, resistance, and even fear in northern England—on the very terrain where the early harm reduction and public health movements had enjoyed their first successes. Legislative changes were not fully promulgated; many remained unaware of them despite editorials in respected journals and seemingly endless rounds of talks and trainings. As the next chapter shows, the 2005 change had much more significance for Scotland than for England.

The NTA had mobilized service users through National Service User Forums, which had energetically debated different positions on drug policy and treatment, including THN. "Experts by experience" received a boost from the 2001 and 2008 NICE Clinical Guidelines; peer education migrated from mental health and into addiction treatment.[74] As a "technology of solidarity," naloxone concretely embodied the agreement that people should not be dying preventable deaths—but left up for grabs how to effectuate ongoing tensions between active users, service users in recovery, and those governing and seeking to govern. Lamenting the NAT's loss of "teeth," Dale-Perrara anticipated increased inequality and feared that the "localism agenda, putting the power down to the community, works against marginalized and discriminated against populations."[75] The ACMD did what any neoliberal body might do—took audit culture at its word and campaigned for better monitoring systems based on public health–oriented KPIs.[76] The ACMD bent the system toward monitoring timely and consistent DRD reporting, which was necessary for harm reduction priorities to take hold and become naturalized. Harm reduction drug policy naturalized the idea that governments might do something strategic to reduce DRDs; performance monitoring rendered outcomes discernible.

JUST DO IT! HANDS-ON AT SOUTH LONDON AT MAUDSLEY (SLAM)

Naloxone training and supply was implemented by Strang and colleagues in South London and Maudsley Hospital drug treatment services. Attributing the impetus to Dan Bigg's impatient admonition to "stop talking about THN and just do it" back in Y2K Seattle, Strang ensured that the effects of wider naloxone provision were documented.[77] Once the harm reduction nursing service that implemented the SLAM program got the "bit between our teeth and started doling naloxone out to everybody who walked around,"[78] they kept a roster of lives saved and collected accounts that testified to what users experienced when administering naloxone to peers:

He was straightaway half gouching, really sleepy, his eyes were glazed. Then his lips started going mauvish and he was starting to fall asleep. He was slumping forwards where he was sitting. I grabbed hold of him, his breathing was really shallow. I told someone to get my rucksack with the (naloxone) kit in it. He was going mauver and mauver. His breathing was terrible and he made a noise like a death rattle. I took the kit out and put a needle on. I put it in the front of his thigh, right through his jeans. I pulled back a bit to make sure it was ok and then pushed it down. I undid his belt a bit so that he could breathe more easily and sat him up and lifted his head to straighten his airway. Then the colour started to come back into his lips and he was alright again. I wanted to call an ambulance, but he refused and said no (Patient #2).[79]

Posted on bulletin boards and circulated in newsletters, this account served as a reminder that Patient #2 should have given the ambulance a ring anyway, due to the short-acting duration of naloxone and the possibility that the overdose might return. Overdose had acquired an aura of the routine, but the narratives recognized that those who overdosed rarely expressed gratitude for having their lives saved.[80] These posts aimed at affirming and reinforcing the actions of the person administering the naloxone, rather than those of the recipient.

As also occurred in the United States during the early days of bystander naloxone, ambulance crews were initially surprised to encounter peer-administered naloxone. Some peers were reluctant to give naloxone because they feared interactions with police and ambulance crews. A patient who was not herself an injector described feeling compelled to administer naloxone despite initial reluctance:

We rang an ambulance.... The ambulance came and they said it was a good thing to have given the (naloxone) to him. He was taken into hospital and kept in overnight. The ambulance crew were shocked that I had the naloxone; I said that because I had it, I had to use it and they told me it was the right thing to do. He (the man who went over) said he was glad I had it on me. He said it was a bit of bad gear and a couple of other people have gone over when they took it (Patient #3).[81]

However, tensions were also recorded between peers who administered naloxone and peers who received it:

This man had a dig…. Straightaway you could see his eyes roll back. His lips started going a blue colour. We couldn't wake him up. We knew he'd gone over. I slapped him a couple of times and nothing happened. He was making a groaning noise. I had the naloxone in my jacket zip up pocket where I usually keep it so it's handy. I know a lot of people who use, everyone I know uses, so I need to have it with me. I got the thing out of the box and put the needle on. I put [it] in his behind; I thought that's the safest place. It worked and he came round in about two minutes. At first he was saying, "I was alright, you didn't need to do that," but I did need to, he was going to die. I felt great and the other people who were there with me said I did the right thing. I saved his life and hopefully, he won't be so stupid next time. I don't know him well, so I don't know if he had a family, but say he had a Mum, she'd have been destroyed if he'd died (Patient #4).[82]

Another patient had administered naloxone when someone collapsed in the kitchen and went "totally unconscious, his lips went blue and he started fitting and his eyes rolled back. He was completely impossible to rouse and was a dead weight." Annoyed that other bystanders did not assist because they had warned the overdose victim not to do such a high dose because he had recently been released from prison, he had injected naloxone as trained—despite never having been an injecting drug user himself. Another reported that when his girlfriend overdosed, he used "our naloxone kit … on the mirror above the mantelpiece."

These anecdotes indicate something of the mundane prevalence of overdose in this South London drug-using community, and the quotidian way naloxone has been incorporated into it. Harm reduction nurses keep track of lives saved also as an outlet for the emotional toll of the naloxone service. One wrote:

Thirteenth time lucky! Today for the thirteenth time one of our Service Users came in and reported that a life had been saved because Naloxone was available. This was the second time that (Patient X) had overdosed in just over two weeks. He is very worried that what was sold as heroin contained something else, but he is unable to find out what it is. He has reported that he suffered from a serious blackout for more than twenty-four hours after the first overdose and feels strange and uncoordinated after this second one. The first time he overdosed his partner did not know that he had naloxone at home, but she did call an ambulance. She told him that the crew had to give him three doses of naloxone before he came round. When it happened again, last night, she knew now what to do.… He is very glad to be alive and very relieved that he had shown her what to do. He has got replacements for the kit that was used and we have said that his partner is welcome to come in to see us and get more training and advice. All relatives or friends who want to know more about overdose and naloxone can come in to see us. We are very happy to train them.[83]

The stories convey basic signs and symptoms of overdose, necessary actions to be performed, and the importance of knowing where one's kit is at all times. Narratives like these promoted the naloxone service, and were distributed because of their parable-like nature.

Naloxone training of the sort hinted in the narratives was initially available in England through harm reduction-oriented treatment services; over time it began to be delivered by service users themselves through the NTA-supported Service User Councils, and such groups proved important for identity formation. In an interview with an "expert by experience" active on a Service User Council, I asked what had propelled him toward becoming more involved:

My involvement played a major part in changing my life. I was using opiates for many years and treatment was there, it was optional, but really it was kind of take it or leave it, there were some options for OST. So you'd come along, you'd pick up your prescription, you'd have a counseling session, you'd think about some areas of your life, practical areas like housing. It was all very helpful, don't get me wrong. But identity plays a part here, my identity. I found myself at that time, what kind of person I was, where did I belong in wider society, and for me, that was where I fit in, from being involved in user activism, user awareness. It was about, where do I fit in, how does society see me?[84]

Drug users pass through profound shifts in identity when they become involved in the peer education networks through which the harm reduction movement reproduces itself. Recounting how little attention he had previously paid to harm reduction, the speaker quoted above is now a trustee of the Aurora Project in Lambeth, a peer mentoring project and registered charity set up in 2011 by and for people with lived experience of addiction services and addictions. "Our mission is that people in recovery—and I guess that means everybody because there really are very few people who are not trying to make some aspect of change within their life—our mission is that we believe, profoundly, that people with lived experience are not a constant burden, but in actuality an asset."[85]

Asset-based approaches are used to "challenge the stereotype that drug and alcohol users are a constant burden on society" and to promote the "belief that people in recovery from their addictions have much to offer their local communities."[86]

In a conversation along with Martin McCusker and George Porter, two "experts by experience" at the Lorraine Hewitt House in Brixton, South London, harm reduction nurse Reuben Cole noted that naloxone had now been embraced by a culture of positivity:

People's lives are often very challenging and difficult, aren't they? What we provide is a short amount of time when people can feel good about themselves. They can really change—they can save someone's life.… I've never met anyone, well, I remember one training session with one person who was completely cynical, out of hundreds. There's often quite a negativity-based narrative in drug treatment, how much you using? How often have you been arrested? I'm being over the top, but naloxone training is a very positive-based narrative. It's about what you can do for the community, what you can do for another person, another human being.[87]

South London services generally regard THN as standard care to such a degree that people must opt out rather than opting in. Social expectations are that individuals will carry naloxone and know how to use it, results seen as supporting drug user empowerment.

NALOXONE AS A "WONDER DRUG" IN WALES AND NORTHERN IRELAND

Other parts of the United Kingdom also responded to increased DRDs. Wales developed WESTROS, a testing and monitoring system that links toxicology screens with other documentation from local death inquiries, thus freeing death certification from the variable knowledge, skills, interests, or competence of individual coroners. "It's only been in place a year. We saw a little of the initial data at the ACMD and it will be great. It will give excellent, textured data going forwards."[88] The Welsh experience with naloxone was hybrid. Wales mounted naloxone pilots in nine community sites and six prisons in July 2009 and rolled out a national naloxone program in 2011 that had distributed over ten thousand kits by 2016. Welsh drug users embraced naloxone as a "wonder drug,"[89] by contrast with sites such as Glasgow, where drug users initially viewed naloxone with suspicion. Wales also built capacity in the testing services, so that substances themselves could be more accurately identified.

Northern Ireland had experienced a lower rate of DRDs than Scotland, England, and Wales, but starting in the mid-1980s it had slowly begun to climb.[90] According to social worker Chris Rintoul, activist pressure from social workers, drug and service users, colleagues, and friends led to naloxone becoming available across Northern Ireland:

Sadly, some of those people are no longer with us. Naloxone was not available in time for them. But as a result of that pushing and that activism, we eventually got heard, and we eventually got taken seriously, and we eventually got the money to back not just the purchase of naloxone, but also to provide training to the various services that needed it. It is a story for me about human rights and about a baseline harm reduction and prevention that you extend to anybody at risk of overdose, and it also says something about "We care for you, we do give a shit whether you live or not, and we do give a shit about whether you can help somebody else who could potentially die as a result of an opiate overdose, and we may be able to do something about that."[91]

THN implementation went smoothly in Northern Ireland,[92] where Community Addiction Teams and the Northern Ireland Prison Service commenced a pilot program in 2012 after beginning to distribute naloxone through the NHS in 2011. The pilot evaluation summarized participants' universally enthusiastic responses as "developing a sense of empowerment, building or strengthening therapeutic relationships, successfully reaching and engaging those who need it, saving lives, and providing useful education to lower overdose risk."[93] Users spoke of gaining self-worth and life skills, leading to perception of THN as a "community asset" with positive effects beyond those of saving lives.[94] Training was delivered across the country by the Council for the Homeless Northern Ireland because it was in homeless shelters (called "hostels") that most overdoses occurred. The profile of actual risks and potential harms in Northern Ireland made heroin comparably more expensive in Belfast then in Dublin, Glasgow, Edinburgh, or North America. Social costs in Northern Ireland also extended beyond those on drug users themselves due to social attitudes toward drug users as exhibiting "weakness" when "clocking" ("dope sick" or in withdrawal).

Although the Northern Ireland Statistics and Research Agency tracks DRDs, social pressures create a micropolitics of overdose that must be taken into account when thinking about how the various places that comprise the United Kingdom respond to DRDs. Rintoul indicated that social pressures made coroners "kindly to families. If you think about it, to have to record a cause of death of your son as a heroin overdose can be quite a difficult thing to bear in the midst of grieving a lost child. So it is sometimes thought that there is another cause recorded there, or we're not sure of what the reason was."[95]

Thus does the politics of overdose in the United Kingdom vary with the local character of heroin-using subcultures, as well as interactions between them, governing authorities, and the broader society. Naloxone was at stake in debates over whether harm reduction and recovery were seen as polar opposites or on a spectrum with overlap. Twenty years after Thatcher's premiership ended, and well after harm reduction had settled into routine in most of the United Kingdom, naloxone still "stuttered" in England.[96] Barriers to rolling out a national harm reduction strategy in the United Kingdom were such that naloxone's uptake had been patchy; "groups of brave clinicians and user groups ... plough a lone furrow without national guidance, support or coordination."[97] But some of the protagonists deduced that Scotland contained fields more conducive to naloxone taking root, as described next.

10

"GROWING ARMS AND LEGS":
The Scottish National Naloxone Program

Scotland grapples with high levels of poverty, inequality, and social deprivation. Since the 1980s, excess death disparities between Scotland and other European countries have widened, a high proportion of which can be attributed directly or indirectly to drug and alcohol use. The Scots have such a long and conflicted relationship with opioids that hardly a family in Scotland has not experienced some form of drug-related harm.[1] Physician Roy Robertson has seen thousands of people whose mortality and morbidity stems from entrenched cultures of drug use:

It's just part of the community. Everybody's got a brother or a sister whose died of drug use, a mum. I've got one family where this woman came in and showed me a photograph of her granddaughter and said, "You remember my daughter, she died of a drug overdose." I said I remembered her. Then she said, "And you remember my mum, don't you, she died of a drug overdose." And I said, "Yeah, I do remember her. Oh, gosh, that's four generations." And she said, "My grandfather, he was your patient as well, and he died of a drug overdose." That's five generations. It's all over the community. Our grip on it, our association with it is so tangible, so day-to-day that it ceases to become a surprise. Drug users, we're all drug users, aren't we?[2]

Depending on where and how drug users live, the forms and effects of social stigma play out in their everyday lives:

Stigma at the highest levels comes all the way down. In our community, in our waiting room, people will ask for an appointment tomorrow and the receptionist will say, "Well, doctor is a bit busy tomorrow," and the next thing they'll say is, "If I was a junkie, you'd give me an appointment." I've got this great picture of the front door of my practice with graffiti all over it saying "Junkie Paradise."[3]

Glasgow and Edinburgh especially gained reputations as "Junkie Paradise," a term that reflects the exceptional integration of opioid use into the social fabric—in ways that stand out in world perspective. As the memorably irreverent Mark Renton said while shooting up in Irvine Welsh's novel *Skagboys*, the protohistory of *Trainspotting*,

If being Scottish is about one thing, it's aboot gittin fucked up…. Tae us, intoxication is just a huge laugh, or even a basic human right. It's a way ay life, a political philosophy…. Whatever happens in the future to the economy, whatever fucking government's in power, rest assured we'll *still* be pissin it up and shootin shit intae ourselves.[4]

Renton might have shaken his head at the futility of the Scottish government trying to reorient drug users toward just saying no, but he might nevertheless have embraced the distinctive Scottish National Naloxone Program (SNNP) that put his country on the harm reduction map. Scotland was unique in straddling both recovery and harm reduction in an era when these were largely viewed as divergent and contradictory. The SNNP was one of the signature programs to emerge from the first Scottish National Party (SNP) government. With only a minority government formed in March 2007, the SNP created an outcomes-based National Performance Framework that required local authorities to specify goals and then demonstrate they were being achieved. The existing Scottish drug policy, which had been set out in *Tackling Drugs* (1999), had not been achieved and in fact had led to "no measurable progress."[5] The SNP crafted a new national drug policy in 2008 and announced the SNNP in the fall of 2010. Thus did Scotland adopt one of the world's most progressive harm reduction strategies, while simultaneously phasing in an outcomes-based, recovery-oriented framework based on personal responsibility well characterized as neoliberalism in action. This chapter tells the story of the various "arms" and "legs" comprising the SNNP, situating it within the Scottish-style opioid crisis, the international harm reduction movement, and the partisan politics of neoliberal governance.

METRICS OF RECOVERY: MORTALITY AND
MORBIDITY IN SCOTLAND

Scotland's social, economic, and political tensions make governing akin to navigating a ship through rough seas. Although differentiated from its populous southern neighbor, Scotland remains subject to the Crown and bound by UK laws. In 1997 a popular referendum devolved health, safety, and welfare responsibilities into the hands of a partially autonomous Scottish Parliament, which was constituted the next year. The drugs mortality problem as dramatized in the movie *Trainspotting* (1996) merged with the country's overall mortality problem, which goes by the name of the Scotland Effect or the Glasgow Effect.[6] Scotland—particularly its second-largest city, Glasgow—is a "public health outlier" in departing from overall declines in mortality and morbidity in most of Europe:

The divergence of Scottish mortality from that of its European neighbors from 1950 onwards was largely due to a slower reduction in mortality from cardiovascular disease, stroke, respiratory disease, and cancer. It was not until the 1980s that alcohol-related deaths, DRDs [drug-related deaths], suicide, and road-traffic accidents became drivers of higher mortality rates. There has been less research into the divergence of Scottish mortality from 1950, and the precise nature of the causal pathways in this earlier period is, therefore, far from certain.[7]

Whatever their causes and drivers, these became fully Scotland's problems.

Scottish drug policy diverged from the UK Conservative government policy as set forth in the late 1990s.[8] Soon after, the UK Home Office adopted a Drug Harm Index, with indicators to measure various "costs" stemming from drug misuse and crimes associated with it.[9] Harms indexed were individual but socially cumulative, each differentially weighted such that the Drug Harm Index overestimated the amount of crime that was drug related, and underestimated the social cost of mortality and morbidity:

The cumulative economic costs of the innumerable crimes that are committed by drug users may be very high. But do they really belong to the same order of cost as the devastating impact of losing children, siblings, or parents to DRD? The health harms of drug use (such as HIV and hepatitis C) may form a small proportion of the national burden of disease, and a larger part of the burden of crime. But can we combine these figures into a tool that appears to tell us that it is more important to reduce the criminal than the health harms of problematic drug use?[10]

Measurable harms can be enumerated; those that "count" are those that can be quantified. Quantification and statistical analysis often make sense in an evidence-based policy world. Yet as work on "psychic numbing" shows, the "more who die, the less we care"—large numbers create paradoxical cognitive effects that figure into failures of moral intuition and empathy.[11] Although unfeeling may be somewhat overcome by appeals to identifiable individuals, there is evidence that presenting the needs of a charismatic case alongside large-scale quantifications erodes individuals' belief in the efficacy of their actions to reduce harm.[12] This effect, often called "compassion fatigue" in the literature on humanitarian aid, is compounded where social stigma shapes responses.

Harm reduction drug policy appeals to concretely identifiable effects of harm that cannot be easily separated from less tangible harms such as grief, loss of relationality, or social suffering.[13] Although such harms may be reduced to matters of individual consequence for purposes of representation, they are more broadly and deeply social matters. How a society deals with uneven distribution of preventable harms says much about that society's values and priorities. During the process of reimagining British drug policy as recounted in the previous chapter, Scotland emerged as an exceptional site where an endemic pattern of "excess deaths" had become normalized. From 1997 to 2000 alone, DRDs rose by 25 percent in England and Wales but by 150 percent in Scotland, far outstripping any other country in the world. Although it took nearly a decade for the Scottish government to craft a drugs strategy that would begin to come to terms with its exceptional status, *The Road to Recovery: A New Approach to Tackling Scotland's Drug Problem* (2008) was billed as a "paradigm change in thinking about drugs policy in relation to those with a serious drug problem in Scotland."[14] This document aimed to reorient Scotland's stance on drug treatment away from the harm reduction measures it had adopted in response to HIV/AIDS, and toward recovery.

Treading this new path, Scotland's government pursued its own approach to recovery, joining a current of broader UK forces moving toward "outcomes-based" recovery as policy and practice. Much ink has been spilled characterizing the recovery movement, which has been especially popular in the United States, where most publicly funded treatment programs incorporate elements drawn from the Minnesota Model,

Alcoholics Anonymous, or Narcotics Anonymous. Recovery is a mutual aid movement within which many former drug users and alcoholics maintain abstention and manage relapse; it has sometimes been cast as a form of spirituality that offers ideological resistance to science and evidence-based practice. The Scottish version of recovery—which was not indigenous to Scotland—moved beyond the governing mentality entrenched almost everywhere that made harm reduction and abstinence mutually exclusive.[15]

A social concept of recovery has recently come into productive tension with neuroscientific theories of relapse and chronicity (embedded in the largely scientific and US-driven redefinition of addiction as a chronic, relapsing brain disease).[16] Tenets of long-term recovery were woven into the institutional fabric of domestic US drug policy in the 1990s—despite the science-based orientation implicit in the evidence-based practices discussed in the introduction. By contrast, the United Kingdom lacked a well-established recovery movement. Driven by government bureaucrats, the nascent UK recovery movement shifted away from addiction experts' approach, which was painted as "pathology-based and intervention-based knowledge of addiction as the organizing centres of policy and service activity."[17] The recovery emphasis downplayed the previously robust role of expertise and harm reduction as drivers of UK policy and practice.[18] Many UK medical professionals, including general practitioners (GPs) providing specialist services to treat drug users and those participating in methadone maintenance, think about the conditions their patients face in nuanced, deeply contextualized ways.[19] However, tensions over the desirable degree of medicalization often come out in discussion of the political agenda of recovery.

Recovery advocates de-medicalize drug treatment, but there its similarity to harm reduction ends. As ongoing, potentially lifelong processes, neither recovery nor harm reduction fit well with UK audit culture. Recovery "outcomes" had to be made measurable, with "metrics" for assessing them; in practice, this imperative moved recovery closer to abstinence because that was a specifiable outcome that could be counted. My interviewees were mostly skeptical of outcomes-based approaches, preferring to mark milestones rather than destination points on pathways to recovery. Moreover, these professionals tended to resent the representation of

The Road to Recovery as a "consensus" document based on a new work-ing paradigm because they believed it was neither new nor consensual. The recovery agenda was contested in ways that dramatized whatever fragile agreements might have existed. Shortly after the Scottish report, "Suffused by the Recovery Lexicon," recovery was adopted UK-wide as bedrock for all decisions made by local authorities.[20] This initiative was widely viewed by UK interviewees as a hostile move that threatened to displace or subsume harm reduction.

Emphasizing abstinence as the ultimate goal of the recovery-oriented agenda was taken to diminish the value of harm reduction. While the relationship between recovery and harm reduction is often treated as contradictory in political discourse, the Scottish Drugs Forum (SDF), a voluntary sector organization established in 1986 to coordinate Scottish expertise on drugs, considered it a "false dichotomy":

For too long, debate in Scotland has centered on whether the primary aim of treatment for people who use drugs should be harm reduction, or abstinence. We fundamentally disagree with the terms of this debate. We do agree with the United Nations Office on Drugs and Crime, which said in a recent report that "harm reduction is often made an unnecessarily controversial issue, as if there were a contradiction between treatment and prevention on the one hand, and reducing the adverse health and social consequences of drug use on the other. This is a false dichotomy. They are complementary."[21]

It was not so much the logic or illogic with which critics took issue, but the degree to which relapse was—or was not—recognized as integral to recovery. Ideologies and policies that made no room for relapse risked denying social and health services to drug users.

In importing concepts of mutual aid, the Scottish government empha-sized self-directed care and individual empowerment—but sidelined professionals whose expertise had long dominated drug treatment. The Scottish Drugs Recovery Consortium commissioned US recovery advo-cate William L. White to review the evidence base for recovery so as to implement a recovery agenda as a basis for governance in Scotland.[22] *Research for Recovery* acknowledged that almost all evidence for recov-ery had been produced by alcohol researchers in the United States, who had suffused the federal treatment infrastructure with recovery-oriented approaches to health and human services.[23] Recovery was presented as the United Kingdom's "new paradigm," the real agenda being to displace Euro-style harm reduction by depicting harm reduction as a barrier to choice—as if people would be so enmeshed in harm reduction that they could not elect recovery and were instead trapped within the realm of drug and alcohol use. Home Secretary (and later Prime Minister) Teresa May stated that "a fundamental difference between this [recovery] strategy and those that have gone before is that instead of focusing primarily on reducing the harms caused by drug misuse, our approach will be to go much further and offer every support for people to choose recovery as an achievable way out of dependency."[24] Going "further" meant building "recovery capital" within each individual.

The idea of recovery capital aligned just enough with the harm reduction movement's emphasis on empowerment that a middle ground became visible to some. In this conflicted context, it is all the more remarkable that in 2010 the Scottish government set out to undo the normalization of high rates of DRD. There evolved a unique community-based effort to saturate the country with naloxone targeted toward those at highest risk of experiencing or witnessing overdose.

THE SOCIAL CONDITIONS OF SCOTTISH EXCEPTIONALISM

By the early 2000s, harm reduction had crept into Scottish pharmacies and treatment services, although access to naloxone itself was uneven due to Scotland's mixed terrain.[25] The Scottish government undertook a national investigation into DRDs, led by Australian justice and forensics scholar Deborah Zador.[26] Due to social cohesion among Scottish drug users, the investigation was able to situate each death within the context of the six months leading up to it.[27] This "social autopsy," in the words of Stephen Malloy, who later became Scotland's first National Naloxone Coordinator, included very detailed information about the particular circumstances of the last days of each decedent.[28] The study revealed that several had died within days of prison release; that most were accessing services in a "chaotic" manner; that not all were injecting; and that polydrug use—particularly involving opioids (heroin) and benzodiazepines—was prevalent. But most importantly, the study noted missed opportunities to intervene in time to save lives. Although naloxone was not specifically mentioned in this 2005 report, the need for overdose education was central, and the report set baseline expectations against which future interventions could be measured.

When naloxone emerged as the solution to reducing DRDs in Scotland, it showed up against this backdrop.[29] Official strategy was spearheaded by public health-oriented physicians, pharmacists, harm reduction nurses, and other professionals. Except for Malloy, who later worked for the SDF, few became activists for human rights or health equity as did their counterparts in the United States. The Scottish government set forth a plan called "Taking Action to Reduce Scotland's DRDs" (2005). Once naloxone joined the list of life-saving prescription drugs that could be used by people other than physicians under Article 7 of the UK Medicines for Human Use Act in 2005, the stage was set. Recommended by the Advisory Committee on the Misuse of Drugs in 2000, this official change galvanized thinking in Scotland.

According to Carole Hunter, lead pharmacist of Glasgow Addiction Services, the 2005 change was "what got us even thinking about naloxone in the first place. It was a number of things coming together—awareness of drug deaths, an increase in drug deaths, and John Strang's editorials. John's just the guru for everything."[30] Hunter wrote a Patient Group Directive (PGD) so that pharmacists could supply naloxone over the counter when customers requested it—despite its continuing status as a prescription-only medication under UK law. The PGD was a legal mechanism—akin to a standing order in the United States—that allowed naloxone to be supplied without a doctor's prescription. As pharmacist Amanda Laird explained,

What we had was a PGD which allows nurses or pharmacists to provide that medication without the need for a doctor to issue a prescription.... [A PGD is] quite a regular format of providing medicines in a specific circumstance usually within specific professions such as nurses or pharmacists or paramedics.... [B]ecause we had [government] reimbursement for our kits, we had physical kits to provide to the individual being trained at the point of the training, so we were able to hand them over using the PGD.... For naloxone the PGD only relates to supply, as administration of naloxone is allowed in an emergency situation as per the Medicines Act.[31]

The distinction between who could legally supply naloxone, and those who could legally administer it, was confusing in practice as well as theory.

After 2005, naloxone could be used "by anyone for the purpose of saving life in an emergency," in order to "prevent death from heroin overdose without specific medical instruction."[32] But it could only be supplied by professionals working under the legal umbrella of a PGD until after the 2015 change discussed below.

Social cohesion between professionals and the drug-using community had to be cultivated, although eventually there developed perhaps the tightest culture around naloxone anywhere on the planet. This was still palpable a decade into these events when I interviewed SNNP protagonists. For example, Hunter observed that "a lot of this has happened because of committed individuals in a country like Scotland, that's small, just 5 million people, where within services we all communicate."[33] Professional camaraderie did not, however, guarantee everything went smoothly. Despite familiarity with opioids, the Scots were no exception when it came to ignorance about addiction. One GP described medical training on the subject as "very biomedical, interventional, and fairly transactional—and there was nothing around the human or social factors, how to understand people's context, or how to engage and get positive outcomes."[34] Worse, health care providers had sometimes used naloxone to "punish people who are 'undesirables' attending hospital settings and wasting time."[35] Merely by upping the dose, patients could be thrown into precipitated withdrawal. Almost every interviewee mentioned this no longer predominant practice. Still, the word of mouth regarding harshness undermined naloxone's initial reputation among drug users:

You had attitude from people working in emergency departments in particular. This was not a norm, but this was undoubtedly a drug that gave some practitioners power, and the power wasn't always about bringing somebody back. It was also about ending their kick, ending their trip. So in the wrong hands, it was something that could be misused.[36]

Naloxone far above the dosage needed to reverse respiratory depression can precipitate such abrupt and intense withdrawal that waking is combative.[37] Thanks to some two hundred interviews fortuitously conducted by sociologist Joanne Neale between 1997 and 1999 in five Accident and Emergency hospital departments in two Scottish cities, it is clear that more than a few Glasgow patients were "overantagonized" (given excessive amounts of naloxone).[38]

The drug users whose experiences were recounted in Neale's 1999 paper indicated that they were treated like second-class citizens by clinical staff.[39] Most either were not informed or had no memory of the exact treatment they received: "Although very few interviewees had heard of naloxone, many recognized the opiate antagonist from its description as a drug to reverse the effects of heroin. Among these individuals, naloxone had a very negative reputation and was generally considered a treatment that should be avoided if at all possible."[40] That reputation was based on "hearsay evidence from other drug users or on a previous negative experience of treatment."[41] Drug users treated for overdose were rarely informed about what drugs had been used to treat them. Drug users who overdosed in 1990s Glasgow were treated with naloxone:

[Yet] you never really knew a great deal about it, apart from when it was administered to you. You woke up sharpish, because at that time there was no discussion with the young hands and people in the community. It was always in a medical setting that it was administered. It would be fair to say that in most people's experience, and in my experience, it was not that nice.... Wakin' up in hospital quite sharp and not feeling the greatest. Well, at the time, no, you never realized it saved your life.... Because at that time, maybe [overdose] was viewed by some medical staff as self-inflicted, and you should just be grateful for that they saved your life.[42]

Jason Wallace's experience dovetailed with other drug users' recollections of the information dearth that followed an in-hospital overdose reversals in Glasgow:

Thinking back, I don't recall any information given to me following the reversal of an overdose in terms of what the medication was that was used, or how long the precipitated withdrawal might last, or anything like that. Discussions with the clinicians and the team were very firmly focused on the psychological/psychiatric intentionality behind the overdose, what had been involved, and very quickly discussions of whether I had been in treatment.[43]

Naloxone acquired its negative reputation in Glasgow (as elsewhere) when excessive doses were administered. If naloxone is not administered judiciously—even adapted to the situation—acute onset of withdrawal practically propelled people out the door, even if clinicians made serious attempts to engage them in treatment:

There wasn't a lot of time for discussion because you were craving it and anxious to leave as soon as possible.... They were not able to detain you. It was totally up to you. Maybe if it had been explained to you, actually what had happened to you, the process, maybe you might have stayed. But when you wake up in that environment with acute withdrawals, it's not the greatest environment because you don't feel part of it, you're already feeling the stigma, people really can feel that, and the first thing on your mind is to get out of here.[44]

Medically trained professionals were constrained from expressing negative attitudes toward drug users by codes of professional ethics. Yet it has been documented that health care professionals experience drug users as a "difficult patient group who wasted hospital resources and whose behavior and lifestyles were difficult to understand."[45] Withholding explanations or the name of the drug administered may seem like a relatively small omission in an emergency situation, but many patients experienced it this form of indignity.

Emergency medical staff themselves felt they had little understanding of how to manage people living with addictions, even in the aftermath of an overdose:

The whole idea was could we manage them overnight and get them out the next morning without much damage being done. There was no sense of a disease model in our head. We would never think that someone with shortness of breath or chest pain could just be let go, [that] the strings could be cut the next morning at nine a.m. But you'd make your rounds the next morning and we could let [people with drug problems] go the next day without any sense of follow-up. That seemed to be normative for drug users.[46]

Ambulance services had also administered naloxone punitively. "Historically, that was a problem that we had to overcome initially. We had to go to smaller doses to bring people around without putting them into withdrawal. We had to sell [the idea] to users as well because they didn't know how much of that myth was true, how much that perception was there."[47] Indeed, in the years following these early experiences, the naloxone-related education of both users and professionals has been extensive. Modelled on protocols learned from a field trip to the Chicago Recovery Alliance (CRA), combined with a "Train the Trainers" cascade model, peer education and the involvement of peers in research proved effective in Scotland because drug users themselves became enrolled in it.

An important organizational change occurred when the Scottish Executive formed the NFDRD, a broad-based group of people who analyzed routes for reducing the impact of drug-related harms in Scotland, and narrowed in on naloxone:

[Naloxone] wasn't top of the list. Like everything else, you have a basket of projects when you're looking at drugs policy, don't you? There are Safe Injecting Rooms, Medication Assisted Treatment, and a variety of different initiatives. We wanted better data sets, we wanted to look at inequalities, we wanted to look at poverty and its impact on drug dependence, we wanted to look at drug services. Honestly, I still think the biggest issue is that we don't get enough people into treatment.... I think you need to have a wide range of things on offer for this group of patients because they're not homogeneous.... You have ... young kids at risk of overdose, violence, trauma, the old guys of overdose and dying from deteriorating pulmonary function, poor liver function, just prematurely very old, much older than they should be.[48]

Although Robertson speculated that the Scottish government supported naloxone as a quick fix to a long-standing and well-publicized problem, the NFDRD moved with speed to put it into place. One member who preferred anonymity mentioned that the SNNP "actually gives a false impression.... [The] first nationalist government [was] looking for some good news and something that would be unique and national.... [A] national program appealed to them because Scotland would have it and England wouldn't and it would make Scotland that much more distinctive and unique." Moreover, naloxone united Scottish nationalists and internationalists; Labour and Conservative party platforms; and recovery-oriented approaches with harm reduction practices. Many political fissures were covered by focusing on naloxone.

SCARE ME: A SCOTTISH FIELD TRIP TO CRA

Although the NFDRD grew from Scottish soil, it was influenced by a site visit to CRA. Chicago, Illinois, was an urban area where naloxone was being supplied directly to drug users at a scale comparable to what the Scottish group envisioned. Seven Scots went to Chicago on September 17–21, 2006, to speak with CRA clients and learn from their experiences. "Every client we spoke to wholeheartedly endorsed the scheme, and urged us to adopt a similar service. We met many individuals who had either administered naloxone (many on several occasions), had received naloxone (again some on many occasions), or who had done both."[49] By this time CRA was delivering a spectrum of harm reduction health services, including vaccination; testing for hepatitis A and B; oral HIV detection; and opiate substitution treatment from sixteen mobile vans and six storefronts. By the 2006 site visit, CRA had given out ten thousand vials of naloxone and documented that at least 5 percent of these had been used at overdose scenes. There literally was no one else in the world as knowledgeable about running that scale of naloxone distribution than Bigg and Maxwell.

While visiting CRA, the Glaswegians met with Karen, a heroin user who had been revived by naloxone some twenty times by her boyfriend Andy and her mother. The couple "always injected together and kept naloxone convenient for quick access, keeping a supply in the house, car and near injecting apparatus."[50] While Karen recounted sometimes feeling "dope sick" after receiving naloxone, she "generally ha[d] no recollection of overdosing or being revived" because Andy adjusted the dose "depending on the apparent severity of the overdose."[51] At the time, Chicago was experiencing a rash of fentanyl deaths. When Karen overdosed on fentanyl-laced heroin, Andy had enough tacit knowledge to adjust the naloxone dose upward. This anecdote, which the Scottish site report recorded faithfully, presented two drug users whose caring for each other was mediated by naloxone. Within the context of an urban heroin market like Chicago's, Karen and Andy acted not as a "parallel couple,"[52] but as an intimate partnership for whom naloxone was a technology of solidarity.

Although the Scots observed little family support in Illinois overdose prevention efforts,[53] they noted that the CRA Community Advisory Groups had been meeting since the 1990s. According to Dan Bigg, "The reason [our programs] work is that we've listened to them. We've got them together, we've paid them, we feed them, we respect their expertise, and we've listened to them. That really is the cornerstone of CRA's success."[54]

The "myths" that dogged naloxone distribution were apparent to the visiting team, which dismissed concerns that widening naloxone availability might make users take more risks: they noted that the "drug users we spoke to imply that naloxone did not make them take more risks with the amount of drug used."[55] The group discussed dangers such as needle-stick injuries to children; confiscation and destruction of naloxone vials by police; and the concern that ambulance calls would plummet if people had naloxone.[56] The latter concern was less relevant in the United Kingdom, where ambulance care was not as costly as in the United States. Of all concerns expressed by those giving or receiving naloxone, the most meaningful in the Scottish context concerned "aggressive" or "punitive" responses of paramedics.[57] The visiting team happened to learn that even in 2006 Chicago paramedics had a reputation for administering such heavy doses of naloxone that recipients became "dope sick."[58]

Most memorable for the visiting team was the inspiration of Sarz Maxwell, a maternal figure for naloxone access who "always gravitated toward the patients that nobody else wanted."[59] Subsequently invited to Scotland, Maxwell became

very influential with some of our National Forum members. When they came back, they were totally converted. [Pharmacist] Duncan Hill, somebody from Scottish Police Service, and somebody from SDF. They came back and they were very positive at the National Forum level, and also, because some people from our local Health Board went, they were very strong advocates for a low-threshold program. So it changed from a very controlled service-based program to, "We should be getting this out there, we should be reaching numbers, we shouldn't be trialing this, it's unethical, we should be getting this out there."[60]

Rather than emulate CRA's user-oriented underground delivery system, the NFDRD worked inside the Scottish government, through the SDF, the Scottish National Health Service, and the Scottish Prison Service. After all, they were themselves "insiders" within these systems. Upon their return, members of the NFDRD, the SDF, and the Scottish Network of Families Affected by Drugs went to the Local Council to make a bid to start a pilot program to supply three hundred drug users in Glasgow with naloxone kits:

I think it was a great thing, but it wasn't an easy decision at that point. Now everybody knows what naloxone is, but at that point, in 2006, they didn't know what naloxone was. The principle of somebody administering it to somebody else, that was totally new. We had a chairman of the relevant committee of the Council at that time, that was Jim Coleman, and what he said has always stuck in my head. He was a traditional Labour councilor, and what he said was, "These are our people and they're dying, and we have to do something." So they gave us the money for our pilot.[61]

The Glasgow pilot, which took place in one of Scotland's largest regional health boards, NHS Greater Glasgow and Clyde, began shortly after the return from CRA in February 2007. The Scottish approach had to follow

European Union guidelines mandating that cardiac massage, first aid, basic life support, and injection technique be included in naloxone trainings. These were steps taken to secure credibility and legitimacy, as were visits from internationally prominent persons. A GP specialized in addiction treatments services, Saket Priyadarshi organized "Our Shared Care" meetings quarterly in Glasgow addiction services and invited Maxwell to speak:

She of course arrived with about ten naloxone kits in her handbag, very energetic and charismatic. She came on a bit of a promotional tour that was very successful and she was really converting people's minds from this controlled delivery system that really needed to be quite a bit lower threshold.... Now you've gone from people sitting around a table discussing whether this was a good idea to a tipping point into a whole movement. But it's not a successful movement yet, because we haven't yet convinced the Scottish government to fund it. That would take Minister Fergus Ewing.[62]

Origin stories abound concerning how and when Scotland's Minister for Community Safety, Fergus Ewing, became interested in promoting naloxone as a technological solution to the problem of DRDs in Scotland. A lawyer by training, married to a now deceased Accident and Emergency physician, Ewing is rumored to have visited Lisa Ross's naloxone distribution pilot program in Inverness (discussed below); to have been persuaded by a local pharmacist with whom he spoke on a train; to have been convinced by sharing the stage with Strang at an Edinburgh conference convened by the Ministry; and to have been looking for a short-term, high-impact government initiative to put Scotland on the map. Regardless, Ewing opted to use his influence to increase naloxone distribution as Scotland's signature response to an issue that had been one of its defining hallmarks since the movie *Trainspotting*:

Scotland has had a substantial and acknowledged problem in terms of drugs-related deaths, and Scotland had a tendency, which I think is a good tendency in public health terms, of focusing on its problems and trying to sort out the things that are the problems where Scotland is at the top of some league or other, and that included drugs-related deaths.[63]

The NFDRD regarded naloxone as so unique that "we really had to develop things that were not there before."[64] Scottish pride in doing things differently from England was evident in the crafting of the national naloxone program:

I think that's always in Scotland. It doesn't matter what field you're in, there is always that. We're a much smaller country as well, and we all talk to each other. We probably know everybody involved in addictions in Scotland.... That is a missing bit in England. It's just the scale and the size of it. They're very envious of us having a Scottish naloxone program. I sit on the Orange Guidelines Working Group, which is the Drug Misuse Prescribing Guidelines, Sir John Strang's group. People round that table—there's five of us from Scotland on that group—the people round that table are actually quite envious of Scotland's National Naloxone Program.... I think they would like to do that in England, but they don't have the structures to do it."[65]

Priyadarshi, lead medical officer within Glasgow Addiction Services when the SNNP launched in early 2011, attributed Scots' compassion toward the unfortunate, which played a role in choosing naloxone as a "Scottish solution for a Scottish problem." "We were prepared to be a bit bolder and to do things differently. That suit[ed] the Scottish National Party agenda to a certain extent.... You have to understand that dynamic in order to advance things, and measure the risks and benefits of what you're suggesting to that agenda. At that time, I don't think we were so calculating, ... we were probably more naive and caught up in it."[66]

Few NFDRD members identified politically with the SNP but most saw why naloxone appealed to Scotland's first nationalist government. The SNNP gained five full years (2011 to 2016) of central support without devolution to local authority. Even after the 2016 migration from national to local decision-making, the Scottish government continued to support two staff members hosted by the SDF in Glasgow.[67] Advocacy by the SDF, a Scottish government charity that was a member of the National Forum, led to such user involvement as there was in the SNNP. Additionally, the SDF organized key national conferences and helped plan the Glasgow pilot with NHS Greater Glasgow and Clyde.[68] The SDF coordinated a Glasgow City Council planning committee; evaluated the Glasgow naloxone pilot launched in 2007; supported approvals for the Lanarkshire pilot (which also launched in 2007); and organized Scottish Parliament Cross Party Group on Drugs and Alcohol meetings in 2009 and 2010. These activities fed into the SNNP, which was implemented via the National Naloxone Advisory Group (NNAG), chaired initially by John Somers and thereafter by Hunter. NNAG rolled out the national program with the Scottish Drugs Policy Unit. Thus did Scotland leverage structural

advantages stemming from central support and political culture constituted through debate, alongside the cultural commitments of a "fairly
practical people."[69]

PILOTS AND PROTOCOLS: FITTING PRACTICE TO POLICY AND POLICY TO PRACTICE

Touted as an exemplary translation of evidence into policy,[70] the SNNP
developed via collaborative efforts and concurrent pilots; in 2009 a third
pilot joined from the Scottish Highlands. Soon after the 2005 change
to the UK Medicines Act, which invigorated the Scottish pilots, Andrew
McAuley was asked to undertake a naloxone pilot in Lanarkshire, where
he had grown up: "Until I'd taken that post, you were always aware
of drug misuse, but you weren't aware of how acute it was, and how
localized it was to certain communities you were quite familiar with. So
without being too sentimental, it felt as if you were doing something
worthwhile for people you knew not directly but maybe indirectly."[71]
Surprisingly, local officials in the mining district of Lanarkshire had previously explored naloxone as a solution to a spike in DRDs experienced
there in the early 2000s. They had gone so far as to request a "letter of
comfort" from the Lord Advocate in 2004.[72] As that request was denied,
there was no legal path forward. Not until the legal barriers lifted did
Lanarkshire look again at take-home naloxone (THN).[73]

Aware that planning for a naloxone pilot was underway in Glasgow,
McAuley recalled friendly competition. The Lanarkshire pilot differed
from Glasgow in that it involved the Scottish Ambulance Service, the
Royal Pharmaceutical Society, and the chief pharmacist. It initially used
a mini-jet naloxone product packaged in flimsy cardboard, which did
not withstand the rigors of drug-using lifestyles. Illustrating the bootstrap
nature of the endeavor, McAuley combed the town looking for "something that people could carry it in that would make it more discreet so
it wouldn't arouse suspicions."[74] He purchased fifty eyeglass cases, into
which the kit could neatly fit. His fledgling program was tiny:

Even to get those nineteen people felt very challenging at the time. It felt as if
we were doing something ground breaking. We knew if even with those nineteen we could get through at least six months and demonstrate some evidence

of effect in terms of the process, that you could train people, that you could give it out, they would manage it responsibly, they wouldn't attack their friends with it, they wouldn't take more drugs, they would still have it on their person after six months, all these very practical things that people were nervous about.[75]

By contrast, the Glasgow pilot took place in one of Scotland's largest regional health boards, NHS Greater Glasgow and Clyde.[76] The Scottish Network of Families Affected by Drugs had helped pressure the local council into funding the training of three hundred drug users, who were outfitted with naloxone kits when the program launched in February 2007. Participants were recruited from the Glasgow Drug Crisis Centre and trained using CRA content modified for the local audience. Both the Glasgow and Lanarkshire pilots received publicity on the BBC and in *The International Herald*, the *London Sunday Times*, and the *Scotsman*. Reporting ranged from strong endorsement to expressions of concern that drug users might be unable by themselves to manage "critical incidents."[77]

A third pilot commenced in the Highlands region in 2009 under leadership of harm reduction nurse Lisa Ross. One of the world's first rural naloxone programs, the Highlands program also distributed kits to police and people released from Porterfield Prison in Inverness.[78] The Inverness pilot trained 170 people, including 68 prisoners, in the first year; resulted in a total of 64 successful uses of naloxone over three years; and contributed valuable expertise to the conversation about what would be necessary to saturate Scotland with naloxone. The Highlands pilot is credited with "training more people quickly and saturating the population with naloxone. They didn't have to get through as much red tape as we did, because [the Glasgow pilot] had effectively trailblazed that for them."[79] Considered impressive for working expeditiously despite dispersed a population and rugged terrain, kit distribution in the Highlands was "far higher per hundred problem drug users than anyone else achieved."[80] Comparisons of the pilot programs revealed that from 2008 to 2009, naloxone was used to reverse overdose twice in Lanarkshire and eleven times in Glasgow—but thirty-seven times in the Highlands.[81]

Once a fully national naloxone program appeared to be moving toward reality in 2010, the NFDRD created a subcommittee called a Short-Life Working Group (SLWG). Their task was to develop a standardized way to legally supply naloxone in Scotland, and to provide evidence-informed advice on products, training, target groups, and outcome measures.[82] Some SLWG members did double duty with NNAG, constituted to monitor the National Program. Despite emphasis on protocols to please the neoliberal governing party, the aim was to make naloxone provision a norm throughout Scotland. At first, this was nearly impossible, as naloxone could only be supplied by members of designated professional groups to a named patient, and thus remained inaccessible to some of the people who needed it most. An obvious goal was to persuade relevant authorities to widen the categories of people who could be supplied it beyond those who were active or former drug users themselves. In a move hearkening back to the maps showing that overdoses occurred primarily in single-room occupancy hotels in San Francisco, the NFDRD realized that many Scottish DRDs occurred in hostels serving the homeless.

Hostel staff had been placed in a legal bind: although they could legally administer naloxone under the 2005 changes to the Medicines Act, they could not legally keep a supply on hand in the homeless shelter:

That was put to the Lord Advocate [Elish Angiolini], the senior legal person in Scotland, and she agreed that she would make an exemption that allowed us as services to supply service managers to hold their own stocks. It was still a prescription-only medicine and that means you have to supply it to a named patient. What the exemption said was that NHS staff who were supplying a Prescription Only Medicine would not be prosecuted for making that supply directly to hostel staff—outside the normal rules where the supply must be made to a named patient.... That also made our English colleagues a bit jealous because it was a very sensible, pragmatic decision. Where should naloxone be? It needs to be where people are around having overdose. In that way the hostels were classic.[83]

Although the Lord Advocate's 2011 exemption expanded the reach of naloxone, it still applied only to "services in contact with those likely to experience an overdose." While the exemption enabled lives to be saved in such settings as hostels and needle exchanges, the distribution of the drug remained a game of chance for some because of the Lord Advocate's restriction. Only those at risk could access a supply; family members or others who did not themselves have a drug history could not obtain a supply of naloxone. Indeed, the NFDRD was advised not to include family members in the petition that gained access for hostel staff as it might make success less likely. But family members desire to obtain naloxone was stimulated by "anecdotal reports from people who've used it and they become empowered. Anybody who's used it can see the effects and they become empowered. To a lay person who's seen that work in front of their eyes, you're not only providing a service, you're also stimulating demand."[84]

Such experiential learning helped rehabilitate naloxone's negative reputation in Scotland. The medication gained an image as a "lifesaver," which multiplied its advocates. According to Amanda Laird, it was difficult early on to train family members because the trainers were legally barred from supplying kits at the end of sessions. In response to this hurdle, Family Addiction Support Services commissioned former addictions worker Ann Mathieson to create a play called "A Chap at the Door" (the term "chap" refers both to a person and to the knock at the door).[85] Performed in 2012 in the Glasgow City Chambers and before the Scottish Parliament, the play dramatized the plight of a family in the aftermath of their daughter's overdose death. The performance highlighted the legal double bind that family members found themselves in—they could be trained but not supplied with naloxone:

> The play was very harrowing, and it was about a family member. There was a death, a very personalized death, and it was about what the family was experiencing and a wee bit about kinship care.... What we found was that the families were quite afraid to access the training because they didn't know what memories would come up. They were afraid they would see pictures of dead people, ... but the small drama group came along to training and they enjoyed it very much and learned that their fears were unfounded.[86]

An initial attempt to enfranchise family members by widening the definition of who could be supplied naloxone failed, but in October 2015 family members and friends who were not themselves drug users did finally gain legal status to be supplied with naloxone across the United Kingdom.

RESTRUCTURING THE LANDSCAPE OF SCOTLAND'S NATIONAL NALOXONE PROGRAM

Scottish law widened again in October 2015 to offer a new framework within which naloxone could be supplied.[87] Hunter again found herself writing PGDs so that friends and family could request a supply of naloxone to "prevent deaths in the critical minutes before specialist care arrives."[88] Scotland devolved the National Naloxone Program to local authority after the first five years (2011–2015). The new emphasis on recovery undid some of the social cohesiveness and predictability with which naloxone was supplied, despite evidence of the effectiveness of THN continually produced within and beyond Scotland.[89] In 2016, the SNNP transitioned to what is called "the framework" in which, according to Hunter, "There's no Naloxone Advisory Group, no Drugs Deaths Forum. There's a new Partnership Action on Drug Strategy [PADS]—that was previously the Drug Strategy Delivery Commission in Scotland. The PADS is the overarching group and one of the subgroups is the Harms Group, which incorporates all of the work of the Drug Deaths Forum and the Naloxone Advisory Group, plus all the other harms, the BBVs [blood-borne viruses]."[90] The Scottish government continued funding the two SDF positions associated with the SNNP—the post of a National Naloxone Coordinator and the post of a National Naloxone Peer Training and Support Officer.[91]

Restructuring created tension between the Drug Policy Unit and NNAG, which wanted continued central funding and a sustained national focus on preventing DRDs. Policy analysis reveals that centralized governments are more likely to originate and implement public health policies leading to mortality decline in response to active social movements that produce "culturally credible constructions of risk to the public's health."[92] In the Scottish case, there was no broad social movement for naloxone; harm reduction was in place and no political agitation appeared. Despite the SDF doing much to open peer education, and the novelty of some of its methods for doing so, the SNNP was never peer-led; it was implemented by a tight-knit cadre of professionals.

Nor did Scotland contend with health insurance issues because the harm reduction-oriented NHS was available everywhere. However, expansion brought in groups who were less enthusiastic than the protagonists involved in the pilots, protocols, and programs recounted in this chapter. Via a new framework that replaced the old systems of PGDs, the SNPP devolved authority to local health boards starting in 2016. Where members of health boards were knowledgeable about naloxone, as in greater Glasgow and Clyde, local Alcohol and Drug Partnerships continued to purchase Prenoxad, Martindale's five-dose naloxone product, despite steep price hikes:

> Once the government stopped funding the national supply aspect, [the recommendation] was that the Alcohol and Drug Partnerships should support it at the level of the local authority. It's where the city or the town, and the Health Board come together with the police, the prisons, and other partner agencies and groups, including families and those with lived experience.... Basically, what's happened is that the funding for the kits has been devolved to the local areas and they're not all making the same decisions, but the government is still collecting the data so they can monitor supplies.... The data recording in Scotland is far superior to the rest of the UK, particularly around drug deaths. You've got to have good data to plan things. Now naloxone reporting is part of official statistics, so they have to keep that effort up.[93]

How have local authorities done now that they are in charge? They mostly have maintained the benefits begun under the SNNP. Scotland continued serving as a model for naloxone distribution to such a degree that nations may be indexed and compared according to how much "Scotland-equivalent power" they achieve in terms of naloxone distribution.[94] These Scottish protagonists estimate that there must be twenty times the number of naloxone kits as there are deaths from overdose in a given area to mitigate the risk of dying.

BEYOND A CUP OF TEA AND A BISCUIT: PEER EDUCATION AND ENGAGED RESEARCH

No longer buffered from the forces of neoliberal market governance, costs will no doubt rise as decentralized naloxone programs "grow arms and legs" and come to maturity as peer engagement also scaled up. During the SNNP, the SDF built peer networks within all but two of the fourteen territorial Health Boards in Scotland. There was some inertia and resistance because "people started making up fantasies in the early days about peoples' behaviors, and saying, what people are going to do is they'll be doing tons of heroin and using naloxone to balance that out. That just goes against common sense. People can't afford to do that."[95] Harm reduction actively contests judgments about drug users' lack of self-governance and their assumed irrationality, selfishness, and lack of care for others,[96] by representing users as "peers" and "witnesses," and including them as "experts" in peer training and naloxone supply. Evolving attitudes toward drug users may be seen as an indication of both large and small changes that have occurred with the SNNP and its aftermath.

One major effect has been a shift in the locus of credibility from medical professionals to those who gained knowledge by living it: "Actually now, when people are telling their story, it's about their peer-ness, if you like, and the contribution that their experience makes to their work and their insight.... I think naloxone has helped drive that change."[97] Peer naloxone educators in Scotland gained credibility for naloxone by producing what might be termed "evidence-based experience." "Naloxone's probably the one and only thing that has happened in the whole of Scotland that has been able to transcend all views and bring people together and wanting to be in partnership."[98] But they also gained credibility for knowledge derived from connections between experience and evidence. Wallace interpreted his move up the hierarchy of credibility as being not only about naloxone, but also about the loosening of bias toward drug users:

Over the years of meeting with staff about this intervention, it appears that I sound more and more credible by how I represent myself and the quality of training I deliver. In Glasgow quite a lot of the training in community-based and residential rehabs will be delivered by peers now. Peers are delivering a lot of the recovery type stuff and even with the recent HIV outbreak in Glasgow, they've been looking for the peers to be delivering HIV briefings because peers are now viewed as more credible and have instant access to the target group, which is a great thing.[99]

Respected "experts by experience" were carefully educated to initiate their circle of drug-using associates into the habits of harm reduction. Once identified as someone at high risk for dying an overdose death—from his embodied experience and his emotional response to the overdose death of a near relative—Wallace was also trained in a business evaluation technique pioneered in Glasgow called "mystery shopper service evaluation."[100]

Mystery shopping was used in service-based industries as a form of performance evaluation that leveraged experiential expertise. "Mystery shoppers, we'd go into community pharmacies and needle exchanges as mystery shoppers and we'd have a wee script that we would go through and then make a report on the quality and level of service."[101] The SDF has used this inclusive peer research model since the late 1990s for recruiting and retaining people by training them to conduct research and evaluate services in this way. Former users were also trained in communication skills, data collection, data entry, and other skills—investments in human capital that cultivated new forms of expertise alongside those derived from years in drug-using social worlds. The peer education model developed in Scotland addressed the lack of knowledge about overdose, as well as the formerly discussed documented cases of overantagonism in Glasgow.[102] Had peer education or mystery shopping evaluation been used earlier, overantagonism and ignorance about dosage, mixing, and drinking would surely have come to notice sooner.

Naloxone dosage remains contentious and understudied. Overantagonism results from administering too high a dose of naloxone relative to opioids in a person's system, causing people to feel "dope sick" when abruptly revived. Dosage might be brushed off as a clinical or technical matter, but is actually socially impactful. The undone science of "overantagonism" was difficult to conduct because researchers could not ethically put human subjects into overdose in uncontrolled settings. Few papers documented the qualitative experience of administering or receiving naloxone,[103] a knowledge gap that flowed from what might be called "quantitative privilege." Temptation to fill in that gap on what it felt like to receive or administer naloxone led NHS analyst McAuley to do a qualitative dissertation:

When you look specifically at lived experience, there [were] no papers in the UK [that] document[ed] experience of naloxone administration and how that felt to the individual and what the aftermath was for the individual. For us, this was a huge gap. Even applying a theoretical lens, there were very few papers.... [Y]ou need different types of individuals at different stages of the process. You need early adopters and then you need people with influence in the peer network to get the kits, to get that credibility within the network, but that actually can be challenged by the credibility of the individuals themselves that you target early on. [So] ... it's remarkable that we are now 21 years since Professor Strang's *BMJ* paper and how little there is on the experience.[104]

Why was there so little qualitative research on overdose and its aftermath? Engaged research relies on ethnographers witnessing what actually happens to people, finding out how policy and clinical practice guidelines affect how people live their lives or die their deaths. One might assume such knowledge would be considered crucial for interpreting the social value of an intervention. Instead, evaluation (not interpretation) has been almost entirely quantitative—to the point of almost becoming itself a form of epistemological ignorance. The devaluation of qualitative and ethnographic methods in the age of evidence-based medicine sometimes reverses; epidemiologists and others turn to those who work with qualitative methods when they cannot explain what they are seeing in the numbers—when phenomena are emerging or "when paradigms crash."[105] McAuley's own conversion to qualitative research "humanized" those he had been studying by opening his eyes to "some unintended consequences that are not documented in the literature.... There's very little about individuals and relationships and credibility and this kind of stuff. For me at the end of the whole process, I had found something new. The whole journey had been worth it. I didn't get to the end of it and say, "So what?" ... I came to respect and enjoy qualitative work."[106] Such work conveys a broader range of experiences and unsettles overgeneralizations, but quantitative studies are taken to provide more definitive evidence; hence they are considered more persuasive in cultures where data is held to standards of political neutrality. Qualitative approaches may be regarded as "partisan" or "biased" despite sophisticated measures to reduce bias. When too rigorously upheld, such standards squelch experiential knowers, who in this case mobilized those among them who had experienced overdose personally to give evidence that naloxone worked.

To counter the delegitimation of harm reduction as compared to evidence-based approaches, the Eurasian Harm Reduction Network and the US-based Harm Reduction Coalition created a campaign called "I AM THE EVIDENCE." There is a nontrivial risk of a narrowing or "silencing" effect that can flow from the reification of evidence-based policy, and doing policy by the numbers has become central to contemporary relations of ruling.[107] The policy climate itself could even be construed as a barrier to widening the scope of "peer-based" or "carry-along" naloxone."[108] Yet multiple forms of evidence are crucial in any complex endeavor with diverse partisans and a broad range of perspectives. "I AM THE EVIDENCE" tapped into the emotive and experiential register as a direct counter to the reductions necessary for quantitative, data-driven studies made at the population level. The Naloxone Advisory Group England embraced the normative stance that naloxone availability should not be a gamble or, in British parlance, a "post-code lottery." Embedded in the "child's play" of figure 10.1 are larger stakes of human rights, social justice, and loss of life to overdose as a loss of human potential. The text on this advertisement comes from a once "chaotic" drug user, who might easily have numbered among the dead without access to naloxone: "My life is full now. I am passionate about my work in advocacy.... Naloxone saved my life and kick-started my recovery."

This chapter has recounted how various forms of evidence and testimony running from vernacular theater to state-supported peer education networks to public and private-sector efforts led to policy change in Scotland. Growing from the combined efforts of physicians, public health professionals, and clinical and social researchers, the need for previously undone science to be done became a social and political necessity

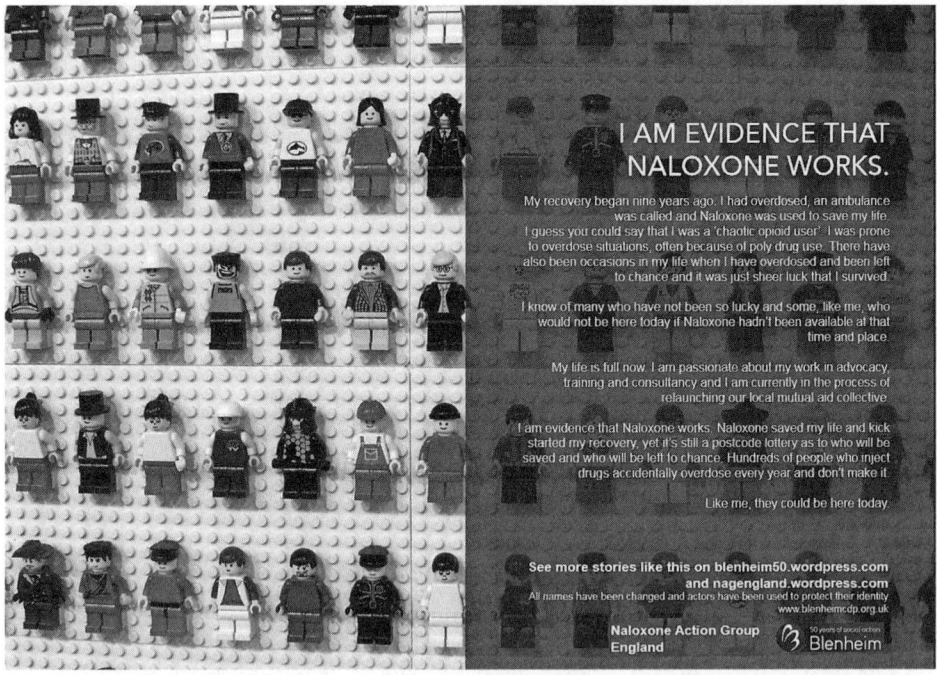

10.1 "I am evidence that naloxone works." Crafting a playful "body count" composed of Lego figures from all walks of life, the testimonial made the point that naloxone availability in England was the "sheer luck" of a "postcode lottery."
Used with permission of Blenheim CDP.

in Scotland, and in the United Kingdom more generally. There remain stories to be told about Scotland's role in the global movement to place naloxone in the hands of those who need it most, some of which appear in the next chapter. One event often credited with tipping the Scots toward the SNNP has not been fully explored in this chapter, which has presented the SNNP's story from the perspective of "insiders" who pragmatically took up the opportunity to saturate Scotland with naloxone to the greatest extent possible. Their actions challenged the world with a model of what the normalization of naloxone looks like in a place that regrettably continues to generate a high DRD rate despite many protagonists' efforts. Multiply characterized as a "miracle medicine," a tool of "rescue" and "reversal," "lifeline" or "lifesaver," naloxone was a technology around which professional solidarity was built. Depending upon who is doing the characterization, which historical and cultural contexts are invoked, and which targets are chosen at what levels of action, naloxone went on to become significant for many places. Scotland stands as an exemplar—just as the SNP hoped it would.

11

EVIDENCE FROM PILLAR TO POST:
Researching the Varieties of Overdose Experience

Sometimes lethal but more often survived, overdose is patterned but not predictable. Overdose evades the usual modes through which science tames social worlds to reduce bias and confound. It tests the limits of researchers' capacities to deal with "dirty" data. *All* data is dirty in the situated knowledges of overdose. Historians, too, deal with dirty data, considering it a duty to examine the confounds that clutter evidence-based policy worlds and to document the multiple realities that complicate lessons one might draw from history. This chapter examines some of the complexities encountered as the United Kingdom considered how to best respond to rising incidence of overdose. Often brought about by blending research with advocacy, new infrastructures changed both "overdose experiences" and "outcomes" in the knowledge politics of overdose. Such research entwined moral and ethical questions with science, magnifying the logistical difficulties of recruiting vulnerable populations and studying interventions into risky or barely legal activities.

Given that is difficult to pre-arrange to witness potentially lethal overdoses, how can they be subjected to a "randomized" design? Yet in the evidence-based world, randomized clinical trials (RCTs) count as the gold standard.[1] Randomization is crucial for those who wish to command the serious high ground at the top of the evidentiary pyramid. Clinical trials rest on an "unparalleled standardization of everything,"[2] yet are viewed as necessary for unassailable claims-making. Even in the context of emergency medicine, randomized trials are increasingly used as political tools.[3] The tightly bounded form of standardization that occurs in RCTs overspills the bounds of any given research situation.[4] Yet standardization must also be remade in specific situations. The very logic of evidence-based "intervention" assumes that effects may be known, measured, and compared over time and place. RCTs nevertheless proved both epistemologically and politically unacceptable to many activists and researchers in the overdose arena.

Research designs based upon holding experimental conditions steady, randomization, and quantitative demonstration—all came into direct conflict with activist worldviews that "naloxone works." After all, participants in social movements seek to remake their worlds, and documenting and maintaining records of "saves" fit with that as a first step. But then came imperatives to integrate data with that of paramedics and other emergency responders, which some saw as tantamount to collusion with authority. Engaging with neoliberal concepts of the "preventive self"[5] was another imperative on which drug users and their advocates held quite different views from those of professional rescuers. Activists often had "authority issues," holding worldviews in stark conflict with the professional commitments of police, sheriffs, prison administrators, and members of state and local governments.[6] But all constituencies were needed if the rules, regulations, and social conditions within which naloxone access was to occur were to be expanded. Surprisingly often, those working to democratize naloxone proved pragmatic and flexible when it became apparent that they had to "act with" authorities—with whom they might otherwise not be caught dead.

DISTURBING NUMBERS: IN WHOSE DATA DO WE TRUST?

Knowledge problems abound in the making of statistical claims regarding drug-related deaths (DRDs). Time lags in the slow administrative processes through which deaths are certified; toxicology reports must be made, inquests held, records forwarded, cases analyzed. For those tracking DRDs or naloxone reversals, epistemological holes dot the effort like Swiss cheese. Methods to count and certify overdose deaths differ in different places, and even within a jurisdiction there can be conflicting methods. In England, the Office of National Statistics (ONS) bases determinations on "drug poisoning" data from inquests, using the date that the death was registered after inquest rather than the date when it occurred. In contrast, the National Programme for Substance Abuse Deaths (NPSAD) counts any death involving illicit substances, includes relevant detail about the person's life, and goes by the actual year when the death occurred. The two routes differ in how they define DRDs, even though coroner's offices and drug treatment programs feed data to both.

In an example provided by general practitioner Judith Yates, a keen advocate of community-based naloxone programs in England, ONS might "count an elderly lady who was dying of pneumonia and given some tramadol" as a DRD, but not "somebody who ... jumped off a car port roof high on MDMA." Only drug poisoning counts: "They won't count somebody who committed suicide by hanging with heroin or methadone in their system. But the NPSAD will."[7] Due to pressure from the National Forum on DRDs (NFDRD) described in chapter 10, Scotland developed a national protocol that included toxicology at forensic autopsy and ensured deaths are registered rapidly (within eight days of death). In England, one coroner's office might track drug poisonings via autopsy, but another might interpret mere presence of drugs at the scene as sufficient for the determination of DRD. Sometimes reporting is in the shaky hands of undersupported officials. Yates experienced intervals as long as two years between the date of death and determination of cause in Birmingham. "Our [old] coroner got very tired and stopped doing inquests in 2013.... [When] our new coroner came along and she did inquests from 2011, 2012, and 2013 in 2014, ... the Office of National Statistics thought we'd had a huge spike in drug misuse deaths in Birmingham in 2014."[8]

Seeking to reconcile systems by creating and maintaining her own, Yates recognized that many patients she knew to have died DRDs appeared in neither database. At the time of my interview with her in March 2017, she had been manually going through all inquest files at the coroner's office in Birmingham to maintain an accurate database of all DRDs since 2009: "Nobody was doing it in Birmingham and I kept asking the commissioners to set up a local inquiry group, because I knew that most [decedents] were not known to drug treatment services. It was clear that somebody should be looking at this. Then in came 2012 and we started our naloxone scheme. I wanted to know if we were making a difference, so I kept going in, and have carried on going in."[9]

While this physician's personal dedication to tracking DRDs is laudable,[10] her experience illustrates how overdose deaths are certified through social as well as technical processes. In "Why People Die: Social Representations of Death and Its Causes," Lindsay Prior and Mick Bloor explored how a community of medical experts—hospital personnel in Edinburgh, Scotland—fit individual explanations for any given death into a culturally mediated system of rules and representations such as actuarial "life tables," mortality reports, and death certificates. Despite variation in infrastructure and process, causes of death are integrated into "culturally structured system[s] of 'social representations,'" the rules and practices for making deaths from particular causes visible.[11] Intensified pressures to standardize causes of death led to the genesis of the category "DRD." Although Priori and Bloor were not studying DRDs, their interviews with "death certifiers" in central Scotland in the mid-1980s spelled out the rules as follows:

1. Death is a product of pathology.
2. Death is a physical event.
3. A cause of death is a physical thing ... susceptible to detection.
4. A cause of death is always a singular event (though it might exist as part of a sequence).
5. A pertinent cause of death is usually one that is proximate to the event.
6. A cause of death is always to be determined by its context.[12]

Deaths departing from these rules are transformed from the "natural" to the "unnatural" category. Social imperatives to make death visible—and arguably meaningful—have intensified at the same time that the uncertain and speculative forensics of DRDs have come into focus.

Death certification narrows and stabilizes a contested set of social representations. With DRDs, there is almost always doubt about pathology, proximate cause, intentionality, and the precise set of drugs in play. Rule-governed practices push "cause of death" toward some categories and away from others; DRDs are by definition always already "unnatural." Drug deaths stem from social inequalities and health disparities; singular deaths cannot be attributed to macro-level social structures or indirect forces. "The entire notion of "natural" death serves both to obscure and to naturalize the effects of the social and economic inequalities which are related to mortality differentials in the advanced industrial world."[13] The normalized categorization derives from the social construction of overdose as resulting from unnatural, individual excess. As Prior and Bloor noted, death certifiers have conflicting moral views: they found that those for whom the primary purpose of death certificates was "amelioration of grief"[14] listed neither alcoholism nor drug overdose as causes of death due to the social stigma carried by these conditions. Causes of death were chosen for their "socio-cultural aptness, rather than their empirical accuracy."[15]

National-level overdose death statistics obscure local patterns of use, in effect simplifying the cause of death. Individual cases of overdose seem so obviously to have been caused by *drugs*. The term "overdose" implies a more precisely calibrated "correct" dose, less potent or less adulterated, that might have led to a less lethal outcome. The individual might have done something differently—abstaining, obtaining supplies from more legitimate or knowledgeable sources, otherwise reducing harm. Instead, the individual became, as we say, "a statistic." Such cases comprise the data about which claims are made at local, regional, national, and international levels. A state of "data denial" prevails concerning the accuracy of DRD statistics, yet these are used to make meaningful comparisons between patterns.

While policy makers might question where DRD statistics come from, their actions grant interpretive legitimacy, especially when statistics are used to draw public attention to overall trends. The European Monitoring Centre for Drugs and Drug Addiction (EMCDDA) has kept close watch on overdose deaths since 1995—during which what counts as Europe has changed; methods for determining, reporting, and "ascertaining"[16] drug-"related" or drug-"induced" deaths have changed; and the visibility of national-level death tolls has changed. Differences in criteria, categorization, and variations in scrutiny of deaths by gender, race, ethnicity, age, or immigration status have occurred. Countries that count DRDs are countries where they matter to people in charge; statistical offices tend to count what matters for political purposes. Thus evolving overdose death tracking systems make opioid overdose deaths matter by extracting them from other adverse drug events. Rather than dwell in the "arithmetic sublime,"[17] I point out that large numbers of adverse drug events that do not involve opioids—including excess deaths from alcohol, tobacco, or prescription anticoagulants[18]—annually kill more people than do opioid overdoses. Yet anticoagulants are not subject to the moral-panic politics of drug crisis rhetoric. There is no "anticoagulant epidemic" (arguably there should be more awareness that anticoagulants kill, how they do so, and how to prevent them from doing so). Deaths from alcohol and tobacco have different temporal horizons measured in decades; are indirectly linked to a variety of diseases; and are not so tightly associated with stigmatized populations.

Wrinkles in our ways of knowing abound in the overdose and DRD arena partly because overdose varies in its visibility as a political artifact; that variability is in itself a risk factor. Overarching claims are made—approximately one-third of overdose deaths in Europe occur in the United Kingdom. The EMCDDA has "counted more than 140,000 drug overdose deaths" in Europe since counting began. Nineteen lives per day are lost in Europe and close to two hundred lives in the United States. Such confabulations congeal and circulate as "fact."[19] The cultural work that it takes to produce such narratives and artifacts is accomplished by epidemiologists who understand that grave epistemic holes riddle the factual edifice—but do so anyway thanks to their good intentions of preventing preventable deaths.

"GATE-HAPPY": STAYING ALIVE AFTER PRISON RELEASE

Official outcomes for the Scottish National Naloxone Program (SNNP) were defined by biostatistician Sheila Bird, lead of the Medical Research Council's Biostatistics Unit in Cambridge.[20] Deeply involved in HIV/AIDS epidemiology, she had worked with the Scottish Prison Service (SPS) to quantify transmission risk and impact of incarceration on mortality and morbidity of drug injectors. An initial record-linkage study was published in the *British Medical Journal* in 1998 by Seaman, Brettle, and Gore (Sheila Bird's previous name),[21] showing that risk of DRD was eight times higher during the two weeks following prison release than at other times at liberty. Based on a sample of HIV-positive male injectors, these findings were later shown to hold for all men ages 15 to 35 years who had been released from Scottish prisons from late 1996 through 1999.[22] The findings seemed incongruous given the "harm reduction era" in the United Kingdom and particularly in Scotland:

We had assumed ... that we would be looking at one drugs-related death per three thousand recently released days (i.e., in the first fortnight) rather than one per one thousand at the time of the Seaman study, which was 1983 through to 1994. We expected that the drugs-related deaths had actually decreased because of harm reduction measures.... But in fact the validation study of 19,500 releases found 57 drugs-related deaths in the first twelve weeks, 34 of them in the first two weeks.[23]

Thinking about how they might intervene in and so interrupt this statistical surprise, Bird and Hutchinson (2003) proposed a "prison-based, randomized controlled trial of naloxone, to see if this would reduce the drugs-related death rate soon after release from prison."[24] Their idea was that all prisoners could be given a naloxone kit upon release—but only some of the kits would actually contain naloxone. Biding her time until after the 2005 change to the UK Medicines Act, Bird began laying groundwork for the trial with the SPS.

The SPS worked from a basis of compassionate custody and knowledge of the typical lives and deaths of modal prisoners. Since 1996, SPS had told prisoners that they faced a higher risk of DRD in their early weeks at liberty. "The guys are going out of there, what they call 'gate happy,' they're met by their dealer, off to the pub, and there are lots of risks for them immediately upon release, not just the loss of tolerance."[25] Tearing a page from the book of harm reduction, an informational booklet warned gate-happy prisoners not to use alone, and to avoid returning to their previous dosage. Framed as a problem of elevated risk and based on epidemiological reasoning, the solution was practical advice and information about individual risk reduction that might or might not be read, absorbed, and acted upon by any given prisoner. "As we know in public health," Bird drily put it, "information is necessary but seldom sufficient to change behavior."[26]

When researchers replicated the Scottish prison study with larger numbers in the English prison environment, they also found a "strikingly acute" concentration of DRDs in both men and women the week after prison release.[27] In one of few recognitions of gender differences in the naloxone literature,[28] National Addiction Centre researchers found that while "male prisoners were twenty-nine times more likely to die during the week following release," "female prisoners were sixty-nine times more likely to die during this period."[29] Female mortality rates were higher than male rates for an entire month after release.[30] Despite experiential knowledge and statistical evidence that DRDs were higher in the short window following prison release, and despite knowledge of naloxone's use for preventing such deaths, there remained legal constraints on naloxone (addressed in previous chapters). The question was how to prove beyond all shadow of doubt that naloxone supplied to prisoners upon release would reduce their chances of drug-related death.

Statistically speaking, death is a relatively rare event. Massive numbers of study participants would be necessary to show that naloxone definitively reduced deaths.[31] But the overdose death rate was so high among the recently released that a naloxone intervention targeted at this group might stand out.[32] Ghoulish as it might sound—and everyone involved in structuring the NALoxone InVEstigation (N-ALIVE) trial and pilot trial understood that some might find it so—that elevated post-release risk period represented an "extraordinary opportunity for a definitive research trial."[33] The situation at prison release was understood as something that could be designed into a structured experiment, an opportunity to produce new knowledge at a higher level of certainty and thus scientific credibility and legitimacy.

Historians of science have considered the so-called natural laboratory as one among several different types of experimental situations, including those using animal models, in vivo or in vitro tissue, and laboratory-based scenarios.[34] With the ascendance of RCTs as the gold standard for global public health, "intervention-cum-research projects in global public health often use the notion of a 'natural laboratory,' suggesting that the social world can generate laboratory-like data."[35] Such natural history experiments might be called highly unnatural, yet they offer opportunities to learn something of benefit. Strang's "tendency as a scientist [was] to look out for natural history studies and to look out for contexts and windows of opportunity where special studies can be done.... What we had with the prison system was that it created that experimental situation."[36] As Strang explained,

So the thinking went, Hold on a moment, if you're going to do a trial, to prepare for hostile times where people don't believe this works, you've got a problem because what you're trying to prevent is death, and death is a rare event, even when it's happening much more often, it's still rare so you need massive numbers. The only way to do it, the smart way to do it from a scientific point of view, is to choose a time when you know you have a horrific concentration of deaths.[37]

Once the early Scottish prison studies consolidated the claim that the first two to four weeks after prison release presented a high-risk window for DRD, this finding opened the door to an RCT designed to demonstrate—definitively—naloxone's effectiveness for preventing deaths. After a baseline death rate was established, a core group of RCT enthusiasts sketched plans and explored logistics with an eye toward a protocol involving the Scottish prisons. In the UK's evidence-based audit culture, they understood that until an RCT was conducted, naloxone advocates would command neither rhetorical nor political high ground.[38] Bird knew that naloxone was the "heroin antidote used in an emergency" from Strang's editorials in medical journals and from the Advisory Council on the Misuse of Drugs (ACMD) 2000 report.[39] As she put it, "[Naloxone] was, to me, the direct response to the problem that our studies had quantified"[40] and proposed in Bird and Hutchinson (2003). She informally approached Dr. Janet Darbyshire, Director of the Medical Research Council Clinical Trials Unit, and waited for the 2005 change to the UK Medicines Act to enable broader use of naloxone.

Once legal barriers were gone, the path was cleared for an RCT to achieve closure[41] on naloxone distribution as harm reduction. Bird formally approached Darbyshire's successor, Mahesh (Max) Parmar, to broach a UK-wide, prison-based trial of naloxone on release.[42] Parmar joined Bird and Strang, a trio that came to be called the Three Musketeers. Each member of the trio added different dimensions: Parmar was a clinical trialist experienced with trials of 100,000 or more subjects; Bird was a biostatistician with experience in prisons, record-linkage studies, and DRD statistics; and Strang was "immersed in the addictions field, but also coming at it from how do you do clinical trials in a patient-orientated way."[43] Having called for THN since the early 1990s, Strang's "strong clinical hunch was that what we needed was evidence from such a definitive trial if THN was ever to become standard clinical practice."[44] He himself sought the "equipoise zone" where there was enough benefit to justify the effort and the involvement of patients in an experimental treatment:

It is only justifiable to do a trial of this sort if you and/or others are not 100 percent certain about whether there is benefit—and that's why you've got to do research. Once you've got to the point where you are 100 percent sure (and especially if the system has reached the point where it has decided to implement an intervention as standard practice), then you have missed the window of opportunity when you could and should have done the study. So I think we have a responsibility to spot those times when you could do a study.[45]

The Three Musketeers faced the question of how to represent their intent to randomize to various gatekeepers. Recalling the 2008 meeting when the NFDRD first heard about the proposed trial, Priyadarshi stated,

That was the first time I met [Sheila Bird]. She came up on behalf of N-ALIVE and basically presented the evidence base for naloxone and asked for our support. We had a very lively, very lively, session because there was a strong feeling locally and a strong consensus in Scotland that we felt there was enough evidence base—it wasn't absolute, gold-standard, RCT-type evidence—but we felt there was enough evidence internationally and that we had enough local experience from our pilots, to say we didn't need a trial of this, we needed to get this out.[46]

Resistance to randomization was based primarily on ethical grounds, but also on the fact that naloxone was an old drug with an established clinical track record. Hunter explained how the N-ALIVE proposal troubled her:

I think initially we thought this wasn't really ethical because we felt we knew it worked, and why would we need to prove what naloxone did. I think we came to the conclusion that it wouldn't be right to do this in Scotland because there were so many confounders. We were already supplying it in Scotland, so if you did it in the prisons, and half had the information and half had the kit, the other half would still have access to it immediately as they went out.... So the whole thing would be very confused."[47]

The NFDRD construed randomization as withholding a treatment demonstrated to save lives. Naloxone was used on an emergency basis in institutional settings (ambulance, hospital, and SPS), but community distribution was small-scale even in the Scottish pilots that had begun in 2007:

In Glasgow, which is the biggest area, and we have Barlinnie, which is the biggest prison [in Scotland], it was already out in the community. Staff already knew about it, so if you had people at risk and you had something you could offer them, why would you not do that? We felt, aside from any ethical issues, it wouldn't be possible [in Scotland]. Then we looked at … England, where they didn't have anything. So we thought that if they don't have any means of getting it out there, getting 56,000 kits, which was a huge amount proposed at that time, in my head, that was a means of getting some of that out there. If you look at it deeper than just the pure, some get it, and some don't, it's more than they had to start with. Not everybody gets it, but at least some do. If Scotland had been in that position, I think we might have done it, but we were at different starting points.[48]

Staggered starting points and confounding factors were the reasons offered for rejecting Scottish participation in N-ALIVE. These rationales readily translated into statistical and methodological reasons for not pursuing the study as planned.

Rather than participate in the N-ALIVE trial, the NFDRD recommended that the Scottish government simply supply naloxone to those who needed it. "Presumably, the government could choose to accept advice or not. The [NFDRD's] recommendations are generally agreed upon. There was discussion but no dissent on the final recommendation. By then we were all invested in just making supplies in Scotland."[49] Consensus was the outcome of a social process in which the NFDRD took the position that naloxone's efficacy was crystal clear. Vulnerable people were at high risk for death after prison release, and NFDRD members feared such individuals would be offered information about overdose without being supplied naloxone:

That was the conceptual thing that we just couldn't get our heads around and accept. We totally understood it from a science point of view. And this is where we enter some interesting territory, because there's an emotional element to that as well. So I think there was a feeling that the evidence suggested there was a benefit in doing this—just getting it out—and also, there doesn't seem to be any harm. So why would we want to deny people? Would we not want to get the kit to as many people as possible and try and measure some form of evaluation? Try to do some measurement to figure out its effectiveness rather than try and do a randomized, controlled trial? Because in any RCT, you're saying that for 50 percent of people who are vulnerable to some kind of opiate overdose after leaving prison, you may not release something that potentially could save their lives.[50]

Some participants in these meetings recalled more shared vocalization of the intertwined ethical and emotional concerns, and others remembered less. However, it is fair to say in retrospect that critical responses to randomization were not based on ignorance, nor were they interpreted as such by the still-hopeful researchers:

John and Max and I went up to Barlinnie Prison with a colleague from SPS [Karen Norrie], and talked with a set of eight to ten guys about the N-ALIVE trial. I was trying to explain, without using the term "randomization," what it meant to be randomized, and a prisoner turned to me and said, "You mean, you'll randomize us?" We all just fell about laughing because I was at great pains to avoid using this word.... Well done, lads![51]

The design aspects of N-ALIVE extended beyond the discursive framing of randomization. Creative effort went into designing a credit card–sized naloxone kit that could fit into a wallet:

The prisoners said they really wanted the pack to be, ideally, the size of a credit card. Initially, we had a false start of trying to get something that ... would literally have been the size of a credit card. We lost at least six months going down that route.... A lot of effort went into the choice of the kit. We had to use something that was basically a standard kit, because we couldn't afford, we hadn't got the resources, to go through the regulatory approval for a new sort of kit. So we had to use approved product, but the credit card thing inspired John [Strang] to come up with the idea of the N-ALIVE wallet. One of our research team at the time, her partner was an eighteen-stone rugby player, and so he put it in his jeans pocket and tested it by sitting on this thing.[52]

But the major design problem that dogged N-ALIVE could not have been anticipated. The N-ALIVE pilot trial, undertaken to work out the kinks of prison provision, was initially supposed to occur in all Scottish prisons and fifteen out of roughly 120 British prisons. Stemming from previous experience, Bird thought highly of the SPS Medical Director's capacity to

cascade things across all of the prisons and monitor whether it was being done, as they did with the universal offering of hepatitis B immunization by Alan Mitchell. So I knew the patch in Scotland and I knew they could deliver, whereas the service in England had had more of a repressed history in terms of research. If they didn't like what the research said, it would be held up for as long as possible and there were more hoops to go through in England.[53]

Scotland had been crucial to the original conceptualization and planning for the N-ALIVE trial. All eligible prisoners who gave informed consent for randomization received information about how to administer naloxone—all learned how to assemble and administer naloxone and all received an N-ALIVE pack. The pack's external appearance was identical whether the subject had been randomized to the control group or to the group that received naloxone. Neither N-ALIVE's prison-based staff nor correctional officers knew the assignment. Only upon opening the N-ALIVE pack on release did the subject know whether they had a naloxone kit or not. That was a big "if." Harm reduction advocates had already argued that naloxone should be the standard of care; in an ideal world, every prisoner with a history of opioid use would be provided a kit upon release. But naloxone distribution programs could not meet that need.

Three years into discussion and after Medical Research Council funding had been secured for the N-ALIVE pilot trial, colleagues in the SPS alerted Bird in early 2010 that they were going to go a different direction because "the minister was minded to make take-home naloxone a public health policy in Scotland."[54] Subsequent implementation of this policy with the support of Fergus Ewing, Minister for Community Safety (2007 to 2011) shifted the entire N-ALIVE pilot to England. The Three Musketeers understood their opportunity in Scotland was foreclosed once the SNNP was in place. The experimental situation had changed; Scottish prisons could no longer be considered the ideal site:

We originally conceived of the N-ALIVE trial as taking the pure question, "Does provision of take-home naloxone reduce the death rate?" So we said to ourselves: "Let's take a prison release population who we know are at high risk who are not currently being given any naloxone, and let's see whether the addition of naloxone reduces deaths." That first bit was crucial to us. Nobody was getting any naloxone, so we were not withholding it from anybody.... [A]t around the time that the funding was agreed, Scotland at a political level decided they were going to initiate take-home naloxone. They'd essentially listened to the argument we'd been making from an advocacy point of view and decided to implement take-home naloxone nationally. So, with Scotland having made the policy decision to introduce THN as national policy, it was no longer ethically acceptable, in our opinion, to continue plans for conducting the N-ALIVE trial in Scotland.[55]

As recounted in chapter 10, the exact motivation for moving toward national naloxone distribution remained unclear. Some credit Strang's vociferous THN advocacy; others Bird's "knowledge of the patch"; still others the results of the Inverness naloxone pilot. Despite differing opinions, the consensus was that Scotland had become an untenable site for the trial:

From the moment Scotland said, "We're not going to run a trial, we are going to just do it," Scotland became a no-go zone for the N-ALIVE trial. You could only do a randomized trial in a country where there isn't provision. I think people mistakenly thought we were not giving naloxone just to do the trial. I wouldn't even apply.... I think that would be discriminatory.... We had approval for England and Scotland. The moment Scotland had the chance to go for a universal policy, our view was that Scotland's just not eligible.[56]

However, it was important to the Scots that the SNNP be evaluated along the lines proposed by the N-ALIVE team's letter, which emphasized prioritizing high-risk populations.

Minister Ewing asked Bird to conduct the evaluation as the statistician to the National Naloxone Advisory Group (NNAG), which was constituted to guide the national naloxone program. The agreed-upon outcome measure was the proportion of opioid-related deaths with a four-week antecedent of prison release. Bird influenced that choice:

She was the driving force behind the outcome measure not being the overall number of DRDs, nor the number of opiate-related deaths because there's a large aging cohort effect going on in DRDs, not just in Scotland and the UK but in Europe as well.... [T]hat aging effect is largely consistent across all the northern European countries, the Scandinavian countries and the UK countries. It's very, very strong. So Sheila was very insistent that we shouldn't focus on that because that aging effect is so strong—and would be persistently strong for a number of years to come based on her own previous research—that it would confound any of the naloxone impact and what we had to do was we had to target the outcome measures to something that was going to be much more sensitive to the groups that were going to be at the very highest risk of DRD and those [in] which the aging cohort effect was less pronounced.[57]

Recently released male prisoners with histories of injection drug use were at highest risk of DRD. Bird told the ministers that if they were "minded to have a policy or program of take-home naloxone, prisoners should be first, not last. So that was the first thing—we needed to protect the prisoners. There wasn't going to be a trial, but if the policy didn't protect the prisoners, the people who were most at risk were going to lose out."[58]

Despite ongoing preparation for the N-ALIVE pilot trial in England, Bird agreed to evaluate the outcomes of the SNNP:

We also warned that Scotland's drugs-related deaths had been on a rising trajectory, … so the deaths could continue to go up, might continue to go up, and yet the policy would have been successful because they could go up less than they otherwise might have done, but we couldn't tell him what might otherwise have been because there were no projections of drugs-related deaths in the same way as there were AIDS projections against which you could do calibrations, as it were.[59]

Variations in deaths also occurred because heroin market dynamics affect the heroin supply. Because of that, Bird recommended against the primary outcome being an across-the-board reduction in the number of opioid-related deaths:

What they should look at, because there was a sort of research history or research basis, was to look at the proportion of those opioid-related deaths that had occurred within four weeks of prison release. Because if the program was working, because these were the people at the highest risk, then they should benefit first and the proportion should go down as a proportion of however many opioid-related deaths there were.[60]

Despite the high-risk period following prison release, it was tricky for Bird to determine how to ensure that a relatively rare event like death would show up in the three-year time frame initially set by the Ministry. To accomplish this, she quantified the effect that naloxone distribution should have had, relative to baseline data from 2006 to 2010 on the numbers of naloxone kits distributed optimally in the Scottish pilot programs and by Scottish prisons. Extrapolating, she predicted a 20 to 30 percent reduction in opiate-related death following liberation from prison after three years. Once results became calculable, they exceeded initial predictions:

Three-years into the program, we'd achieved a 30 percent reduction. Then by five years, we'd achieved almost a 50 percent reduction in that group. It's a fantastic achievement in terms of that population, and Sheila, Carole, myself, and Sam Perry published a paper in *Addiction* (2016) which looked at that drop and applied the Bradford-Hill criteria to it to ensure that what we were finding wasn't a spurious finding and it wasn't as a result of chance. We concluded that the decrease was associated with the national naloxone program.[61]

Instrumentalization of statistical data operates as a "technology of truth" for imparting to the diagnosis of problems a place within the "symbolic economy of knowledge self-evidence."[62] Rates, figures, numbers, maps, patterns, and estimates "function as necessary technologies, which by abstracting from their own process of production simplify and thus dramatize the symbolic value of a societal issue that is meant to demand biopolitical action."[63] I have argued elsewhere that "confabulations," compressed narratives about abstractions proceed not as "simple discussions of 'fact,' but instead assess the moral and symbolic value of particular paths and patterns of risk and blame" that are accomplished through the rhetorical and discursive effects of realist claims about the scope of a biosocial problem like drug use.[64] Such confabulations bundle "facts" that become normalized as accepted common sense; these bundles are then used to "settle" issues of social import. Settlements secure and simplify knowledge, reflecting the power of reason.[65] Not coincidentally, confabulations can be crucial for shoring up institutional structures that trade in them. The production of biostatistics requires interpretive skills and sensitivity to the political, social, and economic contexts in which such claims are made. But policy arenas tend to undo these interpretive bundles, selecting those aspects deemed most useful for action. Naloxone distributed through the SNNP was credited with decreasing DRDs in Scotland, an accomplishment that had implications for the N-ALIVE pilot trial then underway England. That "social fact" gained a life of its own—with implications for the N-ALIVE pilot trial.

RESTRUCTURING IN *THE HOUSE OF THE DEAD*: N-ALIVE AND HEROIN MAINTENANCE IN ENGLAND

The N-ALIVE pilot trial was conducted in fifteen of England's 120 prisons. Randomization began in 2012, ending two years later once the results of Bird's evaluation of the SNNP became clear. The pilot study involved a twist of Dostoyevsky's "The degree of civilization in a society can be judged by entering the prisons" from *The House of the Dead*, a novel set in a Siberian prison labor camp. N-ALIVE was designed to follow up on what happened to prisoners with a history of heroin injection just after release. Undertaken to establish the feasibility of a much larger, full-blown N-ALIVE trial, which never came to fruition, the pilot was set up to follow what happened to randomized prisoners in terms of the prevention of overdose deaths; did they, for example, return to heroin use by injection?

Did they carry the naloxone kit they had been given? Did they witness or act upon overdoses? But above all, did they themselves live or die?

Measuring the effectiveness of prevention—a nonevent—preoccupies any field of "prevention science," which attempts to apply "the scientific method" to areas of social life deemed individually risky and/or socially dangerous. But what evidence can be produced to show that death has been prevented? According to Strang, the N-ALIVE trial was borne out of a normative, pragmatic, and interventionist mind-set designed to advance the field, albeit on slimmer evidence than he would have preferred. Recalling that one of the commentaries written in response to his 1996 editorial derided "such big conclusions from such weak data," he mused that this did not mean that "there's anything wrong with the logic [of THN], it looks as if it's correct."[66] In playing the "modest witness" whose actions are metonymic with those of his field,[67] Strang's laboratory life[68] has sought to fill in evidentiary gaps in the undone science of overdose. But this modest witness was on a quest.

Designing clinical research situations to track overdose required scientific ingenuity. When the British undertook the policy experiment of maintaining forty or so heroin injectors on pharmaceutical heroin, Strang saw it as a chance to study acute overdose from the clinical vantage point within an unprecedented "natural laboratory":

We've done some studies with respiratory physiologists measuring what happens when people inject heroin. In the UK there's a small number of people who are prescribed maintenance injectable pharmaceutical heroin and we've written papers about the heroin clinic, which existed for about eight years but doesn't exist anymore.... The total population at any given time was about forty people for the whole of the United Kingdom. We had three clinics. Ours had about fifteen to twenty people in it and the others were even smaller. It was incredibly intensive and incredibly interesting.... [W]e had an extraordinary research opportunity, quite apart from the heroin treatment, because it also meant that we could study what happens when an intravenous heroin injector injects pure diamorphine intravenously or intramuscularly, with no contaminants or anything else. It allowed us to study what actually happens physiologically.[69]

Conditioned as such experiments are by the material conditions and social relations of laboratories,[70] few have been possible since closure of the US Public Health Service's Narcotic Farm Addiction Research Center in the mid-1970s.[71] Strang's heroin maintenance studies were conducted in a special ward outfitted to control respiratory depression; naloxone was near to hand (as it always is in operating rooms). Researchers explored exactly how the depth of oxygen deprivation changed with heroin purity or dosage: "Obviously, you can't do that in an unsafe environment. So we began doing studies where we said, 'Look, you've got to come into, essentially, an intensive care ward.' A few years ago, we opened up a clinical research facility, which is a dedicated unit just for this sort of work."[72] British heroin maintenance joined heroin-assisted treatment trials across Europe.[73] At the time of writing, Strang and colleagues continued to develop different naloxone formulations, work that requires precise measures of bioavailability.[74] The social, material, and legal circumstances of the NAC allowed overdose to be investigated ethically in ways once foreclosed.

While the above clinical studies proceeded, the pilot N-ALIVE study in English prisons was as close as the Three Musketeers ever got to an RCT. Designed to enroll 5,600 subjects, the pilot proceeded until just under 1,700 subjects had been randomized as planned: half were provided THN upon release from prison; the N-ALIVE pack for the other half did not contain a naloxone kit. The pilot ended in direct response to Bird's evaluation of the SNNP at the three-year mark: SNNP's primary and secondary outcomes monitoring data for the first three years (2011–2013) were published by the Scottish Information Services Division in October 2014. As in the N-ALIVE pilot trial, Scottish prisoners who were released with naloxone and sought resupply reported more overdose reversals on behalf of "others" than on behalf of themselves. After discussion with the N-ALIVE Data Monitoring/Trial Steering Committee (see figure 11.1), the pilot trial ended and it became apparent that the full-on trial would not take place.[75] Although Scottish and North American onlookers might feel vindicated by the premature ending of the N-ALIVE pilot in England, the decision was wrenching for those who sought to put doubt to rest.

Researchers involved in the pilot trial salvaged lessons from their years-long efforts, the main one being that large prison-based trials were possible in Europe, of which the N-ALIVE pilot trial was the largest up to 2015. Despite debate over informed consent in prisons, research in prison settings could be accomplished in an ethical manner. As Strang summarized, "I would view it as unethical, and any ethics committee today would just throw it out if you didn't have evidence that clearly there wasn't a coercion one way or the other. This one had a very clear

An unscheduled interim analysis of the feasibility outcomes of the N-ALIVE pilot trial [18] and (Bird *et al.*, in review) was prompted by imminent release on 28th October 2014 of the 3rd year results from Scotland's National Naloxone Programme [19].

This interim analysis showed that the N-ALIVE participants who received naloxone were more likely to use it to save another person's life rather than use on self in a ratio of approximately 3:1 (Another: Self (A:S), 15:5 then; 16:5 now); which was corroborated in broad terms by the Scottish data for NOR (A:S, 21:12) [19]. This dual finding had major implications for the main N-ALIVE trial, which was designed around individual randomization.

First, mortality for the majority of those to whom naloxone was administered would not be captured as we were following up only the N-ALIVE participants.

Second, individuals in the control group of the trial could potentially receive the naloxone, thereby diluting the size of any effect of naloxone in the trial context.

Even if, at best, half the administrations of NOR were on self (17/53, 99% CI: 15% to 49%), the size of the trial (with individual randomization) needed reliably to detect the required effect-size of 20% to 30% reduction in DRDs would be infeasibly large.

As the main trial could not go ahead as planned, it was not appropriate to continue randomizing participants to the N-ALIVE pilot trial. This decision was unanimously agreed by the TS-DMC. This decision was also approved by our appointed Essex 2 Research Ethics Committee.

The finding that Scotland's percentage of opioid-related deaths with a 4-week antecedent of prison-release had decreased from 193/1970 (9.8%) in 2006–2010 to 76/1212 (6.3%) in 2011–2013 ($P < 0.001$) led the TS-DMC to advise that, when randomization to N-ALIVE stopped, all participants who remained in custody be offered naloxone on their release, including those who had been due to receive a control pack.

11.1 Explanation of the decision to end the pilot trial of naloxone-on-release, conducted in English prisons. From Mahesh K. B. Parmar, John Strang, Louise Choo, Angela M. Meade, and Shelia M. Bird, "Randomised Controlled Pilot Trial of Naloxone-on-Release," *Addiction* 112, no. 3 (2017): 502–515.
Used with permission of corresponding author Sheila M. Bird.

example. There wasn't a coercion either to get involved with it or not."[76] Although the grand-scale N-ALIVE trial might be looked at as the naloxone RCT-that-never-was, the smaller-scale pilot yielded enough information to affirm that naloxone distribution upon release was a good idea, especially when conjoined with the results of the SNNP evaluation.

LIVED EXPERIENCE AND NALOXONE RESEARCH: PEER EDUCATION IN SCOTLAND

Knowledge about how naloxone worked pharmacologically was considered settled. Obstacles to getting naloxone out included the social, political, and economic intricacies involved in democratizing a technology formerly confined to medicine. Prenoxad, the Martindale product available in Scotland, nearly doubled in cost during the first year of the SNNP. Yet advocates still had to modify the product by indicating doses with magic markers and wrapping the box with yellow tape so that police could immediately perceive the package as undisturbed. As advocates negotiated technical matters such as dosage, bioavailability, and naloxone's fleeting effects, other barriers appeared. These "unsettlements" included lack of knowledge about the social contexts into which naloxone was being inserted, and unknowns about how naloxone would be received in the communities where DRDs were increasingly present. Also problematic were the prevailing relations in conventional clinical settings:

It's the translational knowledge and the utilization of that in managing overdoses in the home or nonservice areas that is revolutionary. I think that perhaps our views about our doctor/patient or service/service user relationships were quite paternalistic. We were not perhaps as trusting a partnership as we are now. If you go to our meetings now, for example, our meetings involve key members who are people in recovery, people who have traveled through our services, who are sitting in meetings, who have been through all of this. We now have a strong component of service user input. In those [early] days, that wasn't so developed. It was a bit more of a divided system. Perhaps we just didn't trust handing this out to people. We were nervous about, would they do the right thing? Could we trust people to do things properly? If they didn't, who would get into trouble?[77]

These questions required social and political, rather than technical, answers. As with naloxone access elsewhere, counter-publics opposed to the democratization of naloxone stymied progress by raising such questions. Generalized opposition to harm reduction—often by those who think that abstinence-based recovery should be the goal—is frequently expressed as a specific objection to naloxone.

Researchers have also asked how having a supply of naloxone affects individual behavior; in particular, some feared that if people had naloxone on hand, they might not call for an ambulance (see figure 11.2). Studies have shown that what prevents people from calling ambulances was that they feared housing loss, rather than some putative increase in risk-taking behavior because naloxone acted as a safety net. Structural conditions such as poverty and homelessness shaped calls for assistance. Andrew McAuley, the National Health Service analyst who had run the Lanarkshire pilot, designed and conducted a study to find out if possessing naloxone deterred people from calling out the ambulance. This concern had been voiced in Scotland even before the Scots visited the Chicago Recovery Alliance.

11.2 "I Made the Call That Saved a Life." This Save Some Naloxone poster emphasized naloxone's short duration of action and emphasized calling the ambulance early
in the critical event.
Used with permission of David Liddell, Scottish Drugs Forum.

In the site visit to CRA, the visitors had observed that seasoned drug users did not mention calling emergency services, even though "call for help" was the second step of the SCARE ME protocol (discussed in chapter 7). They also recalled having met people who had received or administered naloxone on multiple occasions, including one couple who used it five times in one night and others who used it dozens of times. The visiting team met with early adopters who enthusiastically embraced naloxone's presence in their community, were proud of their "saves," and had so much familiarity with naloxone that they did not always call for emergency assistance.

By character an optimist, Hunter foresaw that misinformation would have to be overcome in order for the Scottish program to reach the mainstream. She recalled headlines that accused her and her colleagues of "putting drug-filled syringes into chaotic households." Such attitudes diminished once family members became involved, and social acceptance of naloxone increased with the strength of peer networks. Although people with active drug use histories have long worked as front-line treatment professionals, the emphasis on peer education mobilizes expertise through a cascade model called, as in many fields, "training the trainers." In the "the new public health, ... the 'imperative to participate' is cultivated as a practice of self-discipline."[78] The SNNP and the post-SNNP framework supported people with lived experience of addiction. Their participation "presupposes a whole range of personal attributes, skills, attitudes, and commitments as well as detailed work upon the self. Few individuals—without a great deal of 'free' time, personal inclination and commitment, and specialist training—would be able to follow through with more than a small proportion of this agenda, even if they had time to read and assimilate the mass of published material."[79] And yet they did follow through and seek to integrate peer-reviewed literature.

Scottish service user participation worked in concert with advocacy driven by "key individuals [on the NFDRD] who were experts in the field ... [and not by a] groundswell of advocacy and demand from potential service users. It's through management and expertise that these things are being developed. It isn't the same elsewhere in the world."[80] The Scottish Drugs Forum (SDF) pioneered a novel route to user participation by involving peers in research aimed at quality improvement. As Smith put it,

Most of the services in Scotland are provided for by the state and the state has over decades become more interested in service user involvement. If you're going to provide these services, if you're going to improve the service, let's ask people. So that level of service user involvement has been promoted. In most services, that's absolutely fine. It's really straightforward. If you're running a service for older people, you get a group of those older people together and you have a cup of tea or coffee and a biscuit, and you discuss how your service could be improved. The problem with drug users, service users, is that we've got a group of people who struggle to engage around a cup of tea and a biscuit because they've got other things going on in their life.[81]

The SDF specifically sought individuals who embodied a peer education model that would work for drug users—for whom, it was understood, a cup of tea and a biscuit would simply not work. They used community-based research as a way to bring about the current model:

So what used to happen in the early days of user involvement was that people would say, "Well, if we can get somebody along who's used our services in the past, but is a bit more stable, because most users are in a bit of a state of chaos, we'll just ask them about their opinion and experience in the past." So what you'd get was their opinion and their past experience. So the idea was if we did research, we would be able to interview them to try to get the experiences of maybe a hundred people. And as Jason says, it's also quite a powerful experience if you can get people with quite a lot of experience to speak on behalf of those hundred people, and the findings, and also appropriate their own experience and their own interpretation of that data.[82]

Smith attributed this model to Alex Miekle, a Glasgow drug service worker at the SDF. Such involvement originated with the need to instill evidence-based practice throughout a community of participants—but was very different from what I had observed in the New York State process from 2000 to 2005.

When the SDF began offering Critical Incidents Prevention and Intervention Training in 2004, "festive periods" and prison release were emphasized as the most dangerous moments for users. Avoiding the crisis-ridden register of moral panic, the SDF flatly informed users about an unusually broad array of risks assumed by those injecting opiates, including anthrax. Their materials emphasized that individuals may be most endangered when using *less* (i.e., abstinence decreases tolerance). Rather than insist on abstinence, they assumed that individuals might well continue doing drugs and sought to involve a handpicked group in a structured three-hour naloxone training that reviewed evidence that directly linked naloxone to saving lives. For Jason Wallace it was first time anyone had spoken knowledgeably about how overdose worked and how the risk of dying might be reduced. He characterized this as a revelation to his peers, each of whom thought they were "well-schooled and well-experienced with probably seventy, eighty years' addiction experience between us."[83]

So here you were in a training with four others linked directly to saving lives. For me, and I can only talk on my own behalf ... here I was first time in my life after twenty plus years of addiction where somebody was actually explaining to me what the mechanisms of an overdose was and what polydrug use was.... You find out that if you used methadone and benzos one day, they'll still be in your system on the next. Because if you spoke to people who used drugs, they couldn't have told you that. Mixing drugs meant that you took them in the same day. What you took yesterday, was gone, was yesterday. So if you have a training and you learn that the methadone's going to affect you up to thirty-six hours, benzos up to two days, this was a revelation.[84]

Practical knowledge of lived experience was respected and balanced with evidence-based research. The SDF recruited career drug users and helped them learn communication and research skills needed to carry out community surveys. Most were entirely unfamiliar with evidence-based practice. As Austin Smith, who was hired in 2001 to coordinate these groups, observed,

People would come in who had life experiences and they understand the services and what the quality was and what the issues were. They might not have paid work experience. There might be some exclusion of people who didn't have literacy skills. Beyond a certain point, they hadn't attended school, and had problems with literacy to the point they couldn't complete questionnaires, you see, but it was fairly inclusive. People worked with peers so they would always support each other.[85]

According to Jason Wallace, the induction experience married an evidence-based mind-set to harm reduction values:

For me what it was all about, what we were doing here in the early days was you were bringing your lived experience and what you were bringing [points to the SDF office generally] was evidence-based practice that we never knew about. Some of us had twenty-plus years of experience of being on services but never understood we were on the short end. What the SDF was ultimately doing was marrying up your lived experience with evidence. It makes a far more coherent and stronger argument if you're arguing the evidence and not just arguing the experience.[86]

The knowledge transfer process was inductive in Scotland. The peer education group generated insights that were then compared to published research in advance of the group setting out on its own data collection process.

Our wee group was pretty tight. There were about six of us in the group who were core volunteers involved for several years. I was here for maybe five, six years as a volunteer. For us what happened was we would begin doing scoping what we were going to look at. What's our experiences with services A, B, and C? Why do you think that happened on a service? We'd have our ideas and then look at the evidence, the legislation, the plans for getting that pinned down, and get up a patch of what's actually researched about why stuff happens.[87]

Premised on user involvement, this model was then institutionalized for the purposes of improving treatment quality.[88] Stable service users conducted surveys and developed a "representative overview," rather than allowing any of their individual perspectives to hold sway.

Such work emerged in the context of the radically inclusive international harm reduction movement, which steered clear of technocratic or bureaucratic entanglements in favor of self-organization based on the diversity and heterogeneity of drug users. Rather than forming a "constituency" that issued a unified voice for "speaking truth to power," the seemingly fragmented associations organized by drug users within specific countries have been described as "much more introvert, defensive, and vulnerable" organizing efforts by contrast to the "powerful collective actors traditionally described as social movements."[89] Knowledge production is often an explicit or implied aim of drug users organizations, reflecting a social movement approach convergent with those seeking to extend civil and human rights to historically marginalized social groups; women's, gender, and sexuality studies, black and ethnic history, and queer studies are full of examples of such groups. Although much research is conducted on drug users, very little has been conducted by drug users. Harm reduction researchers began to realize that including drug users in shaping research or framing research questions made for less tokenized and more meaningful experiences.

Having observed some of the early Practice-Research Networks funded by the federal Center on Substance Abuse Treatment back in the early 2000s, when the New York State Office of Alcoholism and Substance Abuse Services (OASAS) was attempting to infuse evidence-based practices throughout the drug and alcohol treatment system by involving practitioners and clinicians in research, the contrast with the peer research model used at the SDF could not have been greater. The SDF recruited drug users with ready access to peer networks, whereas OASAS rarely involved so-called consumers, concentrating all resources on treatment providers. Both processes involved cultivating increased trust in research processes and products that comprised the evidence base. However, the SDF enfranchised former users with multiple motivations for becoming involved:

For me, I got cleaned up again by losing somebody very close to me [to] ... an accidental overdose.... He never touched substances but he was drunk in the pub one night, and somebody gave him a swig of methadone, maybe 20–30 mg, and it killed him. I kinda thought, there must be something you can do to stop this. That was Christmas Day, 2005, and then I was involved with SDF in 2006. I just came at the right time, when they were talking about possibly going to visit Chicago to look at the naloxone program. I was very much at that time one of the high-risk individuals, at risk of dying.[90]

By 2017 the SDF had built thirty peer networks providing more than 180 individuals a path to employment and sustained recovery through an annual Addiction Worker Training Project.

Delivering naloxone into the hands of "peers" was different from giving it to "laypersons," "bystanders," or "witnesses." Trying to capture what made naloxone special in Scotland, Smith said:

So the big change is actually a much more subtle thing—and it's about the notion of the struggle for credibility in Jason's story, which used to be told like this, "I'm not a doctor, but I do have this experience," or almost as a revelation, "Here's something you have to do." But actually now, when people are telling their story, it's about their peer-ness, if you like, and the contribution that their experience makes to their work and their insight. I don't know if you see that. [Gestures toward Jason, who nods and says, "Aye."] I think naloxone has helped drive that change, more necessarily than the recovery agenda, because they kind of talk down and that kind of stuff.[91]

SDF gradually created a support system for peers to develop hybrid expertise about overdose prevention with naloxone—hybrid in the sense that their tacit, experientially based knowledge was integrated with the finding of peer-reviewed literature. Despite lacking academic credentials, Wallace embraced experience as understood through the lens of scientific and clinical studies. "For the peers in Scotland, I think partly, in a sense, a big part of what I teach them is my own experience and the importance of coupling my experience with the evidence, ... we've got a model that means, we've got a place for you, ... we've got something that you can do."[92]

Operating in a high-overdose death toll country, the SDF undid the binary opposition between harm reduction and recovery. Their 2012 campaign "Save Some(one) with Nalox(one)" emphasized the *temporary and contingent* nature of reversals: "Naloxone can temporarily reverse the effects of opioid overdose." Figure 11.3 establishes a knowing distance between the subject of the photograph—who knew what to do—and the risk-taking user, who blundered into mixing drugs and alcohol. Such differences are coolly presented in a nonjudgmental tone, as a matter of *knowledge—knowing what to do and what not to do*.

Emphasis fell on the performative knowledge of the person who did the saving, who is portrayed instead of the person saved (in the fertile terms of situational analysis, persons saved are "implicated actors").[93] Viewers, too, are implicated—called upon to acquire the requisite knowledge. In figure 11.4, a pensive mother reflects on her boy's life, having gained the knowledge to save it. It is her knowledge, rather than the mere technology, to which actionable power is attributed.

Because drug users are often presented as ignorant or unknowing, the matter-of-fact presentation of them as knowing subjects made the SDF campaign unusual. In another poster (see figure 11.5), a young woman concluded her reflections on her own overdose with her intention to "keep myself safe" after she came round to find her boyfriend crying above her and holding a naloxone syringe.

The posters invoked intimate bonds between the person who saves someone, and the person who needs saving. No longer confined to gritty documentary realism, overdose was represented within the lexicon of scientific realism and the new public health version of safety and prevention.

11.3 "I Saved My Best Friend's Life." One of four posters made to signal that the law had changed to allow family members and friends to access naloxone. "He was always taking risks … mixing drugs and drinking."
Used with permission of David Liddell, Scottish Drugs Forum.

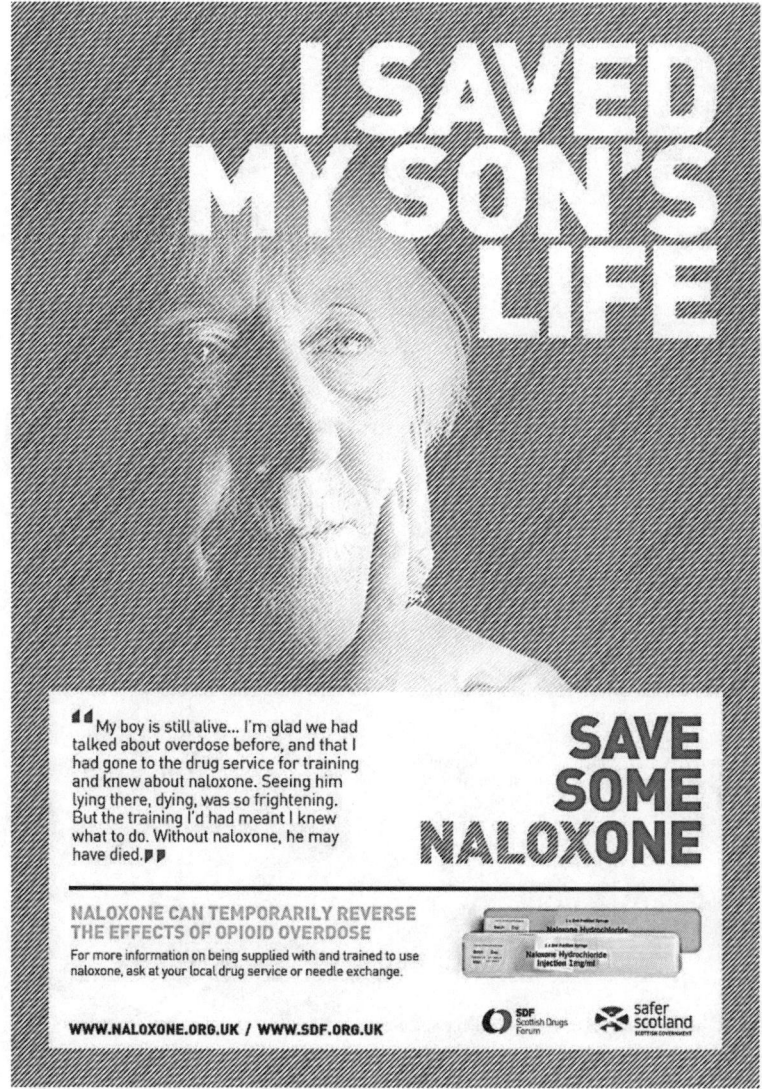

11.4 "I Saved My Son's Life." One of four posters made to signal that the law had changed to allow family members and friends to access naloxone. "Seeing him lying there, dying, was so frightening. But the training I'd had meant I knew what to do. Without naloxone, he may have died."
Used with permission of David Liddell, Scottish Drugs Forum

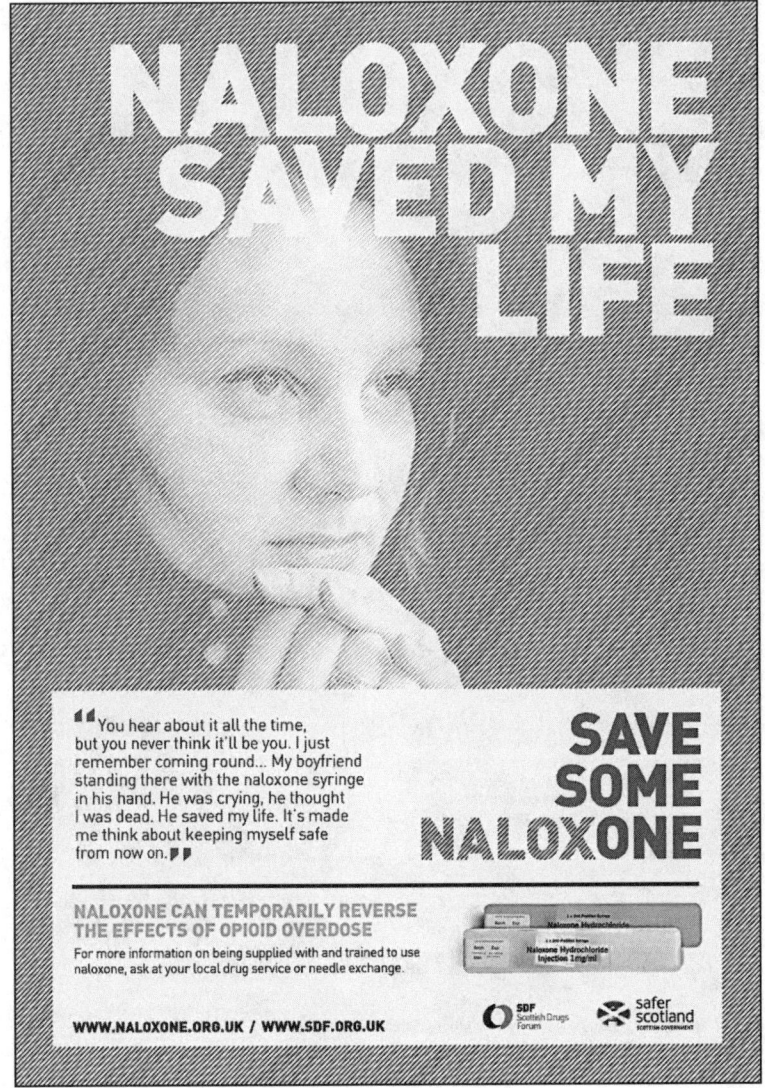

11.5 "Naloxone Saved My Life." One of four posters made to signal that the law had changed to allow family members and friends to access naloxone. "You hear about it all the time, but you never think it'll be you. I just remember coming round…. My boyfriend standing there with the naloxone syringe in his hand. He was crying, he thought I was dead."

Used with permission of David Liddell, Scottish Drugs Forum.

"SERVICE USERS" AND "EXPERTS BY EXPERIENCE" IN ENGLAND

"Service users" is a British term applied to those enrolled in addiction treatment "services," including OST. Widespread use of the term may be attributed to the consumerist patient advocacy movements of the 1990s and New Labour emphasis on democratic participation.[94] Widely used in the United Kingdom, the term routed away from the "client," "consumer," "customer," or "survivor" mental health service model common in the United States.[95] "Service user" is an identity based on claiming and using social service benefits and based in a recognizable knowledge or mode of expertise derived from the experience of drug treatment and associated services. Experts by experience develop specialized knowledge by overdosing personally and witnessing or responding to overdose; administering naloxone or having it administered to oneself; and being trained or training others to use naloxone in the event of an overdose.

A letter to the editor of *Addiction* appeared in 2017 from the Addiction Service User Research Group (SURG) at Aurora Lambeth in Brixton, a group that consults with the National Addiction Centre (NAC). Acknowledging that "naloxone unquestionably saves lives," SURG questioned why THN programs have performed poorly in the face of rising opioid overdose and overdose death rates.[96] The NAC turned to SURG for an explanation in "absence of any current evidence," and with "remarkable ease," SURG members listed more than a dozen concerns that limited naloxone diffusion in England. According to an Aurora trustee, the letter was written in response to a meeting with John Strang, and it was

an extraordinary meeting of the group. He had been before, to talk about different ways of administering naloxone. But for this conversation it was much more about the barriers to carrying naloxone, what people with lived experience, what people in recovery, are concerned about when it comes to carrying or administering naloxone. It was very much about not wanting to condone drug use if you're in long-term recovery, if it means you're considering using again. And there were issues about what the police might do if they found you carrying this, about stigmatization. Certainly, from people in recovery's point of view, these things worry them. The good thing about our work is it gives us a chance to challenge people's views of recovery, and a chance to rationalize some people's views [about naloxone].[97]

Inclusion of "service users" in UK contexts has also allowed new inquiry into the phenomenology of overdose. While few question death as a negative outcome, there is much evidence that the "lived experience" of overdose is not completely negative. Neale and colleagues, analyzing self-reported "overdose accountings" in nonfatal cases interviewed in Glasgow and Dundee in 1997 and 1998, asked, "Specifically, how do individuals feel between the point of drug consumption and loss of consciousness? Are they aware that they are overdosing and could—indeed would—they act to save themselves?" Their findings defied negative stereotypes as their typology included "enjoyable accounts" that offered insight into some-thing rarely acknowledged except in fictional accounts—the pleasures of "going over," the giving of oneself over to "a beautiful feeling" [Hugh]. Each of the three Scottish naloxone pilots also included at least one person who self-administered naloxone when they felt themselves "going over."[98]

Pleasure has been a dirty word in drug studies; banished from science, displaced by "reward," unmentionable within public health.[99] In registering informants' pleasures, ethnographers often relocate them within distinctly unpleasurable structures of inequality and violence. Few tread near pleasure in overdose contexts—framed consistently as a grave public health problem that should be "prevented," "managed," "responded to," and accounted for in policy-friendly terms. In offering plausible accounts of what hap-pens when overdose occurs, there has been little examination of the user-expressed phenomenon of the pleasures of "going over." Unless and until these aspects of overdose experience are considered more fully, there is likely to be ongoing need for overdose studies. Near the end of his 2017 interview with the author, Strang doubted that "we'll ever finish studying take-home naloxone. I think it's wrong to think we'll ever finish study-ing and improving it. You could look at all kinds of other aspects of the take-home naloxone approach. Why don't we look at other antagonists? If we don't constantly challenge and adapt, then I think we're not doing the service we should do to our field and the people we're concerned about."[100]

Need for ongoing discussion includes efforts to unsettle naloxone and work against closure, while strenuously advocating that something be done. There is also a role for experts in drug experience to resist closure, emphasize the many forces in people's lives that go unaccounted, and continue to make naloxone meaningful—as well as accessible—in circuits of opioid consumption.

12

OVERDOSE AND THE CULTURAL POLITICS OF REDEMPTION

Every technology assumes political dimensions, but some artifacts lean toward certain arrangements of power, expertise, and authority. Such highly political technologies transcend impoverished binaries like "intended" and "unintended" consequences.[1] Nuclear power and weaponry, for instance, are not very conducive to decentralized decision-making or communitarian values. Some drug-related technologies embody the prerogative powers of police surveillance and drug control;[2] others such as fentanyl test strips, clean needles, and naloxone embody principles of harm reduction. The politics of a technological artifact depend on what people do with it and where they do it, but also upon the tenacious forms of social solidarity that grow around it and hold it in place. Writing that technologies are "ways of building order into our world," Langdon Winner points to how the earliest "structuring decisions" made about a technology can often fix its fate.[3] Fortunately, there is usually latitude for reshaping along the culturally preferred lines of protagonists alert to potential social and technical flexibilities. Initial decisions made about naloxone confined it to emergency medicine and the surgical suite; breaking it out of those structuring decisions took reimagined commitments to a "resuscitative society" and enabled new forms of sociality. Naloxone's life-and-death politics were reanimated into what I have called a "technology of solidarity."

"BREAKING OUT THE NARCAN"

Emergency medical services have become so expected and ubiquitous that it is difficult to remember that most urban areas had no paramedics until the 1970s, and rural areas even later. Uneven diffusion meant that naloxone often was unavailable in prehospital emergency situations—and it could be a long way to the hospital. According to Pennsylvania paramedic Duane Nieves, naloxone was limited to the most highly trained responders because it was a prescription medication requiring intravenous administration: "Back then, it was Basic Life Support, ambulances, volunteer fire companies. There was no advanced care in the field, no IVs, no medications. I've been involved in the entire history of Advanced Life Support care in this area. [Naloxone has] been paramedic-only in Pennsylvania up until last fall [2016] with Act 139, the expansion of naloxone to other public safety. That includes firefighters, police, and Basic Life Support, EMPs."[4] Prior to that enabling legislation, only licensed paramedics could legally administer naloxone; yet harm reduction activists had been giving it out since 2005 through needle exchange programs in Philadelphia and Pittsburgh. Prevention Point obtained naloxone from the Chicago Recovery Alliance (CRA), which remained the central node of the far-flung underground distribution network described in chapter 7.

Considering that police treated naloxone with almost as much suspicion as heroin itself, risks were pronounced for licensed physicians who prescribed naloxone via standing orders and for activists staffing weekend needle exchanges, speaking to local officials, and starting up naloxone education in jails and prisons. Framing overdose deaths as an injustice, harm reduction activists did more than help motivate "bystanders" to become equipped to act. By encouraging ordinary people to keep naloxone within reach, they implicitly constituted Good Samaritans who knew something about antidote and overdose; they widened the circle of people who acknowledged moral responsibility not only for their loved ones, but also for a larger, less defined and less beloved community. Activists constituted themselves as experts, as authorities in their own right. There is much in the lore of "experts by experience" to indicate that they knew about doing "tastes" or "tester shots," about not mixing opioids with alcohol or downers, and about managing the protective aspects of tolerance and familiarity. By contrast, drug-naïve crusaders promoted abstinence as the only approach with redeeming social merit—and thereby increased risk of overdose and death. Harm reduction widened the aperture and promoted an almost managerial style toward one's habit, a framing that resonated especially in the United Kingdom. Basic harm reduction messages—don't use alone, don't mix drugs, don't drink while using drugs, keep Narcan around—are directed toward managing degrees of risk,[5] itself a form of expertise that goes unrecognized.

Managing neurochemical selfhood is an ever-present challenge for most of us.[6] But drug users especially live in tension with institutional structures of biomedicine, sutured into it when enrolled in treatment, but relapsing to living "wild" or "sleeping rough" when not. This neurochemical selfhood is grounded in cultural, clinical, and industrial practices that have promoted drugs to a prominent role in twenty-first century social life. Being a user means maintaining a metabolic phenomenology of the self, keeping up with the demands of contextual factors such as setting, market, or season. This is in some sense an incredible skill, as anyone managing to comply with a number of legal prescriptions can attest.

But it is important not to miss the social epidemiology that makes evident a strange and uncomfortable fact: most who die overdose deaths have experienced up to ten overdoses prior to the one that killed them. So those in relationship with them have opportunities to intervene. Why, then, has the uptake of naloxone not been even broader and more systematic? One explanation lies in the dearth of media portrayals of naloxone. One of the first in the United States was "Heroin Hassles: Medical and Legal Issues Keep Overdose Antidote Out of Users Hands," which appeared in the *Village Voice* in early 2000. Authored by an insider, Maia Szalavitz, the article began in the apartment of an experienced user, Valerie S., as she was drawing up a syringe of naloxone to inject it intravenously into a college-educated "neophyte" who had overdosed. The article went on to describe early harm reduction efforts, including several mentioned in previous chapters: the Torino Outreach Project coordinated by Susanna Ronconi in Italy; Santa Cruz Needle Exchange; and CRA. Prior to "Heroin Hassles," the popular culture of overdose prevention had been produced entirely by the casual press of the harm reduction movement itself—the *junkphood* collages, the CRA "Any Positive Change" T-shirt, and early versions of the SCARE ME protocol spring first to mind.

The spectacular exceptions were *Trainspotting* (1996) and Quentin Tarantino's *Pulp Fiction* (1994), a film credited in some harm reduction circles with inspiring action despite the inaccurate depiction of Narcan. A stylized overdose event appeared in the movie, scripted by Tarantino but based on *American Boy* (1978). The latter was a short documentary by Martin Scorsese that was based on a fifteen-hour-long interview with Steve Prince,[7] who had run a Manhattan shooting gallery equipped with medical supplies during the 1970s. Most other shooting galleries of the era were spaces that maximized harm by ensuring no one left with drugs, money, or dignity intact. In the documentary, Prince recounted his response to the OD later made famous. Using a medical dictionary, a Magic Marker, and a huge cardiac needle, Prince injected adrenaline— not naloxone—straight into the woman's heart. The "Pulp Fiction" scene paralleled Prince's description in all but one respect: "I didn't get that plunger all the way down before she was up and wide awake. I, however, removed the needle, unlike the movie."[8]

Pulp Fiction and *American Boy* raised intriguing social and moral questions. Educating people about naloxone was hardly the filmmaker's intent. The drama lay in the awakening, the seeming rise from death, and the surprise that even a moment of redemption much less a full recovery was possible in the closed cultural circuits of overdose. The phrase "break out the Narcan" comes from Scorsese's *Bringing Out the Dead* (1999). Set in a fictional ambulance service in New York City, the paramedics worked for Our Lady of Perpetual Mercy (or Misery) Hospital. Paramedic Frank Pierce (played by Nicholas Cage) was tormented by the imagined accusations of those he had revived: "What haunted me now was more savage: spirits born half-finished, homicides, suicides, overdoses, innocent or not, accusing me of being there, witnessing a humiliation which they could never forgive."

Such humiliation was famously dramatized in Danny Boyle's *Trainspotting* (1996). The character Mark Renton literally sinks into a trap door in the floor after shooting up during the section of the film titled "A Visit to the Mother Superior." Against the transcendent backdrop of Lou Reed's "Perfect Day," Renton's head made no sound as his overdosing body was dragged down a steep flight of stairs, across the pavement, and into the back seat of a cab to be dumped outside the emergency room. There he is wrestled onto a gurney, his reverie rudely interrupted by a shot of nalox-one administered by a doctor who does not linger to witness the revived patient gasping for breath. The white ache from the fluorescent lights and Reed's line "You're going to reap just what you sow" unsettles view-ers, reminding them that the wages of sin may be death, but a heroin user's perfect day ends with naloxone.[9]

Bringing Out the Dead centered the unnatural technologies of the trauma service from which Frank derived his livelihood—but which trau-matized *him* nightly. The "Narcan revival" scene was set in a nightclub to which Frank and his partner Marcus[10] had been summoned. When they realized they were dealing with an overdose, they turned to each other, and said, "Let's break out the Narcan." They then enacted an irreverent healing ritual, during which Marcus preached to the nightclub-goers who have formed a gospel choir. Without denying the film's creativity, it is worth noting that activists of that time period had been weaving such fic-tions into moral economies that situated naloxone as a "miracle medicine" that could raise the dead.[11]

Overdose is still represented in contentious ways even as the movement has enjoyed considerable success in changing law, policy, practice, and culture. What physician-anthropologist Helena Hansen terms "wasted whiteness" refers to the promotion of young, white, middle-class women as iconic overdose victims, their promising futures foregone forever.[12] By contrast overdose deaths were already familiar among communities of color by the 1970s.[13] As those working closely with drug users in commu-nities of color are wont to point out, heroin "never went away," as 2017 *New York Times* coverage of the "endemic" put it, but became a "ground of shared misery" instead.[14] But there were differences in attribution of overdose to "root causes" that were more or less redemptive or legitimate. Those who mourned white opioid users attributed their deaths to cultur-ally legitimate causes such as despair,[15] and often emphasized redemptive traits of the deceased—being a loving mother or an award-winning high school athlete. Often written by parents, obituaries in the United States began to mention overdose around 2015—a change in practice so news-worthy that there was reporting and blogging about the explicit naming of the cause of death.[16] Popular representations of candlelight overdose vigils, protest rallies and marches, and even public art projects received more play in public space.[17] The new visibility still emphasized the acute episode, but often highlighted chronic patterns of relapse and the "natural history" of the individual. The young, the white, the once promising, the socially redeemable, but above all, the beloved—these are the portraits continually rehung in the national gallery of the opioid crisis.

Increasingly public memorials and testimonies offer marked contrast to the silence and stigma that haunted overdose deaths when they occurred primarily in communities of color. But naloxone has offered a tool for broadcasting new stories. As I finished this book, the New York City Department of Health launched a campaign to widen social acceptance of naloxone, particularly in communities of color. The campaign drew from the Scottish Drugs Forum education effort analyzed in chapter 11, but it was very much the product of the multicultural boroughs of New York City. The motivational campaign called upon New Yorkers to equip themselves to save the life of a father, a fiancé, a neighbor, or the "one best friend [they] could always rely on" (see figure 12.1).

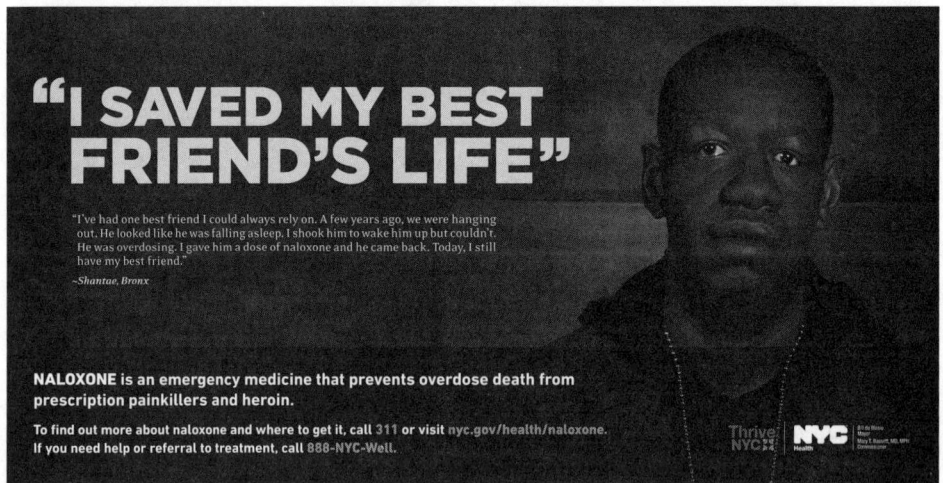

12.1 "I Saved My Best Friend's Life." New York edition.
Used with permission of the New York City Department of Health.

DENATURALIZING EXCESS DEATHS: AGE, CLASS, GENDER, RACE, AND PLACE MATTER

When beginning this research, I never anticipated that overdose death rates would continue to spiral upward throughout, never seeming to peak, plateau, or decline. The most recent figures suggest that 70,237 people in the United States died from overdose in 2017, more than two-thirds from opioids; between 2000 and 2017 close to 300,000 have died, according to the Centers for Disease Control and Prevention (CDC). Despite less ignorance, more prevention, more naloxone, and more integration between recovery, treatment, and harm reduction, "excess" death has become naturalized. The very notion of "excess" death implies that these deaths are preventable or premature. But they are far from evenly distributed, and their patterns reveal much about what a society values or devalues. As David Sudnow showed in *Passing On*, "social death" is reflected in the rationing of resuscitative care according to accountings of social value and moral worth. Some such deaths are discounted—those in lower-income brackets, those who lived in darker skin, those who used drugs. A common, if morally repugnant, interpretation is that those who die "excess deaths" were already leading "excess" lives.

Cultural anthropologies of excess death suggest that people who do not feel needed die early.[18] For instance, Michelle A. Parsons's work on premature mortality, titled *Dying Unneeded*, concluded that in post-Glasnost Russia, men perceived themselves as useless whereas women experienced themselves as having something to offer others. "Being unneeded is a distal driver of the mortality crisis. Being unneeded translates social collapse to bodily death from cardiovascular- and alcohol-related causes. Being unneeded is related to the death of the body, but it is also related to the life of the soul."[19] "Addicts" are often treated as "excess" people who are "wasting" their lives, having already passed into states of "social death."[20] Parents or partners may value the lives thereby lost, but it is not unusual to come across a cruder moral calculus in which drug users are spoken of as "deserving" their deaths. Prompted by parallels between *Dying Unneeded* and "American overdose,"[21] I wonder *what* social, political, or economic orders are coming undone that so many are dying unneeded? An aptly titled study, "Bowling Alone, Dying Together," has argued that higher levels of social capital and community resilience protect against overdose deaths.[22] This article joins other signs that there may be a shift away from an emphasis on individuals carrying naloxone to communities showing care by collectively reducing harm in a public manner. What might it take to rebuild capacities for resilience under such conditions—for individuals, kin, communities?

This new emphasis on community was on display in the New York City campaign (see figure 12.2). Billy recounts a neighbor's boyfriend knocking on his door one night, gesturing toward the role of chance or fate in his having been home and having naloxone on hand. There is something interesting here as the campaign directs the viewer away from close kin, and toward a neighbor whose racial, ethnic, social class, or other markers of individuality remain unclear. In saving a life, these characteristics seem not to matter, as the important thing is the community capacity

12.2 "I Saved My Neighbor's Life." Billy, Manhattan.
Used with permission of the New York City Department of Health.

for saving lives. And there is an ethical remodeling of the "bystander" into a "neighbor"—someone who is involved enough with his neighbors that the boyfriend realizes that Billy would have naloxone. This story, of course, could only unfold in densely populated urban environments.

But the uneven topography of excess death has meant that those who live in rural areas more commonly suffer such deaths.[23] The US opioid overdose death rate is 45 percent higher in rural areas than in urban or suburban areas.[24] Thanks to changes in the business end of illicit drugs, people living in rural areas—and the United States is 80 percent rural at the county level—have access to licit and illicit opioids to a much greater extent than was historically the case.[25] Drug sellers shifted toward largely white small cities and towns in the Midwest and rural Northeast, running operations resembling pizza delivery. Heroin markets sprang up where pharmaceutical opioids like OxyContin, Vicodin, Percocet, and generics had made inroads. When access to prescription opioids was tightened and OxyContin was reformulated, this had the apparently unanticipated consequence of pushing rural and suburban opioid users toward heroin, which was easier to get. Whereas adulterants had usually been inert in the past, adding synthetic fentanyl changed the calculus and multiplied the dangers of illicit markets, particularly in rural areas that lacked access to harm reduction infrastructures such as needle exchanges and naloxone.

Until recently rural paramedics were unfamiliar with naloxone. Studies show that more than 40 percent of people who die from overdose did not receive naloxone.[26] Some died alone, but some died because responders' decisions to "break out the Narcan" depend upon social characteristics of patients and settings. A Rhode Island study of all opioid overdoses in which EMS responders did not administer naloxone ($n=124$) found women much less likely to have received naloxone. A wicked intersection between gender and age showed that older women were four times *less* likely to get naloxone that would probably have saved their lives. Surprisingly, naloxone administration in Rhode Island did not differ significantly by race/ethnicity or social location, but such factors surely help determine whether emergency response reaches an overdose victim in time.[27]

Judgments about people, places, and things shape who takes naloxone home and keeps it ready to hand. Being prescribed an opioid is itself is a risk factor for opioid abuse and overdose; arguably, therefore, prescribers should "break out the Narcan" to ensure that patients' friends and families have the antidote on hand regardless of the source of the opioids in use. The text accompanying figure 12.3 implied that Theresa's fiancé had been prescribed opioids for chronic pain or illness but also used them with heroin: "My fiancé was addicted to heroin and prescription pills. One night I came home from holiday shopping, and found him lying on the bathroom floor. His lips were blue, his skin was gray. I called 911, grabbed my naloxone, and gave him a dose. If I didn't have naloxone, he would have died that night." Here Theresa is revealed to be knowledgeable and realistic, undertaking a lifesaving action out of love and, significantly, not representing her fiancé as unlovable.

Unlike Theresa, those caring for people with addictions or chronic pain patients often do not realize that overdose is a risk for which they need to prepare. Sometimes pain patients die of overdose but are categorized as

12.3 "I Saved My Fiancé's Life." Theresa, Manhattan.
Used with permission of the New York City Department of Health.

"dying in their sleep," so strong is the idea that OD happens to "addicts," not "patients." Yet physicians who treat chronic pain patients sometimes find them unable to access naloxone from pharmacies even where state laws permit them to do so.[28] Some physicians' offices have begun assembling naloxone kits and educating the highest-risk patients and families about how to use the reversing drug. The fact that naloxone still is not routinely coprescribed with opioids in high-risk situations sits awkwardly with the redefinition of addiction as a chronic disease.

A New England epidemiologist who has studied opioid overdose patterns and naloxone programs for two decades, Traci Green,[29] attributed the region's increase in overdose deaths to women managing multiple chronic diseases in midlife. According to the CDC, deaths from prescription opioids increased fivefold among women from 1999 to 2014. In a chilling example of undone science, what is known from National Institute on Drug Abuse (NIDA)-funded prevention efforts typically involves only children and adolescents. There has been almost no research on middle-aged or older adults, particularly those living rural or small-town lives. Situated knowledges are matters of life and death. The politics of overdose render urban women of color hypervisible and overvulnerable to scrutiny, whereas women who look like me—white, middle-aged, housewives, academics, or grocery store clerks—are hypovisible and overlooked. White women are not going to get naloxone, a very safe and effective drug that could save lives, because of who we are and where we live. Nor are many African Americans, who have long confronted dehumanizing legacies in relation to pain and social suffering. Prior to recognition of *over*prescription of opioids for pain to white Americans, there was an acknowledged *under*prescription problem for black Americans. An epidemiology for the people could be employed to clarify exactly how these differences and disparities comprise a multisited social justice issue that may manifest differently in different places.[30]

WARM HANDS: HARM REDUCTION RECOVERY SUPPORT SERVICES

Harm reductionists draw moral equivalency between people living with drug problems and everyone else. By 2017, harm reduction strategies had come in from the cold; the Food and Drug Administration (FDA) was working closely with the private sector to expedite review of consumer-friendly naloxone delivery products. Policy and innovation had changed considerably since the 2012 Over-the-Counter FDA hearing, to which harm reduction activists were not invited. Naloxone was on the way to becoming just another pharmaceutical "product"; "addicts" were consumers; and access issues no longer seemed to have social justice implications. Naloxone may yet become just another drug commodity as it becomes a normalized part of technocratic governance that helps lower the high social costs of premature death. Naloxone may not remain the "miracle medicine" or "lifesaving antidote" of yore. Naloxone Ninjas may be an endangered species. In the waning days of my research, I detected new voices and more nuanced shades of the persistent "othering" that has haunted drug users in the past.

The "average user will continue to use heroin whether harm reduction measures exist or not," despite risks that are "real, repeated, and prolonged."[31] Someone who drove that home to me was fellow Pennsylvanian, Chris Benedetto, who became an intravenous drug user at age 13 and made "dying" a regular activity—"getting hit with Narcan" nine times before finding his way through treatment and into long-term recovery.[32] He recalled once waking up with his jeans soaked in a cold shower, hearing his friends saying, "Don't worry, that's just Chris dying again." However, he and his peer group thought naloxone was available only in hospitals or on the ambulance. Yet it has been *the* drug most commonly administered to adolescents by emergency responders since the early 1990s.[33] Youth began to be depicted as taking up more than their fair share of naloxone, as police equipped with naloxone began to encounter repeat "customers," typically white males in their twenties, veteran drug users who valued the sensation of being pulled toward death. So-called multiple naloxone administrations (MNAs) increased substantially for this category in the Northeastern United States.[34]

Overdosing marks social status within some subcultures,[35] but only a small minority of overdose incidents involve "pushing the envelope." Chris gave me to understand that he sought the sensation of dying, rather than avoiding it or seeking to prevent it. He did not seek that sensation as a romantic concept or an escape from routine, but as a phenomenological extension of heroin use. Dying was just what he did. Now working for the Recovery, Advocacy, Service, and Empowerment (RASE) Project, Benedetto embodied a hybrid harm reduction recovery identity. As he said, "When you're using, you can't imagine anything else.... This [recovering] lifestyle wasn't imaginable, it is outside the scope of an addicted mind."

The RASE Project was founded by Denise Holden, a self-identified former heroin addict now in long-term recovery. Laboring across a parking lot from the Pennsylvania State House, RASE offers recovery support services that are not fully abstinence based, including MARS (medication-assisted recovery services) with buprenorphine or naltrexone (Vivitrol). The RASE Project primarily employs people whose "expertise comes from the fact that we are people in long-term recovery who can build rapport with the people we're working with."[36] RASE exemplifies how naloxone and the knowledge to use it has traveled beyond the urban harm reduction communities where it originated and into the small towns and rural communities of southeastern Pennsylvania. This unanticipated turn of events also indicates how much easing of social antagonisms between abstinence-based recovery and harm reduction communities is currently underway. Although still committed to long-term, drug-free lifestyles, recovery-based organizations now stock naloxone—it was on people's desks when I visited the RASE Project offices—and teach people how to use it. Yet Holden wondered how widely democratized naloxone and the supports to access and use it really are:

The whole epidemic has gotten more attention. Finally, attention is being paid, where before it wasn't a high priority. I would say the majority of the people I see who walk into our programs [are white], so I would say that once again the minority populations are being missed, left there languishing somewhere else. We do a lot of outreach and minorities know of our services, but especially [we miss] the Latino population, because of their cultural norms and mores are not as prone to seek services. Now we have Latino Recovery Support Specialists at most of our sites, and that definitely helps because, obviously, cultures feel more comfortable working within their own cultures or with people from their own cultures. I don't know, but I doubt that Narcan kits are in the homes of the urban black community.[37]

Naloxone has become a resource for many people despite patchy supply and sometimes outrageous pricing (in some times and places, it is free when across town it costs hundreds or even thousands of dollars a kit).

The conversation has widened to encompass the "breadth and depth of the resources you have to stay clean in a nutshell—so it's about clothing and food, emotional and spiritual support, physical, like health and housing needs, all the things that people new to recovery don't have. The more recovery capital you have, the greater your chances are of remaining in recovery."[38] Inclusion of naloxone and overdose prevention training in this broader "recovery capital" approach is nothing short of remarkable; it shows how far the US movement has traveled from previous decades when harm reduction was unmentionable.

TECHNOLOGIES OF SOLIDARITY: SHOULD NALOXONE BE "WILD" OR "TAME"?

Harm reduction went from "unspeakable" in US policy contexts to an implicit but widespread practice, and on to a state of "commodity access" with the actors changing as it progressed through these phases. Institutionalization has alienated protagonists such as Daniel Raymond, policy director for the Harm Reduction Coalition, who argued that naloxone should remain "wild," in the hands of "the unredeemed" who unabashedly use drugs, and are neither in recovery nor seeking treatment. Some places have retained that spirit, such as the People's Harm Reduction Alliance (PHRA) in Seattle, Washington, among the largest naloxone providers in the United States as this book went to press. Following the death of its founder, Bob Quinn, in 2012, PHRA was anchored by Shilo Murphy (Jama), who self-identifies as "someone who enjoys drugs."[39] He started handing out clean needles and naloxone in the early 2000s with a core group of four Needle Pirates:

It was myself, a circus performer, another street kid, and a skinhead who all met and decided that they were going to stand up for their community and we were going to fight. Each one of us brought in people to be distributors. We said the four of us were going to start handing out Narcan aggressively. They started bringing in other people. No training. We trained each other. It was a liquid-based naloxone that was 10 cc. I know that sounds scary to anyone who is not an injection drug user, but getting a liquid into your body is something drug users are really awesome at. It is a skill they have perfected.[40]

Police confiscated naloxone and charged one of the pirates with "distributing pharmaceuticals without a license." Kits acquired a street value close to twenty dollars per kit. Believing that "Narcan is Narcan and it should be given out in every way possible," the group decided to "flood the market, make it so worthless that people have five or six [kits and] tons of people have it all over their house. We've reached that point in some parts of the community.... The street punk who lives in the squat has a good chance to have Narcan. If you're in the University District, there's a good chance you have Narcan. It's getting better downtown, but people are still dying of overdose. Public Health now gives out Narcan, but they're Public Health so their training is thirty minutes. Ours is two minutes."[41]

Given negative experiences with local authorities during his needle pirate days, Murphy was initially suspicious of researchers such as Caleb Banta-Green and Phillip Coffin, with whom he and his organization later worked.[42] He was also thrown in with local and state government, police, firefighters, and pharmacists. Despite working to get first responders on board, he reminded me that naloxone's power lies in its capacity to connect people:

With all the hatred and stigma, I can take care of my brother and I can look after my sister. We can look out for each other and we can be a family and we can unite to keep each other alive, because this is a life or death situation. Drugs don't kill people, stigma kills people.[43]

Naloxone fights stigma and signals solidarity, Jama and other advocates say. The symbolic power of naloxone is important to some who walk on its "wild" side and seek to avoid mainstream scrutiny of their pleasures, dangers, and ways of coping with how they have chosen to live. Public health discourse alienates such users, some of whom identify as "the unredeemed."[44] Drug user unions and organizations like the Harm Reduction Coalition and the People's Harm Reduction Alliance occupy what is at times an uncomfortable space of "protagonism." Naloxone serves these communities as a technology of solidarity, helping secure a place of expertise with authorities they once repudiated—but which first repudiated them with the symbolic and material force of stigma. Whose account of safety will prevail becomes the question.

New-breed naloxone programs encounter a globally commodified space in which overdose is highly mediated and naloxone far better exposed—a "tame" naloxone that may disappoint some, but signals the success of social rehabilitation to others. As naloxone shoulders its way into the well-stocked household medicine cabinet, it is above all a commodity that shows how markets may be used to coordinate social expectations. In the process, formerly stigmatized relationships and activities may be normalized, becoming inserted into the usual cycles of production, distribution, and consumption. Even kinship itself, as recognized in figure 12.4, is subject to remaking. While in the Army, Brian was trained to use naloxone. Once returning to Queens, he was living with his father who had formerly used heroin. When his father fell out of bed and started turning blue, Brian knew what to do and he implied that this knowledge was "life-changing" for both his father and himself.

Relationships between harm reduction activists and local, city, and state governments improved dramatically once public officials realized that visible opioid overdose reflected poorly upon them. Their goals turned out to

12.4 "I Saved My Father's Life." Brian, Queens.
Used with permission of the New York City Department of Health.

converge with those of protagonists. Official support of harm reduction—and harm reduction research—was key to the taming of naloxone. Yet the global game of Whack-a-Mole is unlikely ever to resolve: as soon as one sociotechnical objection has been overcome, another pops up, only to be knocked back down, and so on. Evidence-based policy environments are infinitely productive and relentlessly amnesiac. Old evidence means nothing, even when old drugs are upcycled as in the case of naloxone. Since naloxone had decades of proven efficacy in emergency medicine as well as the operating room, it took considerable sociotechnical expertise to rework the drug for community use.

We might expect to see social movements settle into the staid patterns of interest groups once they institutionalize many of the changes for which they worked. Harm reduction does not quite fit the conventional story, partly because naloxone is a unique form of symbolic capital—"power in a bottle" carries socially transformative meanings. Consider the difference between naloxone and the EpiPen, a product that required little reorganization for integration into regular clinical practice. Prior to the EpiPen's advent, my father, a family physician, recalled drawing adrenalin up into a syringe to inject a patient in anaphylactic shock. By the late 1980s, he cynically recounted, "Somebody figured out how to make money on it." Makers of EpiPens quietly profited; physicians quietly prescribed; and the EpiPen did not become a "technology of solidarity" by drawing together collectivities to assemble kits on weekends, create new distribution channels, or produce zines or YouTube videos. Despite lifesaving potential, neither EpiPens nor the automated external defibrillators that are now the unremarkable wallpaper of airports, gyms, and other public gathering places galvanized people into action or connection.

Naloxone, on the other hand, required extensive remodeling due to the undone science of overdose. "Undone science is not only about absent knowledge; it is also about a structured absence that emerges from relations of inequality that are reflected in the priorities for what kinds of research should be funded. It is about the contours of what funders target as important and unimportant areas of research."[45] Absence has structured the development of knowledge around and about naloxone. Although it may seem too obvious and important an issue to be left undone, the problem of drug-related death had to be reconstructed as an interesting scientific problem. Advocates worked against the notion that overdose was a settled issue to be studied epidemiologically as a simple matter of mortality; they saw it as something that occurred as a result of a wicked combination of social, political, and economic forces.[46] Overdose had to be "undone" as a simple matter of individual mortality if it was to be redone as a matter of social life instead of social death. This process has recently tended toward the sentimental, another sign that its pervasive stigma is finally coming undone.

Many naloxone access activists with whom I have spoken over the years nostalgically recall hand-assembling naloxone kits, inserting cleverly adapted flyers, marking up prefilled syringes so people knew when to stop, or searching local shops for ways to package kits to evade police detection when naloxone was illegal. Since those heady times, commercial entities have entered the harm reduction space, producing everything from naloxone itself to the "Keep Calm and Carry Naloxone" T-shirts with which this book began. As a "technology of solidarity," naloxone's social relations have been catalyzed, cultivated, and calculated by the protagonists to whom I have turned. Naloxone is now poised to become a high-tech commodity; innovative formulations may appear—needleless naloxone patches or "sticking plasters," rapidly dissolving tablets akin to breath fresheners, or even *Star Trek* phaser-like applicators may appear on the overdose market. Who will benefit from these commercial novelties? Who will purchase, use, or teach others to use them? Naloxone, in other words, may not go on being a "technology of solidarity" once tamed as a commodity.

Nontraditional protagonists have become involved in harm reduction. Live4Lali, a group of parents who have lost sons and daughters to overdose, has started up "Needles, Naloxone, and Nice People" events where naloxone is distributed in hopes of saving the children of other "nice people." This harm-reduction-oriented parent movement differs greatly from the parents' movement of the "Just Say No" era. Yet ignorance continues to claim lives, and might even be said to have become epidemic. The flip side of evidence is ignorance—sometimes, we know so much that we can only see and understand rather poorly, a situation considered in the final section of this chapter.

BACKLASH: EPIDEMICS OF IGNORANCE

By the summer of 2017, all fifty US states had changed naloxone access and/or Good Samaritan laws, and stories about "Narcan parties" proliferated.[47] These events were described as gatherings where drug users took greater risks with potent opioids, figuring naloxone would save them from fatality—just as "designated drivers" reduced the harms of alcohol-fueled social gatherings. Authoritative statements issued from police officers and newscasters indicating that drug users were overdosing intentionally in public parking lots near ambulances or police stations.[48] Supposedly, drug users were "pushing it way too far, overdosing to the brink of death at what's being called a Narcan or Lazarus party. They are intentionally trying to find out what it feels like to be brought back to life."[49] Local law enforcement repeatedly quoted alarming numbers of people using naloxone as a safety net that promoted risky behavior.[50] When local news stories are promulgated as a "national" crisis, the amplifying effects resemble those of classic drug panics such as the model airplane glue-sniffing parties of suburban Denver, Colorado, in the late 1950s, with which Edward M. Brecher illustrated his chapter "How to Launch a Nationwide Drug Menace."[51]

Naloxone parties became such a menace; local news anchors pertly pointed to the dangers of safety nets. One aspect of naloxone particularly lent itself to these tendentious interpretations. Because proper naloxone dosage can rarely be ascertained due to the unknowns central to overdose,[52] naloxone must be titrated in ways that "new users and settings" are said to "lack the medical expertise or equipment" to do.[53] These stories resonated with popular conceptions of addiction but skirted the obvious: they did not acknowledge the aversive aspects of getting hit with naloxone. While those who overdose may appear to chase death, the subjective effects of getting hit with naloxone mitigate against that pursuit becoming a regular part of party scenes. An interviewee who had experienced overdose and naloxone on multiple occasions said he and his peers had never put naloxone and Narcan together, much less thought about "keeping Narcan around," as the long-ago CRA T-shirt said. Aversive effects were far more prominent for them:

It didn't come up because we knew if you got hit with Narcan, you were sick, and that's not what we were trying to do, so we never really thought about it like that. Later I had gone to prison, I had gotten out, and I used almost immediately for maybe six to eight weeks after, and I went to [a halfway house].... [D]uring that time, Suboxone had just started to come around, ... so we had started to use that, and I knew that had naloxone in it, and it was the first time I had put Narcan and naloxone together. Because I had never heard that, so it wasn't until Suboxone came out that I knew what was in Narcan.[54]

Similar to Glaswegians' experiences in the 1990s,[55] settings where naloxone is given condition the interpretation of the situation. A polar opposite of hospital and prehospital settings was that of taxi drivers who got high while waiting for fares at Chicago's airports; they self-titrated low-dose naloxone to "wake themselves" out of a nod when passengers arrived unexpectedly due to changes in flight schedules.

The idea that drug users might self-administer naloxone to modulate opioid effects has been cast as deeply disturbing, but resistant practices— uses of naloxone that depart from harm reduction protocols—are not completely unknown. "Noncompliance" is not limited to users of illegal drugs; according to the World Health Organization's Adherence to Long-term Therapies Project, fully half of patients do not comply with instructions given when the medication was prescribed.[56] Opioid users mix drugs obtained through illicit means with those diverted from the so-called grey market or obtained legally. These consumers have no patient package inserts or other expert advice on dosage or effects, despite purchasing drugs that lack quality control or assurance.[57] Harm reduction organizations have stepped into the breach with fentanyl test strips and similar services now that market pressures have intensified incentives for suppliers to add lower-cost synthetic fentanyl to dilute the supply of higher-cost product derived from poppy plants. Fentanyl is now found in upwards of 90 percent of heroin supplies in the United States and was beginning to be encountered in the northern United Kingdom and Ireland at the time of writing. In short there is no real way for responders to ascertain the dosage of opioids used in an overdose they are trying to reverse. They must proceed on the basis of experience with their combined knowledge of all markets.

The community of naloxone knowers cited in this book have shown pretty clearly that naloxone is rarely used intentionally to enable risky practices. Yet outdated reservations about naloxone have proved tenacious in the face of evidence, and protagonists feel compelled to continually disprove them. At the same time, no less a personage than the US Surgeon General advised on April 18, 2018, "Knowing how to use naloxone and keeping it within reach can save a life. BE PREPARED. GET NALOXONE. SAVE A LIFE." This was directed to all "patients currently taking high doses of opioids as prescribed for pain, individuals misusing prescription opioids, individuals using illicit opioids such as heroin and fentanyl, health care practitioners, family and friends of people who have an opioid use disorder, and community members who come into contact with people at risk for opioid overdose."

Consider the Surgeon General's advice against that of the Brookings Institution, a Washington, DC–based policy think tank usually relied upon for evenhanded analysis and strict political neutrality, which supported a dubious econometric study criticizing the naloxone access movement and the public health field more generally for making naloxone more widely available.[58] Economists Doleac, Mukherjee, and Schnell argued that naloxone distribution was keeping drug abusers alive longer, enabling them to take more risks, use more drugs, and/or commit more crimes.[59] Their statistical analysis may have been exemplary, but the premise and causal connections were downright silly to anyone knowledgeable about opiate markets, naloxone distribution, or the lives and behaviors of drug users. As an econometric study performed on pre-2015 panel data on overdoses in each state, there was no determination showing how passage of naloxone access laws affected actual availability of naloxone (despite extant research on modes of distribution[60] or Medicaid or insurance coverage,[61] findings widely discussed in the public health literature). Working in splendid isolation, the economists had no realistic estimates of the number of people carrying naloxone or trained to use it. The distance between the realities that harm reduction pragmatics address and the assumptions of econometrics was well illustrated by the resulting public controversy. Claiming that their findings on "behavioral effects" showed unintended consequences—and indeed the "moral hazard," a term discussed below—of wider distribution of naloxone, they did not actually study what drug users do. What they studied instead was how numbers of *Google searches for "naloxone" and "drug rehab"* changed after naloxone access laws went into effect.

The data and methods used in this study were so abstracted from the "natural experiment" that the economists were claiming to study that the relevance of their results on the consequences of legal changes would strike most non-economists as slim.[62] They got off to a shaky start by defining out of scope two decades of peer-reviewed literature produced by the protagonists and public health governing bodies in Australia, the United States, the United Kingdom, and the European Union. Contradicting this vast archive, the economists relied on "media reports offer[ing] evidence of the[] effects" of making naloxone available.[63] Claiming to be motivated by "worried legislators," Doleac, Mukherjee, and Schnell recycled discredited and warmed-over stories about "naloxone parties," casting public health officials as pushing for naloxone access without evidence that it worked.[64]

It is hard to imagine how anyone could have reached such a conclusion. The harm reduction movement and allied researchers have been evidence-based almost from the beginning. As we have seen in previous chapters, it was public health officials who had to be nudged, cajoled, and pressured—by a motley crew of drug users and advocates, health equity activists, and harm reductionists working inside and outside of government—into easing naloxone access and reorienting public safety towards harm reduction. Researchers who were doing "good science," rather than letting it go "undone," began working on naloxone access despite the risk of damage to careers in the decade before the US NIDA or the UK Medical Research Council jumped on the overdose bandwagon. Instead of engaging with the knowledgeable research and advocacy domain, Doleac disparaged naloxone proponents for proceeding on an "anecdotal" basis: "Our analysis of panel data from across the United States shows that such anecdotal reports reflect valid concerns about the unintended consequences of naloxone. We use the gradual adoption of state-level naloxone access laws as a natural experiment to measure the effects of broadened access, and find that the moral hazard generated by naloxone is indeed a problem resulting in increased opioid abuse and crime, and no net reduction in mortality."[65] Disregarding the evidence base and taking to Twitter, Doleac found it "utterly disheartening that a discipline as important as public health is filled with so many people who collectively have so little understanding of rigorous research methods.

Advocates should acknowledge that many of their strongly held priors are not evidence based. Anecdotes and personal experience are valuable but are not a substitute for rigorous causal inference methods."[66]

Casting naloxone as a morally hazardous material obviously contrasts to my presentation of it as the pro-social tool of Good Samaritans. Doleac and colleagues found that the "moral hazard generated by naloxone is indeed a problem resulting in increased opioid abuse and crime, and no net reduction in mortality."[67] An economic term-of-art, "moral hazard" was used in the actuarial sciences to indicate externalization of the costs of risky behaviors—that is, that costs are assumed by parties other than those behaving badly. The term is used for individual and institutional behavior; banks, insurers, think tanks, or opioid users may have more or less taste for risk. The moral hazards literature is based on the premise that individuals and institutions increase exposure to risk in response to behavioral incentives.[68] For instance, seatbelts or condoms could present a moral hazard by increasing a wearer's willingness to engage in risky behavior. Does the presence of seatbelts or condoms increase participation in risky behaviors? If one's neighbor has seatbelts or condoms, does this affect your use? Does presence of naloxone increase a drug user's risk behavior? No matter how often that question is answered, or how much evidence is produced, it will always be raised again, given the current state of ignorance, amnesia, and strangely unsituated knowledges within which such questions rest.

Despite seeming to be about individual behavior, the social harms and consequences of naloxone have come to matter for the kind of society we want to live in. I hope you will wear seatbelts and condoms—and keep naloxone within reach—because I want to live in a society where it is considered a public good to prevent preventable deaths. And that is naloxone's redeeming social merit as a technology of solidarity and a signal not of moral hazard but of the moral value of life, not death.

CONCLUSION: Harm Reduction Infrastructure—
We Have Hardly Begun to Be Human

People are dying. Naloxone is everywhere. Naloxone protagonists have brought the opiate reversal drug into twenty-first-century conversations and medicine cabinets. More people are living drug-related lives, and all too many are dying drug-related deaths. Finding its way into handbags and backpacks of Good Samaritans and drug users alike, there is little mystery surrounding what drove demand for naloxone.[1] Increased use of opioids, legal and illegal, synthetic and opium based, has made naloxone a household word; it is now standard equipment in many US high schools, shopping malls, UK housing estates, train stations, and homeless hostels, promoted on billboards, T-shirts, and subway signs. But naloxone is not nearly enough to stem the tides of drug-related death in either the United States or United Kingdom. No matter how scientifically marvelous the technological fix, this problem turns out to require more than naloxone. For it has been surprisingly complicated, difficult, and time consuming to "tame" naloxone—to create the social infrastructure for making it more available to those who need it.

Making overdose "simple" proved difficult because of the social structure of drug markets and drug policy, not simply because of the complexities of people dependent on opiates for their sense of well-being and attunement with the world. Despite parallels, drug worlds differ in supply, markets, consumption patterns, and cultural practices between the

United States and the countries of the United Kingdom. For instance, drug availability in Northern Ireland differs enough that Chris Rintoul actively sought to avoid what he characterizes as "the North American disaster":

We have some pharmaceutical opioids but it's nothing like the United States, where everybody got hooked on OxyContin or something like that, and you get escalating doses and then you get cheaper heroin. That hasn't been our experience yet, and I think there has been a concerted effort to see that the North American disaster doesn't happen in Northern Ireland. And it has been a disaster, we are well aware of that. I was recently in Montreal, speaking at the International Harm Reduction Conference about opiate overdose in a hostel in Belfast, and we had the heroin tested [in Montreal], in various batches from all over the world, and all heroin in North America, northern North America, at least, contained fentanyls of some kind or other.[2]

Once fentanyl was identified in Northern Ireland's heroin supply, efforts to promote naloxone training and supply redoubled. Matters of drug supply are intimately connected to matters of demand, a point emphasized by Stephen Malloy, who has argued that the United Kingdom was

protected from fentanyl ... because we have such a robust, well-financed illicit heroin market. That is genuinely part of the reason, and I do believe that because we have such a good heroin market, we are less likely to see the emergence of fentanyl at such a high rate. That said, Brexit and the various other things that we're seeing now will put other pressures on that supply chain, and I think that fentanyl, the more it's reported, the more it's brought up, the more people are talking it up, the more we will see people choosing to bring it in.[3]

As drug markets change, naloxone dynamics will do so as well. These are socially situated problems, and it takes context-specific knowledge to respond well to them.

For the past thirty years, overdose deaths have been rising in the United States, tripling between 2002 and 2014 (counting legal and illegal drugs obtained from both legal and illegal markets, with fentanyl increasingly implicated from late 2013 on). While writing this book, I got much practice as a parent, teacher, administrator, and interpreter of an exponentially unfolding epidemic. When I thought it couldn't get worse, it did. Not until 2017 did a falloff in opioid prescriptions begin in response to publicity on the "third wave of the U.S. opioid epidemic."[4] Periodization often begins with the overprescription of opioid pain relievers in the 1990s, when pain was declared the "fifth vital sign," followed by a "second wave" fueled by "pill mills" capitalizing on the first wave. Shutting these down pushed users of legal and illegal pharmaceutical opioids to heroin, creating the conditions for the third wave. Fentanyl is now found in the vast majority of the US heroin supply, a problem because it often requires even more naloxone to reverse an overdose in which fentanyl is involved. In the face of a daunting failure to make a dent in the overdose death rate, denizens of the North American disaster may well ask, "What's next after naloxone?"

Naloxone availability falls far short of the quantity required to counteract every overdose,[5] and the United States still does not have OTC naloxone as of this writing in the summer of 2019. But deeper questions are still worth asking: What's next when basic human rights and social goods have not been extended to drug users in most places? Why ask what's next when so much about the lives of drug users takes place as a form of precarity? We have hardly begun to be human when it comes to remaking people, places, and things to fit into a mature harm reduction infrastructure that goes beyond the places where it grew up. We ask, "What's next?" when bored, cranky, and tired of what is. Tessie Castillo's article on "post-naloxone outreach programs" took that "What's next?" crankiness seriously—she put her finger on the pulse of new "post-overdose" follow-up programs, then gearing up in the United States and the United Kingdom:

As communities explore options to offer resources to people post-naloxone administration, it is important to note that these programs are just one piece of a solution to a very complex issue. Everyone is looking for the "golden ticket" to do away with the problem of chaotic drug use and tragic deaths from overdose, but none of the proposed solutions—not naloxone, not post-overdose outreach programs, not more inpatient treatment, not injected medications that block cravings for opioids—is a panacea. The situations and needs of people who use drugs vary widely, as does resource availability in any given area. Post-naloxone visits will help some people but they need to be coupled with robust efforts in overdose prevention education, naloxone distribution, diverse treatment options, drug policy and criminal justice reform, and many other interventions. The questions we explore need to go much broader and deeper than "What's next after naloxone?"[6]

I hope that this book has animated past and present in ways that allow us to go broader and deeper into productive dialogue with the question of "What's next?" Inhabitants of resuscitative societies are all "service users" in some sense. But for many of us, neither market choice, nongovernmental advocacy, nor governmental representation works to represent and reflect our interests. All of us, across much of our lives and the technologies we adopt to enhance and prolong them, could learn from the ways clients/users and families were integrated into harm reduction, drug treatment, and naloxone provision in the United Kingdom. Drug users should not have had to push so hard in order to be included in basic human rights to health and health care. We simply must become more reflective about the North American disaster not solely in terms of drug policy, in terms of what is legal and what is not, but in terms of what these distinctions say about social exclusion if we are to wisely answer the question, "What's next after naloxone?"

Two answers to "What's next?" appear to be gathering steam in relation to naloxone and the politics of overdose. One answer is more: more harm reduction, more naloxone, more attunement, more justice, more treatment. According to health law professor Scott Burris, who tracked naloxone's legal progress from its earliest days in the United States, the answer could be "more investment in a public health, harm reduction approach [that] makes sense given the evidence of what works," a kind of harm reduction infrastructure. But he also cautions that "there is every reason to expect that actual policy will be more mixed."[7] Another answer is that enthusiasm for naloxone dwindles and gives way to "mixed" response, and reluctance, backlash, and containment emerge. How overdose is represented matters. Videos of overdoses in public places, cars, streets, or on public transport have been recorded by cell phone–wielding cops or bystanders, going viral as morality tales and positioning naloxone as an unnatural tool of the devil. Whereas naloxone protagonists once heralded lives saved via miraculous resurrections, opponents of the expansion of naloxone emphasize social costs, trauma, wasted resources, anoxia, and even brain damage if the OD is not reversed quickly.[8] As discussed in the last chapter, naloxone has been cast as a "moral hazard." An emerging "is-there-life-after-Narcan" genre shades sadly into tired arguments in which the "undeserving" and "unredeemed" soak up more than their share of public resources. The social movement work of harm reduction in reshaping drug users' self-care and conduct toward others—and with it the notion that we are all drug users, all service users—disappears in these accounts from anti-resuscitative corners of the society.

We cannot pretend that naloxone-for-all is not contentious. The politics of overdose animates ambivalent forms of social solidarity. Where did you start to feel ambivalent as a reader of this book about naloxone as a technology of solidarity? On the one hand, I presented examples in which social solidarity has helped communities and individuals adapt to and sometimes reimagine the social-structural conditions of deindustrialization and neoliberalism in order to prevent people from dying. Did that ring true for you? Or did you resonate more with harm reduction as a "biopolitical therapeutic regime" that makes precarious subjects responsible for managing their own health?[9] Does thinking pharmacologically— thinking and acting as "neurochemical selves"[10] or "thinking with the molecule"[11]—help people inhabit their present lives on their own terms? Or does it render them more vulnerable to further involvement with the carceral state?

Cultural assumptions about drugs and drug users have worn deep grooves in our neurosocial substrate. Harm reduction offers a policy path beyond criminalization and (bio)medicalization, but falls smack dab into the "responsibilization" of new neoliberal public health in ways that place it at odds with the patterns of social solidarity analyzed in this history. Anthropologist Jarrett Zigon characterizes most harm reduction services as premised on the "closed normalization of biopolitical metaphysical humanism," pointing to the "isolated fixity" that he encountered in such sites in places as disparate as Honolulu, New York City, St. Petersburg, Russia, and Denpasar, Indonesia.[12] He contends that the rules and regulations pervading these spaces require drug users to remake themselves into someone they are not: "To be in the center, then, is to be a certain kind of person who can manage being that kind of person."[13] But he did uncover one remarkable space in Vancouver, British Columbia, an organization known as InSite, where he found an exceptionally nonjudgmental ethics of attunement in a "parallel world" that was not "uninhabitable and unbearable."[14] As many places consider "safe" or "supervised" injection rooms as the next harm reduction frontier—from Sydney, Australia, to

Glasgow, Scotland, to Vancouver, British Columbia, and quite possibly to US cities—it becomes important to consider how people inhabit the spaces of harm reduction.

People matter. It matters who is in these parallel worlds and what they are doing there. State-supported harm reduction may change the terms of engagement so substantially that programs risk falling out of attunement with those whose lives they seek to enhance and whose deaths they seek to prevent. The moral sentiments required to deliver nonjudgmental or low-threshold harm reduction services are not easily instilled and maintained in neoliberal social regimes, which assess whether or not individuals meet their responsibilities. The assessments built into audit culture are an exercise of power; living with their rules and regulations—and the technocratic public health discourse from which they flow—can be unbearable. Successful social movements that negotiate changing landscapes of social, political, and epistemic conflict have often had to settle for less in order to gain something "more"—if they choose to become infrastructural in order to live on into the future. People and organizations compromise when they institutionalize services, yes, but there is little future in underfunded enterprises in which the time and energy of volunteers is largely taken up with assessments and reports.

Protagonist Daniel Wolfe, director of International Harm Reduction Development in the Open Society Public Health Program and an expert on the intersection of HIV and drug policy, brought this to my attention early in my research for this book. He and others had to convince HIV funders to attend to overdose and naloxone: "We would say, of course, it is the same people who you are seeking to reach with antiretroviral treatment who are dying of overdose given how concentrated HIV is among injection drug users." Moreover, Wolfe argued that "naloxone was a critical means of strengthening the relationship between outreach worker and client, because you needed to offer something that they needed. We had a lot of anecdotal evidence that when people started giving out naloxone, it would invigorate the HIV relationship because people were excited— and because it is a miraculous experience to feel like you've saved someone's life." Enthusiasm over naloxone as a "miracle medicine" enabled a foregrounding of drug users' autonomy and rights:

From the beginning for us, it was … our understanding that the medicine was largely safe, had very limited abuse potential, and, most importantly, was something that people could do for one another without having to pledge to a largely unrealistic program of lifetime abstinence. That was very powerful for people. I remember talking to a Vietnamese woman who had been at one of our trainings and took back some naloxone as a sample. Then she used it two days later when someone in her network had overdosed, and she said, "I finally have something that I can tell my mother that she can be proud of."[15]

According to Wolfe, anecdotes like this circulated to dramatize how drug users could make "rational choices about their own health even if they were still actively using drugs or involved with people who were. This continues to put us both in alliance and in tension with some of the other naloxone actors. It is a creative tension." Whereas other organizations focused on making it possible for people to get naloxone from doctors or pharmacies, Open Society Foundations focused on putting it "in the hands of peers rather than in the hands of physicians. In the international context, this was tied to an incredible tension between physician-controlled technologies and community-controlled technologies."[16] This tension was due to the division between legal regulatory controls on naloxone and other pharmaceuticals, and the lack of regulatory controls on illegal drugs.

The social infrastructure of harm reduction—and the emerging harm reduction market—owes much to the creative capacity to bend ambivalence toward affirmation and enactment of social solidarity. Naloxone's power to cut through "issue overwhelm," in which too many issues need attention simultaneously, was profound. As good a job as AIDS activists and harm reduction advocates did with needle exchange and "any positive change," they were in no position to address the overall drug problem, mass incarceration, or social injustice. Those who may have been in such a position did not think or act in ways that would have been necessary to head off the steep rise in opiate-related deaths or the emergence of a ferocious fentanyl problem in our lifetimes. But naloxone's cultural work was to narrow the aperture—every social movement needs a list of possible concrete accomplishments to which they can bring closure. While a technological fix can serve the symbolic value of signaling the affirmative goal of life, not death—as naloxone does—it may also lend itself to premature closure. However, it took a surprisingly long time for most governments and even health authorities to become hospitable to naloxone.

Building up harm reduction infrastructure requires skills and resources designed to institutionalize programs while addressing disparities, be they based in age, gender, race/ethnicity, class, or sexuality, or intersections of these. These differ greatly from the skills needed to create and sustain the energy of consciousness raising, movement building, direct action, and outreach. Yet a surprising number of people continue to identify with harm reduction as a social movement, even after it has symbolically settled into a more mundane existence. The movement has used expertise to enliven activities while entraining recruits into acquiring and exercising knowledge and commitment, while amassing evidence on multiple dimensions. This liveliness may be seen in STS terms as animating the creative tensions between showing how a technology like naloxone is socially constructed over time, and the work of enacting a more convivial, fair-minded, and otherwise better reconstruction of the technology. The story of the narcotic antagonists casts into sharp relief earlier framings of how drug users would be treated. Nalorphine was put into the hands of police as a technology of coercion and suspicion without sufficient regard for civil and human rights. Naloxone was initially put into place as an aspect of emergency medicine and in ways that essentially denied access to many who died as a result. In reframing and reanimating naloxone as a technology of solidarity, the movement remade naloxone as far more than medicine.

So why is it that all of this dying is still going on? It is said that those who die of overdose die because of "ignorance, not opioids."[17] This book has tried to show the coproduction of ignorance along with knowledge and experience, and to show that undone science works not simply through repression but through the productive work of the science that is done, who does it, and how and why they do it. Goals and intentions matter. Recovery activist Ryan Hampton put it well in his New Year's resolution, stated on the blog of the US Congress, to make *saving all lives* his harm reduction goal in 2019:

The issue isn't opioids: it's ignorance. What are we actually trying to fix? Are we solving the problem, or simply eliminating the people we don't think deserve help? I've looked at this epidemic from both sides. I know where I stand. I hope others will stand with me, and lend their voices to a movement that includes everyone and offers real, meaningful solutions. This doesn't have to be complicated. Simple measures like making naloxone widely available, offering 24/7 access to safe injection sites with fentanyl test kits and clean syringes, and connecting people with help the minute they need it ensures that they can get through 2019 alive.[18]

What do we make of the ongoing reluctance to integrate social with biomedical interventions for drug treatment? One form of ignorance is the problematic compartmentalization of "social factors" from "scientific" ones. We are more likely to end up with better knowledge of how to save all lives if evidence and experience are purposefully enacted together. Better research questions are not simply better in the abstract. We are likely to end up with more actionable knowledge that can be used to animate the parallel worlds we dream up and bring into being. Asking for better knowledge—more attuned, properly contextualized, situation-specific—is asking for more adaptive knowledge rather than more "basic" research or more RCTs.

Naloxone is an unfortunately necessary artifact that has grown political arms and legs in the United States and the United Kingdom. Naloxone is no moral hazard, but a technology with a politics that embodies a moral stance, an antidote to malaise, a symbol of collective care and solidarity, and a hopeful harbinger of cultural redemption. Recounting their naloxone dreams, this book's protagonists brought naloxone to life by placing it within ecologies of harm reduction that made it infrastructural to drug users' practice. The twin project of producing evidence that naloxone works, while simultaneously making people who overdose matter was well summed up by harm reduction educator Lee Hertel: "Nobody deserves to die because of how they choose to navigate life."[19]

Convinced of that stance, I hope you will stand with me, keep calm, and carry naloxone. But I also hope that I've inspired you to share your answers to the question, "What's next?" Will naloxone become just another component of the first aid kit, an evidence-based product that can be bought and sold like any other, something that most people will have on hand just in case? There are signs that it may yet become so. A columnist for my Pennsylvania hometown newspaper, the *Press-Enterprise*, mused that having a few doses of the "OD remedy" on hand "would be a great thing to have, not only for friends and relatives of addicts, but for addicts themselves. Maybe then we could stop writing about people dying while their friends are afraid to call an ambulance." This encouragement came three years after the Commonwealth of Pennsylvania changed its laws to permit anyone to administer naloxone, and the Pennsylvania Physician General issued a statewide standing prescription that theoretically enables anyone to purchase Narcan. Pricing still varied wildly; when I requested some at a pharmacy near the RASE Project described in chapter 12, the pharmacist asked me why I wanted it. Without even checking my insurance, he told me that a kit would run me $800.[20] Rather than drop with economies of scale, prices have drastically increased since naloxone access laws have gone into effect. Luckily, I knew some harm reduction organizations that could supply naloxone for less!

Even in the Democratic presidential debates televised on July 31, 2019, the price of some formulations of naloxone was stated as $4,000. The story of how an old and once affordable remedy came to command prices high enough to be publicly debated is also the story of under-supported harm reduction organizations overwhelmed by increasing demand—and unable to sustain the energy the protagonists had during their years "underground." State governments have yet to truly pick up the slack—although in the waning days of 2018, the Commonwealth of Pennsylvania gave out six thousand naloxone kits for free in one day at eighty locations across the state. These contradictions illustrate that the world has not yet been remade into one where naloxone is fully present in the lives of those who need it most. Harm reduction is a broad approach that may be adapted to prevent many kinds of social suffering through close attention to the people, places, and things that matter. Starting from what matters most, change the world to reduce harms that we know well. "Any positive change" will make us more human.

NOTES

INTRODUCTION

1. See https://www.overdoseday.com/. Since beginning in Australia in 2001, this event has been marked by the wearing of silver ribbons modeled upon the red ribbons worn since 1988 in remembrance of people who have died of AIDS.

2. Karla Dawn Wagner, Peter J. Davidson, Ellen Iverson, Rachel Washburn, Emily Burke, Alex H. Kral, Miles McNeeley, Jennifer Jackson Bloom, and Stephen E. Lankenau, "'I Felt Like a Superhero': The Experience of Responding to Drug Overdose among Individuals Trained in Overdose Prevention," *International Journal of Drug Policy* 25, no. 1 (2014): 157–165.

3. Although by comparison its population base is much smaller, the United Kingdom annually accounts for approximately one-third of drug-related deaths in countries reporting to the European Monitoring Centre for Drugs and Drug Addiction, which are the European Union plus Norway and Turkey; see http://www.emcdda.europa.eu/emcdda-home-page_en.

4. I was far from the only critical drug scholar watching; see Helena Hansen and Mary E. Skinner, "From White Bullets to Black Markets and Greened Medicine: The Neuroeconomics and Neuroracial Politics of Opioid Pharmaceuticals," *Annals of Anthropological Practice* 36, no. 1 (2012): 167–182.

5. David Hess argues that ignorance is produced through sequestration or repression of knowledge, deliberate avoidance of controversial topics and ethical issues, problems with existing data sets, and the mismatch between the scale at which knowledge is produced and the scale at which regulatory policy is made. David J. Hess, *Undone Science: Social Movements, Mobilized Publics, and Industrial Transition* (Cambridge, MA: MIT Press, 2016).

6. Anne M. Lovell, "Addiction Markets: The Case of High-Dose Buprenorphine in France," in *Global Pharmaceuticals: Ethics, Markets, Practices*, ed. Adriana Petryna, Andrew Lakoff, and Arthur Kleinman (Durham, NC: Duke University Press, 2006), 136–170.

7. Centers for Disease Control and Prevention, "Vital Signs: Overdoses of Prescription Opioid Pain Relievers—United States, 1999–2008," *Morbidity and Mortality Weekly Report* 60 (2011): 1487–1492.

8. Alexander Y. Walley, Ziming Xuan, H. Holly Hackman, Emily Quinn, Maya Doe-Simkins, Amy Sorensen-Alawad, Sarah Ruiz, and Al Ozonoff, "Opioid Overdose Rates and Implementation of Overdose Education and Nasal Naloxone Distribution in Massachusetts: Interrupted Time Series Analysis," *British Medical Journal* 346 (2013): f174.

9. Christopher B. R. Smith, "Disorder and the Body of Drugs: Addiction, Consumption, Control, and City Space," in *Cultures of Addiction*, ed. Jason Lee (Amherst, NY: Cambria Press, 2012), 57–88; Christopher B. R. Smith, "Harm Reduction as Anarchist Practice: A User's Guide to Capitalism and Addiction in North America," *Critical Public Health* 22, no. 2 (2012): 209–221.

10. David W. Maurer and Victor H. Vogel, *Narcotics and Narcotic Addiction* (Springfield, IL: Charles C. Thomas, 1954), 228.

11. Ibid., 229.

12. Emilie Gomart, "Surprised by Methadone: In Praise of Drug Substitution Treatment in a French Clinic," *Body & Society* 10, no. 2–3 (2004): 85–110.

13. Nikolas Rose and Joelle M. Abi-Rached, *Neuro: The New Brain Sciences and the Management of the Mind* (Princeton, NJ: Princeton University Press, 2013); and Natasha Myers, *Rendering Life Molecular: Models, Modelers, and Excitable Matter* (Durham, NC: Duke University Press, 2015), on how protein modelers animate and propagate visual facts with their bodies. Naloxone access activists are science educators who explain receptor-level drug effects with gestures that convey dynamic interactions between agonists and antagonists.

14. Deboleena Roy, *Molecular Feminisms: Biology, Becomings, and Life in the Lab* (Seattle: University of Washington Press, 2018).

15. Susan Cozzens, *Social Control and Multiple Discovery in Science: The Opiate Receptor Case* (Albany: State University of New York Press, 1989); Candace Pert and Solomon H. Snyder, "Opiate Receptor: Demonstration in Nervous Tissue," *Science* 179, no. 4077 (1973): 1011–1014.

16. See Susan C. Boyd, Donald MacPherson, and Bud Osborn, *Raise Shit! Social Action Saving Lives* (Halifax and Winnipeg: Fernwood Publishing, 2009); Steve Epstein, *Impure Science: AIDS, Activism, and the Politics of Knowledge* (Berkeley: University of California Press, 1996); Neil Hunt, Eliot Albert, and Virginia Montanes Sanchez, "User Involvement and User Organizing in Harm Reduction," in *Empowerment and Self-Organizations of Drug Users: Experiences and Lessons Learnt*, ed. Georg Bröring and Eberhard Schatz (Amsterdam: Foundation Regenberg, 2008), 333–354.

17. David Herzberg, *The Other Drug War: Addictive Medicines in American History* (Chicago: University of Chicago Press, forthcoming). Attentive to the racial implications of these color-marked opioid markets, Herzberg calls white markets an "open secret, widely acknowledged but rarely examined" because they fall between drug policy historians, who have tended to study illicit markets, and historians of medicine and pharmaceuticals who focus on the licit side of these conceptual and regulatory divides. Left to drug control and law enforcement, the muddy waters of drug diversion are rarely bridged.

18. Shane Darke and Michael Farrell, "Three Persistent Myths about Heroin Use and Overdose Deaths," *The Conversation*, February 6, 2014, https://theconversation.com/three-persistent-myths-about-heroin-use-and-overdose-deaths-22895.

19. Shimon Prokupecz, Steve Almasy, and Catherine E. Shoichet, "Sources: Philip Seymour Hoffman Dead of Apparent Drug Overdose," CNN, February 3, 2014, https://www.cnn.com/2014/02/02/showbiz/philip-seymour-hoffman-obit/index.html.

20. Ray Sanchez, "Coroner: Philip Seymour Hoffman Died of Acute Mixed Drug Intoxication," CNN, February 28, 2014, https://www.cnn.com/2014/02/28/showbiz/philip-seymour-hoffman-autopsy/index.html.

21. Darke and Farrell, "Three Persistent Myths about Heroin Use and Overdose Deaths."

22. For a provocative account that takes this underemphasized aspect of tolerance as a form of adaptation, see Judith Grisel, *Never Enough: The Neuroscience and Experience of Addiction* (New York: Doubleday, 2019).

23. Shepard Siegel, "Pavlovian Conditioning and Heroin Overdose: Reports by Overdose Victims," *Bulletin of the Psychonomic Society* 22, no. 5 (1984): 428–430.

24. Shepard Siegel and Delbert W. Ellsworth, "Pavlovian Conditioning and Death from Apparent Overdose of Medically Prescribed Morphine: A Case Report," *Bulletin of the Psychonomic Society* 24, no. 4 (1986): 278–280.

25. Shepard Siegel, "The Heroin Overdose Mystery," *Current Directions in Psychological Science* 25, no. 6 (2016): 375–379.

26. Jessie M. Gaeta, "A Pitiful Sanctuary," *Journal of the American Medical Association* 321, no. 224 (2019): 2407–2408.

27. Shepard Siegel, Riley E. Hinson, Marvin D. Krank, and Jane McCully, "Heroin 'Overdose' Death: Contribution of Drug-Associated Environmental Cues," *Science* 216, no. 4544 (1982): 436–437.

28. Siegel, "The Heroin Overdose Mystery," 378.

29. Sarah G. Mars, Jason N. Fessel, Philippe Bourgois, Fernando Montero, George Karandinos, and Daniel Ciccarone, "Heroin-Related Overdose: The Unexplored Influences of Markets, Marketing and Source-Types in the United States," *Social Science & Medicine* 140 (2015): 44–53.

30. Shane Darke, *The Life of the Heroin User: Typical Beginnings, Trajectories and Outcomes* (Cambridge: Cambridge University Press, 2011), 114–115.

31. Ibid., 121.

32. Ibid., 112–113; Helena Hansen, "Assisted Technologies of Social Reproduction: Pharmaceutical Prosthesis for Gender, Race, and Class in the White Opioid 'Crisis,'" *Contemporary Drug Problems* 44, no. 4 (December 2017): 321–338.

33. Darke found little difference between male and female death rates among heroin users. Although women heroin users were at similar risk of death as their male counterparts, they were at much greater risk than non-heroin-using female comparands.

34. Walley et al., "Opioid Overdose Rates." Over 50,000 people were trained and over 10,000 reversals accomplished from 1996 through 2010.

35. Maya Doe-Simkins, telephone interview by Nancy D. Campbell, November 27, 2013. Doe-Simkins took precocious interest in this project and introduced me to many of the protagonists.

36. Arnold R. Soslow, "Acute Drug Overdose: One Hospital's Experience," *Annals of Emergency Medicine* 10, no. 1 (1981): 18–21.

37. Scott Burris, Leo Beletsky, Carolyn A. Castagna, Casey Coyle, Colin Crowe, and Jennie Maura McLaughlin, "Stopping an Invisible Epidemic: Legal Issues in the Provision of Naloxone to Prevent Opioid Overdose," *Drexel Law Review* 1, no. 2 (2009): 273–339; Scott Burris, Brian Edlin, and Joanna Norland, "Legal Aspects of Providing Naloxone to Heroin Users in the United States," *International Journal of Drug Policy* 12 (2001): 237–248; Corey S. Davis, Scott Burris, Leo Beletsky, and Ingrid Binswanger, "Co-prescribing Naloxone Does Not Increase Liability Risk," *Substance Abuse* 37, no. 4 (2016): 498–500; Corey S. Davis and Derek H. Carr, "Legal Changes to Increase Access to Naloxone for Opioid Overdose Reversal in the United States," *Drug and Alcohol Dependence* 157 (2015): 112–120.

38. Rooted in "controversy studies," STS responds to everything from nuclear plants to methadone maintenance by studying who was excluded from development, regulation, and implementation of the technology in question. As STS cross-fertilized with scholars studying effects of slavery, colonialism, industrialism, imperialism, and twentieth-century wars, it renewed attention to questions of power. See Dorothy Nelkin and Laurence Tancredi, *Dangerous Diagnostics: The Social Power of Biological Information* (Chicago: University of Chicago Press, 1989); and Ulrike Felt, Rayvon Fouché, Clark A. Miller, and Laurel Smith-Doerr, eds., *The Handbook of Science and Technology Studies*, 4th ed. (Cambridge, MA: MIT Press, 2017).

39. Nancy D. Campbell, *Discovering Addiction: The Science and Politics of Substance Abuse Research* (Ann Arbor: University of Michigan Press, 2007); Steven Epstein, *Inclusion: The Politics of Difference in Medical Research* (Chicago: University of Chicago Press, 2007).

40. João Biehl, *Vita: Life in a Zone of Social Abandonment* (Berkeley: University of California Press, 2005); João Biehl, Byron J. Good, and Arthur Kleinman, *Subjectivity: Ethnographic Investigations* (Berkeley: University of California Press, 2007); Philippe Bourgois and Jeffrey Schonberg, *Righteous Dopefiend* (Berkeley: University of California Press, 2009); Veena Das, Arthur Kleinman, Margaret M. Lock, Mamphela

Ramphele, and Pamela Reynolds, *Remaking a World: Violence, Social Suffering and Recovery* (Berkeley: University of California Press, 2001); Arthur Kleinman, Veena Das, and Margaret M. Lock, eds., *Social Suffering* (Berkeley: University of California Press, 1997); and Angela Garcia, *The Pastoral Clinic: Addiction and Dispossession along the Rio Grande* (Berkeley: University of California Press, 2010).

41. See Lundy Braun, *Breathing Race into the Machine: The Surprising Career of the Spirometer from Plantation to Genetics* (Minneapolis: University of Minnesota Press, 2017); M. Susan Lindee, *Moments of Truth in Genetic Medicine* (Baltimore, MD: Johns Hopkins University Press, 2005); Melbourne Tapper, *In the Blood: Sickle Cell Anemia and the Politics of Race* (Philadelphia: University of Pennsylvania Press, 1999); Keith Wailoo, *Drawing Blood: Technology and Disease Identity in Twentieth-Century America* (Baltimore, MD: Johns Hopkins University Press, 1997); Keith Wailoo, *Dying in the City of the Blues: Sickle Cell Anemia and the Politics of Race and Health* (Chapel Hill: University of North Carolina Press, 2001); David C. Courtwright, *Forces of Habit: Drugs and the Making of the Modern World* (Cambridge, MA: Harvard University Press, 2001); Joseph Dumit, *Drugs for Life: How Pharmaceutical Companies Define Our Health* (Durham, NC: Duke University Press, 2012); Jeremy Greene, *Generic: The Unbranding of Modern Medicine* (Baltimore, MD: Johns Hopkins University Press, 2014); David Herzberg, *Happy Pills in America: From Miltown to Prozac* (Baltimore, MD: Johns Hopkins University Press, 2010); Andrea Tone, *Age of Anxiety: The History of America's Turbulent Affair with Tranquilizers* (New York: Basic Books, 2008); and Elizabeth Watkins and Andrea Tone, *Medicating Modern America: Prescription Drugs in History* (New York: New York University Press, 2007).

42. E. Jerschow, R. Y. Lin, M. M. Scaperotti, and A. P. McGinn, "Fatal Anaphylaxis in the United States 1999–2010: Temporal Patterns and Demographic Associations," *Journal of Allergy and Clinical Immunology* 134, no. 6 (2014): 1318–1328.e7. According to the American Academy of Allergy, Asthma, and Immunology, death from anaphylaxis is a "reassuringly unusual outcome" resulting in less than 225 deaths per year.

43. Stephen Malloy, interview by Nancy D. Campbell, via Skype, January 23, 2019, 32.

1 POISON MURDERS AND NATURAL ACCIDENTS

1. Stephen E. Lankenau, Karla D. Wagner, Karol Silva, Aleksandar Kecojevic, Ellen Iverson, Miles McNeely, and Alex H. Kral, "Injection Drug Users Trained by Overdose Prevention Programs: Responses to Witnessed Overdoses," *Journal of Community Health* 38 (2013): 133–141. This Los Angeles study cast folk remedies used to revive individuals in the past—injecting milk or salt, hitting, slapping, or rubbing with ice—as "barriers preventing participants from employing recommended response techniques." See Shane Darke, Joanne Ross, and Wayne Hall, "Overdose among Heroin Users in Sydney, Australia: II. Responses to Overdose," *Addiction* 91 (1996): 401–412; David Best, Lan-Ho Man, Michael Gossop, Alison Noble, and John Strang, "Drug Users' Experiences of Witnessing Overdoses: What Do They Know and What Do They Need to Know?," *Drug and Alcohol Review* 19 (2000): 407–419.

2. Hugo M. Krueger, Nathan B. Eddy, and Margaret Sumwalt, *The Pharmacology of the Opium Alkaloids* (Washington, DC: United States GPO, 1940), Part II, Subject Index, LI, LII.

3. I. H. Stearns, "The True Antidote for Opium," *Chicago Medical Journal and Examiner* 47 (1883): 47.

4. The *New York Times* notice appeared on June 30, 1867, ironically on the same page as a report on a druggist censured for mislabeling a bottle of laudanum implicated in the suicide of August Homan, a 38-year-old German immigrant who gulped a whole two-ounce bottle in front of this wife, who attested to his intemperate habits.

5. The latter facts were reported in the *New York Times* on December 9, 1868, when the City Court awarded damages to the widower.

6. In a December 9, 1868, *New York Times* story, "Local Intelligence: The Alleged Poisoning Case in Brooklyn," the reporter visited Kennedy in jail to learn more about the "overdose of morphia."

7. A February 19, 1870, news brief from New Jersey published in the *New York Times* noted that "a verdict of death from natural causes concluded the inquest of Coroner Burns over the remains of Mrs. Mould, who died a couple of days ago from an overdose of opium." Natural causes and overdose were not portrayed as contradictory.

8. Medical testimony was suspect in the nineteenth century; expert dithering did not inspire confidence. Toxicologists' disagreements cast doubt not only on their expertise, but upon the very basis of scientific knowledge. See Mark R. Essig, "Poison Murder and Expert Testimony: Doubting the Physician in Late Nineteenth-Century America," *Yale Journal of Law and Humanities* 14, no. 1 (2002): 177–210; and Mark R. Essig, "Science and Sensation: Poison Murder and Forensic Science in Nineteenth-Century America" (Unpublished diss., Cornell University, 2000).

9. "Records of the Coroner's Office for the Year 1866," *New York Times*, December 31, 1866.

10. "An Overdose of Laudanum," *New York Times*, May 14, 1883.

11. "Accidents in the Household: What Should Be Done in Certain Emergencies," *New York Times*, October 10, 1880.

12. "City and Suburban News: New Jersey," *New York Times*, September 6, 1876.

13. "Morphine instead of Quinine: A Young Man Poisoned by a Mistake in the Medicine," *New York Times*, June 4, 1883.

14. Essig, "Poison Murder and Expert Testimony."

15. Sheila Jasanoff, *Science at the Bar: Law, Science and Technology in America* (Cambridge, MA: Harvard University Press, 1997).

16. David Courtwright, Herman Joseph, and Don DesJarlais, *Addicts Who Survived: An Oral History of Narcotic Use in America, 1923–1965* (Knoxville: University of Tennessee Press, 1989), 103–104.

17. Needle-borne malaria among injecting drug users in New York City is discussed in chapter 4.

18. Clifton K. Himmelsbach, "Malaria in Narcotic Addicts at the United States Penitentiary Annex, Fort Leavenworth, Kansas," *Public Health Reports* 48 (1933): 1465; Alexander G. Biggam, "Malignant Malaria Associated with the Administration of Heroin Intravenously," *Transactions of the Royal Society of Tropical Medicine and Hygiene* 23 (1929): 147–153; Oliver C. Nickum, "Malaria, Transmitted by Hypodermic Syringe," *Journal of the American Medical Association* 100 (1933): 1401–1402; Guy H. Faget, "Malarial Fever in Narcotic Addicts: Its Possible Transmission by the Hypodermic Syringe," *Public Health Reports* 48 (1933): 1031–1037; L. McKendree Eaton and Samuel M. Feinberg, "Accidental Hypodermic Transmission of Malaria in Drug Addicts," *American Journal of Medical Sciences* 186 (1933): 679–682.

19. "Chorus Girls Used to Smuggle Drugs," *New York Times*, January 31, 1922, 14.

20. Her family favored homicide as cause of death; others speculated it was suicide. See "The Mysterious Death of a Fiercely Burning Vaudeville Flame," *The Dead Bell*, January 14, 2015, https://deadbell.com/2015/01/14/the-mysterious-death-of-a-fiercely-burning-vaudeville-flame/.

21. "Mary Warburton Dies in Home Here," *New York Times*, September 15, 1937. Three days later her death was publicly attributed to overdose: "Death Ascribed to Opium: Mary Brown Warburton Found to Have Died of Overdose," *New York Times*, September 18, 1937: 7.

22. "New Year Brings 11 Fatalities Here but US Toll Is under Estimate," *New York Times*, January 1, 1950.

23. Courtwright, Joseph, and DesJarlais, *Addicts Who Survived*, 104.

24. Caroline J. Acker, *Creating the American Junkie: Addiction Research in the Classic Era of Narcotic Control* (Baltimore, MD: Johns Hopkins University Press, 2002); Campbell, *Discovering Addiction*; Everette L. May and Arthur E. Jacobson, "The Committee on Problems of Drug Dependence: A Legacy of the National Academy of Sciences, an Historical Account," *Drug and Alcohol Dependence* 23 (1989): 183–218.

25. Horatio C. Wood, "The Indispensable Uses of Narcotics: The Therapeutic Uses of Narcotic Drugs," *Journal of the American Medical Association* 96, no. 14 (1931): 1140–1144.

26. Everette L. May, "Analgesic Drugs and Drug Dependence," *British Journal of Addiction* 62 (1966): 197–202. May worked for the National Institutes of Health on the nonaddicting analgesics project.

27. Gavril Pasternak and Ying-Xian Pan, "Mu Opioids and Their Receptors: Evolution of a Concept," *Pharmacological Reviews* 65, no. 4 (2013): 1257–1317.

28. Aware that allyl ethers stimulated respiration, Leake suggested that attaching an allyl group might overcome respiratory depression from morphine. Nalorphine is essentially an allyl group attached to a morphine ring.

29. E. Ross Hart and Elton L. McCawley, "The Pharmacology of N-Allylnormorphine as Compared with Morphine," *Journal of Pharmacology and Experimental Therapeutics* 82, no. 3 (December 1944): 339–348.

30. Committee on Problems of Drug Dependence (CPDD), *Proceedings of the Agonist-Antagonist Symposium*, February 14–15, 1983 (unpublished), 90–92 (hereafter CPDD Symposium). E. (Eddie) Leong Way was a Leake graduate student with Hart and McCawley; see interview with Nancy D. Campbell, San Juan, Puerto Rico, June 13, 2004.

31. Hart and McCawley, "The Pharmacology of N-Allylnormorphine." The article thanked Major for preparing the material.

32. Klaus Unna, "Antagonistic Effect of N-Allyl-Normorphine upon Morphine," *Journal of Pharmacology and Experimental Therapeutics* 79 (1943): 27–31.

33. Ibid.

34. William R. Martin, "Opioid Antagonists," *Pharmacological Reviews* 19, no. 4 (1967): 463–521, at 482.

35. James E. Eckenhoff, John D. Elder, and Benton D. King, "N-Allyl-Normorphine in the Treatment of Morphine or Demerol Narcosis," *American Journal of Medical Sciences* 223, no. 2 (1952): 191–197.

36. A 1952 National Heart Institute review identified no full articles—just the abstracts examined above, which were joined by H. Franklin Fraser, G. D. Van Horn, and Harris Isbell, "Studies on N-Allylnormorphine in Man: Antagonism to Morphine and Heroin and Effects of Mixtures of N-Allylnormorphine and Morphine," *American Journal of the Medical Sciences* 231, no. 1 (1956): 1–8, 7.

37. Editorial, *Journal of the American Medical Association* 150, no. 11 (1952): 1121. The drug's value was clearly perceived to be in morphine antagonism rather than substitution. Merck worked the phrase "heavily narcotized patients" into promotional advertising to the medical press, as seen in figure 1.2.

38. Abraham Wikler, "Effect of Large Doses of N-Allyl-Normorphine," *Federation Proceedings* 10 (1951): 345; and Abraham Wikler, H. Franklin Fraser, and Harris Isbell, "N-Allylnormorphine: Effects of Single Doses and Precipitation of Acute Abstinence Syndromes during Addiction to Morphine; Methadone or Heroin in Man (Post Addicts)," *Journal of Pharmacology and Experimental Therapeutics* 109, no. 1 (1953): 8–20.

39. Within this 1,500-bed institution designed to house, treat, and rehabilitate all opiate addicts residing east of the Mississippi River, there was a small laboratory directed by Himmelsbach when the institution opened in 1935. Later directed by Harris Isbell, this laboratory was renamed the Addiction Research Center (ARC) in 1948, when it became part of the National Institutes of Mental Health. See Acker, *Creating the American Junkie*; Campbell, *Discovering Addiction*.

40. M. R. King, Clifton K. Himmelsbach, and R. S. Sanders, "Dilaudid (Dihydromorphinone): A Review of the Literature and a Study of Its Addictive Properties," *Public Health Reports*, suppl. 113 (1935); quoted in Nathan B. Eddy, *The National Research Council Involvement in the Opiate Problem, 1928–1971* (Washington, DC: National Academies Press, 1973).

41. Ian Walmsley, "Coming Off Drugs: A Critical History of the Withdrawing Body," *Contemporary Drug Problems* 43, no. 4 (2016): 1–16.

42. Donald R. Jasinski, Rolley E. Johnson, and Jack E. Henningfield, "Abuse Liability in Human Subjects," *Trends in Pharmacological Sciences* 5, no. 5 (1984): 196–200.

43. There are two extant interviews with Himmelsbach, one by Jon M. Harkness and Gail Savitt for the Advisory Committee for Human Radiation Experiments, November 2, 1994, and another by Wyndham D. Miles for the Oral History Interviews, Accession 613, History of Medicine Division, National Library of Medicine, May 4, 1972.

44. Walmsley, "Coming Off Drugs."

45. Ibid.

46. Lawrence C. Kolb and Clifton K. Himmelsbach, "Clinical Studies of Drug Addiction: III. A Critical Review of the Withdrawal Treatments with Method of Evaluating Abstinence Syndromes," *American Journal of Psychiatry* 94 (1938): 759–797.

47. Walmsley, "Coming Off Drugs."

48. Clifton K. Himmelsbach, "The Morphine Abstinence Syndrome, Its Nature and Treatment," *Annals of Internal Medicine* 15 (1941): 829–830.

49. Walmsley, "Coming Off Drugs."

50. Himmelsbach, "The Morphine Abstinence Syndrome," 829. For a more contemporary take, see Darke and Farrell with Sarah Larney, "Yes, You Can Die from Withdrawal."

51. Subjects included 110 white males and 15 African American males. Later published as Havelock Franklin Fraser, Abraham Wikler, Anna J. Eisenman, and Harris Isbell, "Use of N-Allylnormorphine in Treatment of Methadone Poisoning in Man: Report of Two Cases," *Journal of the American Medical Association* 148 (1952): 1205–1207.

52. Ibid.

53. Ibid., 1206; Abraham Wikler and R. L. Carter, "Effects of Single Doses of N-Allylnormorphine on Hindlimb Reflexes of Chronic Spinal Dogs during Cycles of Morphine Addiction," *Journal of Pharmacology and Experimental Therapeutics* 109, no. 1 (1953): 92–101. At the January 22–23, 1954, Committee on Drug Addiction and Narcotics (CDAN) meeting, Isbell stated his intent to "get some patients pretty depressed with morphine and then come in with the Nalline" to prove its efficacy in reversing overdose (852). At the previous CDAN meeting, he had urged Henry K. Beecher to investigate a nalorphine-morphine combination while ascertaining nalorphine's analgesic efficacy for postoperative pain. According to May and Jacobson, "The Committee on Problems of Drug Dependence," 190, this latter suggestion arose in the late 1940s.

54. William R. Martin, "The Evolution of Concepts," in *The Opiate Receptors* 23, ed. Gavril W. Pasternak (New York: Humana Press, 2011), 7–21, quotation at 9.

55. Harris Isbell and H. Franklin Fraser, "Addiction to Analgesics and Barbiturates," *Journal of Pharmacology and Experimental Therapeutics* 99 (1950): 355–397, mentions that nalorphine's analgesic activity was still untested.

56. See Laura Stark, *The Normals: A People's History* (Chicago: University of Chicago Press, forthcoming).

57. Louis Lasagna, "Historical Introduction," in CPDD Symposium, 6.

58. Only at higher doses did nalorphine act as an analgesic; see Louis Lasagna and Henry K. Beecher, "Analgesic Effectiveness of Nalorphine and Nalorphine-Morphine Combinations in Man," *Journal of Pharmacology and Experimental Therapeutics* 112 (1954): 356–363; and Raymond W. Houde and Stanley L. Wallenstein, "Clinical Aspects of the Evaluation of Analgesics," *American Geriatric Society* 4, no. 2 (1956): 167–171.

59. Louis Lasagna, interview by Jon M. Harkness and Suzanne White-Junod, Advisory Committee for Human Radiation Experiments, Washington, DC, December 13, 1994, 13.

60. Lasagna and Beecher, "Analgesic Effectiveness of Nalorphine." In the 1983 CPDD Symposium on Narcotic Agonist-Antagonists, Lasagna mentioned that a *Journal of the American Medical Association* editor had struck all unexplained "clinical anecdotes" from the 1954 article. He recalled patients who experienced hyperalgesia and "complained bitterly of pain all over as if one had stimulated all their pain receptors simultaneously," 9. Hyperalgesia was "highly unusual and interesting," making nalorphine an interesting drug to think with later in the 1960s when hypotheses about multiple opiate receptors arose.

61. Wikler and Carter, "Effects of Single Doses," 92, 101. Widely used in preclinical pharmacology at the time, the technique is described in Abraham Wikler and K. Frank, "Hindlimb Reflexes of Chronic Spinal Dogs during Cycles of Addiction to Morphine and Methadon," *Journal of Pharmacology and Experimental Therapeutics* 94, no. 4 (1948): 382–400.

62. James E. Eckenhoff, George L. Hoffman, and Robert D. Dripps, "N-Allyl-Normorphine: An Antagonist to the Opiate Drugs," *Anesthesiology* 13 (1952): 242–251.

63. Ibid., 249.

64. Francis J. Marx and Julian Love, "Effectiveness of N-Allylnormorphine in the Management of Acute Morphine Poisoning," *Annals of Internal Medicine* 39 (1953): 635–640.

65. These are addressed via interviews with Edward F. Domino and accounts of the experimental work of William R. Martin, one of Unna's students who replaced Isbell as Director of the Addiction Research Center at the US Public Health Service Hospital in Lexington, Kentucky, from 1963 to 1977.

66. Edward F. Domino, interview by Nancy D. Campbell, Ann Arbor, Michigan, November 9, 2006, 27.

67. Ibid.

68. Published as Edward F. Domino, Edward W. Pelikan, and Eugene F. Traut, "Nalorphine (Nalline) Antagonism to Racemorphan (Dromoran) Intoxication," *Journal of the American Medical Association* 153 (1953): 26–27.

69. Murray Strober, "Treatment of Acute Heroin Intoxication with Nalorphine (Nalline) Hydrochloride," *Journal of the American Medical Association* 154, no. 4 (1954): 327–328.

70. Ibid., 327.

71. Louis Lasagna, "Historical Introduction," in Committee on Problems of Drug Dependence (CPDD), *Proceedings of the Agonist-Antagonist Symposium*, February 14–15, 1983 (unpublished), 7.

72. May and Jacobson, "The CPDD," 192–193.

73. Ibid.

74. Acker, *Creating the American Junkie*; Campbell, *Discovering Addiction*; Eddy, *The National Research Council Involvement*. Until 1977, CPDD was part of the National Academy of Sciences; it then became an independent organization.

75. Quoted in Martin, "Opioid Antagonists," 491.

76. Martin, "Opioid Antagonists," 494.

77. Pasternak and Pan, "Mu Opioids and Their Receptors," 1269.

78. Gavril W. Pasternak, "Opioids and Their Receptors: Are We There Yet?," *Neuropharmacology* 76, Part B (2013): 198–203. There turned out to be multiple *mu* receptors; the ARC clinical research is still noted as providing data on questions of individual variation in *mu* opioid effects.

79. Solomon H. Snyder, "You've Come a Long Way, Baby!," in *The Opiate Receptors*, ed. Gavril W. Pasternak (New York: Humana Press, 2011), 1–6, quotation at 4.

80. Michael Kuhar, interview by Nancy D. Campbell, Orlando, Florida, June 23, 2005.

81. Snyder, "You've Come a Long Way, Baby!," 6.

82. Dennis Paul, Chaim G. Pick, Leslie A. Tive, and Gavril Pasternak, "Pharmacological Characterization of Nalorphine, a Kappa 3 Analgesic," *Journal of Pharmacology and Experimental Therapeutics* 257, no.1 (1991): 1–7.

2   THE "CHEMICAL SUPEREGO"

1. Editorials and Comments, "Rapid Diagnosis of Addiction," *Journal of the American Medical Association*,154, no. 5 (January 30, 1954): 414. According to William R. Martin, the unattributed editorial comment was written by Isbell. The editorial drew on Isbell's earlier report to Merck, "Nalline—A specific narcotic antagonist." However, pharmacologist Chauncey Leake attributed the Nalline test to Eddie Leong Way, "another brilliant graduate student."

2. Safeguards were in place on the clinical side of the US Narcotic Hospital, where detoxification was medically managed with methadone after 1948. The hospital needed to differentiate between types of dependence during admission; patients addicted to alcohol or barbiturates were not admitted. Understanding the patient's habit was necessary for determining safe dosage of methadone. Prisoners transferred to Lexington from other federal penitentiaries did not arrive actively addicted, but those sent there by judges, doctors, or family members arrived in various states of need for medical management. During the 1950s and 1960s Lexington and its sister narcotic farm in Fort Worth, Texas, were well known to drug users as the only places that tried to keep patients comfortable during withdrawal. See Holly M. Karibo, "'The Only Trouble is the Dam' Heroin': Addiction, Treatment and Punishment at

the Fort Worth Narcotic Farm," *Social History of Medicine* (2019): hky069, https://doi.org/10.1093/shm/hky069.

3. I. Young Chen and E. Leong Way, "Studies on Antagonism of Morphine Miosis by Nalorphine as a Diagnostic Test for Narcotic Usage," *British Journal of Pharmacology* 24 (1965): 789–797.

4. "Drug Detector," *Time* 68, no. 26 (December 24, 1956).

5. In "Big Step Reported Taken in Hunt for Drug Addiction Cure," James C. O'Neill reported in the *Yuma Sun* on February 27, 1956, that Terry and Braumoeller also experimented with Thorazine, which "completely eliminates the agony of pain and physical torture suffered by addicts trying to shake the habit." Most reports dropped reference to Thorazine, a so-called major tranquilizer then in vogue in mental hospitals. O'Neill referred to Thorazine and Nalline as a "double-barreled attack on the narcotics addict."

6. Thorvald Brown's book *The Enigma of Drug Addiction* (Springfield, IL: Charles C. Thomas, 1961) was an invaluable primary source because it included all forms used to administer the program; detailed architectural diagrams (later used by critics of the program); and a summary of responses given by fourteen physicians appointed by the state of California to administer such tests.

7. Wikler, Fraser, and Isbell, "N-Allylnormorphine: Effects of Single Doses."

8. Chen and Way, "Studies on Antagonism of Morphine Miosis by Nalorphine," 796.

9. Ibid.

10. Fraser, Van Horn, and Isbell, "Studies on N-Allylnormorphine in Man."

11. Ibid., 10.

12. Ibid., 12.

13. California's attempt to combat its narcotics problem became the basis for a US Supreme Court test case, *Robinson v. California* (370 US 660 [1962]). The Health and Safety Code was struck down as unconstitutional for violating the Eighth and Fourteenth Amendments on grounds that it was "cruel and unusual punishment" to imprison someone on the basis of narcotic addiction.

14. Court of Appeal of California, Second District, Division One, *People v. Jaurequi*, 142 Cal. A. 2d 555, 298 P 2d 896 (1956). See also Edwin C. Conrad, "The Admissibility of the Nalline Test as Evidence of the Presence of Narcotics," *Journal of Criminal Law and Criminology* 50, no. 2 (1959): 187–191.

15. Brown, *The Enigma of Drug Addiction*, 289. Brown quoted four full paragraphs of Vernon's views without attribution. Despite the unclear provenance of Vernon's remarks, they indicate police attitudes towards the text as well as convergence of view between Brown and Vernon.

16. James G. Terry, "Nalline: An Aid in Detecting Narcotics Users," *California Medicine* 85 (1956): 229–301.

17. Bob Piser, "Arizona Officials Observe Oakland Narcotics Control," *Arizona Republic*, December 19, 1960, and "N.Y. Officials Study Oakland Nalline Dope Control Plan," *Oakland Tribune*, August 1, 1962.

18. James T. Carey and Anthony Platt, "The Nalline Clinic: Game or Chemical Superego?," *Issues in Criminology: Drug Use and Crime* 2, no. 2 (Fall 1966): 223–244, quotation at 224.

19. Brown, *The Enigma of Drug Addiction*, 294.

20. Ibid., 297–301.

21. Ibid., 45–47.

22. Leona Ward, "Santa Rita: Model Prison Grows Old in Disgrace," *The Argus*, July 28, 1975.

23. Brown, *The Enigma of Drug Addiction*, 277.

24. Ibid.

25. Ibid., 54.

26. Ibid., 55.

27. Ibid., 16. Santa Rita was cast into disgrace by the Civil Rights and Free Speech Movements.

28. Donna Murch, *Living for the City: Migration, Education, and the Rise of the Black Panther Party in Oakland, California* (Chapel Hill: University of North Carolina Press, 2010), 63–65.

29. Ibid., 64.

30. Brown, *The Enigma of Drug Addiction*, 317.

31. Ibid.

32. Ibid., 291.

33. Jim McAuley, "Drug Used to Detect Addicts: Program Designed to Keep Parolees from Return to Heroin," *Independent Press-Telegram*, October 11, 1959, A-10. Prior to the state pilot program, the Long Beach office occasionally ordered Nalline tests in suspected cases of relapse.

34. Ibid.

35. Brown, *The Enigma of Drug Addiction*, 287, 309.

36. Carey and Platt, "The Nalline Clinic," 225.

37. Karen Namson, "New Drug to 'Sniff Out' Narcotics Users," *Pasadena Star-News*, April 29, 1957, 15.

38. George Robeson, "Drug Pinpoints Dope Addict Violators," *Independent Press-Telegram*, January 30, 1962, B-2.

39. Brown, *The Enigma of Drug Addiction*, 323.

40. Ibid.

41. Ibid.

42. Ibid., 327.

43. Ibid., 286.

44. Ibid., 309.

45. Ibid.

46. At the time that Conrad described Dr. Terry's procedure, the good doctor had administered 2,300 such tests over the preceding three years.

47. See the list in Stewart Weinberg, "*Nalline* as an Aid in the Detection and Control of Users of Narcotics," *California Law Review* 48 (1960): 282–294.

48. "Drug Detection Expert Faces Challenge at Trial," *Oakland Tribune*, November 27, 1957.

49. Conrad, "The Admissibility of the Nalline Test," 190.

50. Stanley E. Grupp, "Addict Mobility and the Nalline Test," *British Journal of Addiction* 63 (1968): 227–236, quotation at 233.

51. Ibid., 234.

52. Ibid.

53. Carey and Platt, "The Nalline Clinic," 232–233.

54. Ibid., 242.

55. Ibid., 243.

56. Ibid., 244.

57. Grupp, "Addict Mobility and the Nalline Test."

58. "Nalline Method of Dope Detection," *Humboldt Times*, December 4, 1960.

59. See "Nalline: New Weapon in Narcotics War," *Humboldt Times*, April 20, 1961; "Nalline and Local Crime," *Humboldt Times*, January 18, 1963, 4 (which approvingly noted a declining rate of "positives" from 40 percent to 2 percent, tying these to similarly declining proportions across the state of California). See also "Dr. Collins' Resignation Could Hurt Narcotics Control Here, Board Told," *Humboldt Times*, October 28, 1964, 5.

60. See an article celebrating the tenth anniversary of the Oakland program, "Officers Use Simple Test as Anti-addiction Weapon," *Oakland Tribune-Sun*, June 20, 1965, 10.

61. Opiate-dependent individuals used codeine-based cough syrups when they could not obtain heroin or morphine; see "Men Use False Name to Buy Medicine Here," *Humboldt Times*, October 28, 1964, 5.

62. Brown, *The Enigma of Drug Addiction*, 318.

63. Ibid.

64. Ibid.

65. Jerome H. Skolnick, *Justice without Trial: Law Enforcement in Democratic Society* (New York: John Wiley and Sons, 1966), showed the great extent to which Oakland police retained low-level discretion particularly in drug enforcement during the 1960s.

66. Harney was a member of Harry J. Anslinger's inner circle. See Douglas Valentine, *The Strength of the Wolf: The Secret History of America's War on Drugs* (London: Verso, 2004), 22.

67. Malachi L. Harney, "Current Provisions and Practices in the United States of America Relating to the Commitment of Opiate Addicts," United Nations Office of

Drugs and Crime, *Bulletin on Narcotics*, no. 3 (1962): 11–23, https://www.unodc.org /unodc/en/data-and-analysis/bulletin/bulletin_1962-01-01_3_page003.html.

68. Conrad, "The Admissibility of the Nalline Test," 188.

69. *People v. Moore*, 70 Cal. App.2d 158, 160 P.2d 857 (1945).

70. *People v. Williams*, 164 Cal. App. 2d 858, 331 P.2d 251 (1958).

71. Weinberg, "*Nalline* as an Aid."

72. John B. Neibel, "The Implications of *Robinson v. California*," *Houston Law Review* 1, no. 1 (1963): 8.

73. Ibid.

74. Facts found in "Record from Municipal Court of Los Angeles Judicial District, City of Los Angeles, State of California, in the case of *People of the State of California, Plaintiff, v Lawrence Robinson, Defendant*," reprinted in the appellate file for *Lawrence Robinson v California*, Supreme Court of the United States, October Term, 1961, no. 554, 1–116.

75. Clerk's Transcript, Docket Entries, Transcript of Record, US Supreme Court Briefs and Records 370 (US 660, 1962), 43.

76. Ibid., 1.

77. "Drug Addiction Ruled No Crime," *New York Times*, June 25, 1962, 1, 18.

78. Clerk's Transcript, Docket Entries, Transcript of Record, US Supreme Court Briefs and Records 370 (US 660, 1962), 12.

79. Ibid., 13.

80. Ibid., 14.

81. Ibid., 16.

82. Ibid., 17, 23.

83. Ibid., 24.

84. Ibid., 17.

85. Ibid.

86. Ibid., 20.

87. Ibid., 26.

88. Samuel McMorris's article "What Price Euphoria? The Case against Marijuana," appeared in the *Medico-Legal Journal* of London and was reprinted in the *British Journal of Addiction* 62 (1967): 203–208.

89. William H. McGlothlin, M. Douglas Anglin, and B. D. Wilson, "Outcome of the California Civil Commitments," *Drug and Alcohol Dependence* 1 (1975/1976): 165–181.

90. "Drug Addiction Ruled No Crime: Supreme Court Voids California Law Penalizing Use," *New York Times*, June 25, 1962, 1, 18.

91. Neibel, "The Implications of *Robinson v California*," 11.

92. Ibid.

93. Ibid.

94. See Julilly Kohler-Hausmann, *Getting Tough: Welfare and Imprisonment in 1970s America* (Princeton, NJ: Princeton University Press, 2017), 40.

95. *Robinson v California*, 370 US 660 (1962).

96. "Drug Case Victory Won after Death of Addict: Appellant in Reversal of California Law Found Dead in Alley Here Last August," *Los Angeles Times*, June 26, 1962, 10. Robinson's premature death appears in John G. West, *Darwin Day in America: How Our Politics and Culture Have Been Dehumanized in the Name of Science* (Wilmington, DE: Interscholastic Studies Institute, 2015).

97. McMorris was later disbarred after suspensions unrelated to his appearance before the US Supreme Court.

98. Brown, *The Enigma of Drug Addiction*, 55.

99. Ibid., 56.

100. Ibid., 57.

101. Ibid.

102. Ibid.

103. Ibid., 58.

104. Ibid., 59.

105. Ibid., 58.

106. Ibid., 59.

107. Virginia S. Lewis, Seymour Pollack, David M. Petersen, and Gilbert Geis, "Nalline and Urine Tests in Narcotics Detection: A Critical Overview," *International Journal of the Addictions* 8, no. 1 (1971): 163–171.

108. Grupp, "Addict Mobility and the Nalline Test," 468.

109. Henry W. Elliott, "The Nalorphine (Pupil) Test in the Detection of Narcotic Use," in *Narcotic Drugs*, ed. Doris H. Clouet (New York: Plenum, 1971), 484–492.

3 DEATHS FROM "NARCOTISM" IN THE MID-TWENTIETH-CENTURY UNITED STATES

1. Maurer and Vogel, *Narcotics and Narcotic Addiction*, 229.

2. Ibid.

3. Established in 1952, the Riverside Hospital program was a court referral program serving drug users who were under twenty-one years old. It was considered a "social as well as medical experiment," as "no similar institution has ever been devoted to the treatment and rehabilitation of the adolescent drug user." See Harney, "Current Provisions and Practices."

4. Ibid. Trussell's remarks occurred at a conference on post-hospital care and rehabilitation of adolescent narcotic addicts convened in January 1960 by the Governor's Task Force on Narcotic Addiction, the New York State Interdepartmental

Health Resources Board, and the New York Board of Hospitals in Albany, New York.

5. David T. Courtwright, *Dark Paradise: A History of Opiate Addiction in America* (Cambridge, MA: Harvard University Press, 1982, enl. ed. 2001), 149–152.

6. Eric Schneider located the letter in Folder 2, 1948–1950, box 37, file 1080 (Juvenile Offenders), accession number 170-74-12, Record Group 170, National Archives. See also Eric C. Schneider, *Smack: Heroin and the American City* (Philadelphia: University of Pennsylvania Press, 2008), 35–40.

7. Ibid., 36.

8. Ibid., 40.

9. Ibid.

10. Terms in this section were those used by the New York City Medical Examiner's Office in this period.

11. Milton E. Helpern, "Causes of Death from Drugs of Dependence," in *The Pharmacological and Epidemiological Aspects of Adolescent Drug Dependence: Proceedings of the Society for the Study of Addiction*, ed. Cedric William Malcolm Wilson (Oxford: Pergamon Press, 1968), 221–233, at 228.

12. Ibid., 225.

13. Ibid.

14. Ibid., 226.

15. Edward Preble and John D. Casey, "Taking Care of Business: A Heroin User's Life on the Street," *International Journal of the Addictions* 4 (1969): 1–24; Schneider, *Smack*.

16. Schneider, *Smack*, 101.

17. Joan W. Moore, *Going Down to the Barrio: Homeboys and Homegirls in Change* (Philadelphia: Temple University Press, 1991), 105–106.

18. Nancy D. Campbell, *Using Women: Gender, Drug Policy, and Social Justice* (New York: Routledge, 2000); James V. Bennett, *Oral History and Delinquency: The Rhetoric of Criminology* (Chicago: University of Chicago Press, 1981).

19. Julilly Kohler-Hausmann, "'The Attila the Hun Law': New York's Rockefeller Drug Laws and the Making of a Punitive State," *Journal of Social History* (September 2010): 71–96; Samuel K. Roberts, "'Rehabilitation' as Boundary Object: (Bio)Medicalization, Local Activism, and Narcotics Addiction Policy in New York City, 1951–1962," *Social History of Alcohol and Drugs* 26, no. 2 (2012): 147–169.

20. Preble and Casey, "Taking Care of Business."

21. I use the pseudonym as it appeared in each of the respective media that presented their story: "Karen" and "Johnny" for James Mills's magazine article, "We Are Animals in a World No One Knows," *Life* 58, no. 8 (February 26, 1965): 66–81, and "Helen" and "Bobby" for his 1965 paperback *The Panic in Needle Park* (New York: Signet, 1965), as well as the 1971 movie of the same title with the screenplay by John Gregory Dunne.

22. Bill Eppridge, interview by Nancy D. Campbell, Milford, Connecticut, October 2006, 23–24. On Doriden, one of the first nonbarbiturate sedatives, see Nils Kessel, "'Doriden von Ciba': Sleeping Pills, Pharmaceutical Marketing, and Thalidomide, 1955–1963," *History and Technology* 29, no. 2 (2013): 153–168.

23. See Ben Cosgrove, "'Two Lives Lost to Heroin': A Harrowing, Early Portrait of Addicts," *Time*, October 3, 2013, updated March 4, 2015, http://time.com/3731579 /two-lives-lost-to-heroin-a-harrowing-early-portrait-of-addicts/.

24. Mills, *The Panic in Needle Park*, 36.

25. Ibid., 37.

26. Ibid., 39.

27. Ibid., 5.

28. John Gregory Dunne's screenplay was one of the first Hollywood depictions of overdose. Although Dunne built on the *Life* story, neither Mills nor Eppridge was involved in the making of *Panic in Needle Park* (1971). *Panic* was Al Pacino's first film; he played "Bobby," the junkie character modeled on "Johnny."

29. See Maia Szalavitz, "These 'New Faces of Heroin' Stories Are Just the Old Face of Racism," http://www.substance.com/these-new-face-of-heroin-stories-are-just-the -old-face-of-racism/7555/, which lists titles—"Heroin's New Generation: Young, White and Middle Class"; "For Heroin's New Users, a Long Hard Fall"; "Heroin Addiction: Problem for Middle Class Also"; "Addiction among Middle Class and Wealthy Found on the Rise"; "The Scourge of Youth: Use of Heroin by Students Is Called Deadliest Fad Ever to Hit Campuses"—and then reveals that the stories are from 2003, 1999, 1982, 1969, and 1970, respectively, and yet *"every single one, going back nearly 50 years, implies or states explicitly that Times readers share the paper's assumption that heroin addiction is a problem of the inner-city poor."*

30. Edward M. Brecher, *The Consumers Union Report on Licit and Illicit Drugs* (Boston: Little, Brown & Co., 1972), 101.

31. Helpern, "Causes of Death from Drugs of Dependence," 228.

32. Edema results from impurities filtered out by capillaries in the lungs, that then present as granulomata. See Pekka Sauko and Bernard Knight, *Knight's Forensic Pathology*, 3rd ed. (London: Hodder Arnold, 2004), 577.

33. Milton Helpern and Yong-Myun Rho, "Deaths from Narcotism in New York City," *New York State Journal of Medicine* 66 (1966): 2393, quoted in Brecher, *Licit and Illicit Drugs*, 107.

34. Ibid.

35. Ibid., 102.

36. Ibid.

37. Ibid.

38. Ibid. The latter point was later confirmed by harm reduction research.

39. Ibid., 105.

40. Ibid., 114.

41. Charles Cherubin, Jane McCusker, Michael Baden, Florence Kavaler, and Zili Amsel, "The Epidemiology of Death in Narcotic Addicts," *American Journal of Epidemiology* 96, no. 1 (1972): 11–22, quoted in Brecher, *Licit and Illicit Drugs*, 112.

42. Robert L. DuPont and Mark H. Greene, "The Dynamics of a Heroin Addiction Epidemic," *Science* 181, no. 4101 (1973): 716–722.

43. Ibid., 722.

44. Ibid.

45. Ibid.

46. Stefan Timmermans, *Postmortem: How Medical Examiners Explain Suspicious Deaths* (Chicago: University of Chicago Press, 2006), 28–29, traces what he called the "four filaments" of the medical sociology of forensic practice: who does what, where, when, and with what consequences; the structural elements of the organizational context reconstituted in and through forensic work practices; the content of categories used to explain deaths; and social norms. My historical thinking is indebted to this sociology of forensic practice.

47. Tetanus showed a gendered pattern because men in the United States were typically vaccinated against tetanus in the context of working construction and heavy industry jobs, whereas women were not. Women thus acquired tetanus more easily through shared injection equipment.

48. "Narcotic Deaths," parts 1 and 2, Emory University and National Medical Audiovisual Center, 1969, National Library of Medicine. Many thanks to Rebecca McDonald for pointing me to these YouTube clips at https://www.youtube.com/watch?v=csKZq65z3y8.

49. Helpern, "Causes of Death from Drugs of Dependence," 230.

50. Ibid.

51. Ibid.

52. Ibid.

53. "Narcotic Deaths."

54. M. M. Glatt, "The Abuse of Barbiturates in the United Kingdom," *UNODC Bulletin of Narcotics* 14 (1962): 19–38; Francisco Lopez-Munoz, Ronaldo Ucha-Udabe, and Cecilio Alamo, "The History of Barbiturates a Century after Their Clinical Introduction," *Neuropsychiatric Disease Treatment* 1 (2005): 329–343; Maurer and Vogel, *Narcotics and Narcotic Addiction*, 231.

55. E. W. Adams, *Drug Addiction* (London: Oxford University Press, 1937).

56. Joel Fort, "The Problem of Barbiturates in the United States of America," *Bulletin of the United Nations*, 1964, http://www.unodc.org/unodc/en/data-and-analysis/bulletin/bulletin_1964-01-01_1_page004.html; Glatt, "The Use of Barbiturates in the United Kingdom."

57. Nicholas Rasmussen, "Goofball Panic: Barbiturates, 'Dangerous' and Addictive Drugs, and the Regulation of Medicine in Postwar America," in *Prescribed: Writing, Filling, Using, and Abusing the Prescription in Modern America*, ed. Jeremy A. Greene and

Elizabeth Siegel Watkins (Baltimore, MD: Johns Hopkins University Press, 2012), 23–45, at 26.

58. US Congress. House Committee on Ways and Means, *Traffic in, and Control of, Narcotics, Barbiturates, and Amphetamines*, 84th Congress, 1st session, October 13, 1955 (Washington, DC: GPO, 1956), 295–491.

59. Rasmussen, "Goofball Panic," 38.

60. Brecher, *Licit and Illicit Drugs*, 256.

61. Ibid., 112.

62. Harold Alksne, Ray E. Trussell, Jack Elinson, and Shermand Patrick, *A Follow-Up Study of Treated Adolescent Narcotics Users, Based on a Survey of Records and a Series of Interviews with Adolescent Narcotic Users Who Were Admitted to the Riverside Hospital (New York City) during 1955*, Report to the New York State Interdepartmental Health Resources Board (New York: The New School and Columbia University School of Public Health and Administrative Medicine, 1959).

63. Quoted in Brecher, *Licit and Illicit Drugs*, 111; Alksne et al., *A Follow-Up Study*, 101.

64. Cherubin et al., "The Epidemiology of Death in Narcotic Addicts," quoted in Brecher, *Licit and Illicit Drugs*, 112.

65. Ibid.

66. A. James Ruttenber, Henry D. Kalter, and Philip Santinga, "The Role of Ethanol Abuse in the Etiology of Heroin-Related Death," *Journal of Forensic Science* 35, no. 4 (1990): 891–900.

67. Ibid.

68. Michael Agar's memoir *Dope Double Agent: The Naked Emperor on Drugs* (Morrisville, NC: Lulubooks, 2006) discusses the sidelining of qualitative social research in National Institute on Drug Abuse.

69. On biomedicalization as a social process, see Adele E. Clarke, Janet K. Shim, Laura Mamo, Jennifer R. Fosket, and Jennifer R. Fishman, *Biomedicalization: Technoscience, Health and Illness in the U.S.* (Durham, NC: University of North Carolina Press, 2010).

70. The study was conducted by Ramon Gardner of the Drug Dependence Clinical Research and Treatment Unit, Bethlem Royal Hospital, Beckenham, Kent, and Maudsley Hospitals in London. See Ramon Gardner, "Methadone Misusage and Death by Overdose," *British Journal of Addiction* 65, no. 2 (1970): 113–118.

71. Ramon Gardner, "Deaths in United Kingdom Opiate Users, 1965-69," *The Lancet* 296, no, 7674 (September 26, 1970): 650–653.

72. Ibid., 651.

73. Ibid.

74. Ibid.

75. Ibid.

76. Ibid., 652.

77. Thomas J. Bewley, "Recent Changes in the Pattern of Drug Abuse in London and the United Kingdom," in Wilson, *The Pharmacological and Epidemiological Aspects of Adolescent Drug Dependence*, 197–220, at 218; see also Thomas H. Bewley and Oved Ben-Arie, "Morbidity and Mortality from Heroin Dependence. 2. Study of 100 Consecutive Inpatients," *British Medical Journal* 1, no. 5594 (1968): 727–730.

78. Helpern, "Causes of Death from Drugs of Dependence," 221.

79. Admissions to the US Narcotics Hospitals, prison-like facilities operated by the US Public Health Service in Lexington, Kentucky, and Fort Worth, Texas, also indexed this increase. These institutions were in transition due to changes in the nation's civil commitment laws in 1966. See Robert W. Rasor, "Narcotic Addiction in Young People in the United States," in Wilson, *The Pharmacological and Epidemiological Aspects of Adolescent Drug Dependence*, 11–26.

80. Darke, *The Life of the Heroin User*, 110.

4   BRINGING OUT THE DEAD

1. Tessie Castillo, "Who Invented Naloxone?," *Huffington Post*, December 15, 2014; updated November 30, 2015, at http://www.huffingtonpost.com/tessie-castillo/meet -jack-fishman-the-man_b_6329512.html; William Yardley, "Jack Fishman, Who Helped Develop a Drug to Treat Overdoses, Dies at 83," *New York Times*, December 15, 2013, 30.

2. Ian Frazier, "The Antidote: Can Staten Island's Middle-Class Neighborhoods Defeat an Overdose Epidemic?," *The New Yorker,* September 8, 2014, https://www .newyorker.com/magazine/2014/09/08/antidote, reprinted in Ian Frazier, *Hogs Wild: Selected Reporting Pieces* (New York: Farrar, Straus, Giroux, 2016), 302–326.

3. According to Walter Sneader, *Drug Discovery: A History* (Chichester, UK: John Wiley and Sons, 2005), 120, naloxone evolved from oxymorphone, a thebaine analog synthesized at the National Institutes of Health by Ulrich Weiss in 1955. Oxycodone is a thebaine analogue (synthesized in 1916 by Freund and Speyer at the University of Frankfurt) that increased the analgesic potency of both morphine and codeine through the addition of a 14-hydroxyl group.

4. Other investigators were also pursuing this line of inquiry: see Max S. Sadove, Reuben C. Balagot, Shigeru Hatano, and Eugene A. Jobgen, "Study of a Narcotic Antagonist—N-Allyl-Noroxymorphone," *Journal of the American Medical Association* 183, no. 8 (1963): 666–668.

5. Eugene Garfield, "Chemical Information Entrepreneurship: A Personal Odyssey," Joseph Priestley Society Symposium on "Knowledge: Our Competitive Weapon" (Philadelphia: Chemical Heritage Foundation, December 9, 2004), 123. Garfield founded the Institute for Scientific Information (basis for Web of Science). Well versed in chemical history and pharmacology, Garfield began his career at the Philadelphia-based pharmaceutical company Smith, Kline & French, for whom he created his first information retrieval system. Garfield wrote columns in *Current*

Contents discussing highly cited research and those who produced it. At the time of his piece on Blumberg and Fishman, who won the John Scott Award in 1983, Garfield sat on the award committee and was highly qualified to outline the priority dispute.

6. Nancy D. Campbell and Anne E. Lovell, "The History of the Development of Buprenorphine as an Addiction Therapeutics," *Annals of the New York Academy of Science* 1248 (2012): 124–139.

7. US Patent 3,254,088.31, May 1966.

8. Garfield, "Chemical Information Entrepreneurship," 125.

9. Some attempted to take clinical advantage of its abrupt precipitation of withdrawal, evolving rapid detoxification methods: see Richard B. Resnick, Richard S. Kestenbaum, Arnold M. Washton, and Doris Poole, "Naloxone-Precipitated Withdrawal: A Method for Rapid Induction onto Naltrexone," *Clinical Pharmacology and Therapeutics* 21 (1977): 409–413.

10. In 1969, DuPont purchased the smaller company and expanded Endo's line of analgesics in hopes of boosting its own pharmaceutical profile. Squabbling among research groups slowed development of new drugs at Endo, which was renamed DuPont Pharmaceuticals in 1982. In 1990, DuPont Pharmaceuticals and Merck formed a joint venture known as DuPont Merck Pharmaceuticals Company. In 1994 Endo reemerged as Endo Laboratories, LLC, the genetics division of DuPont Merck, which renamed the company Endo Pharmaceuticals in 1997. See the DuPont Heritage Timeline at http://www2.dupont.com/Phoenix_Heritage/en_US/1969_b_detail .html.

11. Hans Kosterlitz and A. J. Watt, "Kinetic Parameters of Narcotic Agonists and Antagonists, with Particular Reference to N-Allylnoroxymorphone (Naloxone)," *British Journal of Pharmacological Chemotherapeutics* 33, no. 2 (1968): 266–276.

12. Martin, "Opioid Antagonists," 465.

13. Cozzens, *Social Control and Multiple Discovery*; Pert and Snyder, "Opiate Receptor"; Candace B. Pert and Solomon H. Snyder, "Properties of Opiate-Receptor Binding in Rat Brain," *Proceedings of the National Academy of Sciences of the United States of America* 70, no. 8 (1973): 2243–2247.

14. Garfield, "Chemical Information Entrepreneurship," 127.

15. Kirsten D. Lee, Frederick H. Lovejoy, and James E. Haddow, "Childhood Methadone Intoxication: Be on the Alert!," *Clinical Pediatrics* 13, no. 1 (1974): 66–68. These authors recommended administering naloxone/Narcan intravenously at a dose of 0.01 mg/kg every three to four hours. However, all but two of the fifteen Boston-area children's hospital cases of accidental methadone ingestion on which they reported had been treated with nalorphine or ipecac to induce vomiting. Naloxone was still unavailable between September 1970 and February 1972 in many metropolitan areas.

16. Louis Lasagna, "Historical Introduction," in Committee on Problems of Drug Dependence (CPDD), *Proceedings of the Agonist-Antagonist Symposium*, February 14–15, 1983 (unpublished), 8.

17. Donald R. Jasinski, William R. Martin, and Charles Haertzen, "The Human Pharmacology and Abuse Potential of N-Allylnoroxymorphone (Naloxone)," *Journal of Pharmacology and Experimental Therapeutics* 167, no. 2 (1967): 420–426.

18. Frances F. Foldes, Deryck Duncalf, and Shigeo Kuwabara, "The Respiratory, Circulatory, and Narcotic Antagonist Effects of Nalorphine, Levallorphan, and Naloxone in Anesthetized Subjects," *Canadian Anaesthesiology Society Journal* 16, no. 2 (1969): 151–161, quotation at 160. See also Francis F. Foldes, George M. Davidson, Deryck Duncalf, Shigeo Kuwabara, and Ephraime S. Siker, "The Respiratory, Circulatory, and Analgesic Effects of Naloxone-Narcotic Mixtures in Anaesthetized Subjects," *Canadian Journal of Anesthesia/Journal canadien d'anesthésie* 12, no. 6 (1965): 608–621.

19. Daniel Sledge and George Mohler, "Eliminating Malaria in the American South: An Analysis of the Decline of Malaria in 1930s Alabama," *American Journal of Public Health* 103, no. 8 (2013): 1381–1392.

20. Milton Helpern with Bernard Knight, *Autopsy: The Memoirs of Milton Helpern* (New York: St. Martin's Press, 1977), 67.

21. Ibid., 69.

22. Ibid., 66–71.

23. Ibid.; see also Mara Keire, "Dope Fiends and Degenerates: The Gendering of Addiction in the Early Twentieth Century," *Journal of Social History* 31, no. 4 (1998): 809–822; George Chauncey, *Gay New York: Gender, Urban Culture, and the Making of the Gay Male World, 1890–1940* (New York: Basic Books, 1995).

24. Helpern, *Autopsy*, 69.

25. Ibid., 73.

26. Ibid., 70.

27. Milton Helpern, "Causes of Death from Drugs of Dependence," in *The Pharmacological and Epidemiological Aspects of Adolescent Drug Dependence: Proceedings of the Society for the Study of Addiction*, ed. Cedric William Malcolm Wilson (Oxford: Pergamon Press, 1968), 221–233, quotation at 222.

28. Helpern, *Autopsy*, 71.

29. Ibid., 72.

30. Ibid., 75.

31. Ibid., 74. Talks and slides can be found at https://www.youtube.com/watch?v=csKZq65z3y8.

32. Paul L. Montgomery, "Coroners Worry about Addiction: Narcotics-Related Deaths Rise Sharply Here," *New York Times*, October 20, 1969.

33. Helpern, *Autopsy*, 45–46; Helpern disliked the term "morgue," preferring "mortuary" or "office."

34. "Death of a Child," *New York Times*, December 21, 1969.

35. Barbara Campbell, "Boy, 12, Dies of Heroin Dose in Harlem Bathroom," *New York Times*, December 16, 1969. For analysis of this and the Dishman case discussed

below, see Anthony Sole, "Black Tar/White Powder: Race, Class, Gender, and Heroin in New York and San Francisco, 1966–77" (Master of Arts thesis, University of Buffalo, 2015), https://arts-sciences.buffalo.edu/content/dam/arts-sciences/history/documents/Anthony-Sole-Thesis-Final-Copy.pdf. The Vandermeer case is also discussed in Michael J. Fortner, *Black Silent Majority: The Rockefeller Drug Laws and the Politics of Punishment* (Cambridge, MA: Harvard University Press, 2015).

36. I am grateful to Hannah Lloyd, who volunteered as a research assistant with me in summer 2017, for finding this case and associated newspaper coverage.

37. Paul L. Montgomery, "Barnard Mourns a Heroin Victim," *New York Times*, February 2, 1970.

38. Thomas F. Brady, "St. Luke's Yields on Drug Facility," *New York Times*, January 18, 1970, 56. The location of the sit-in was the Community Psychiatry Division. See "College Student, 17, Dies of Heroin Overdose at Party," *Columbia Daily Spectator*, February 5, 1970, 1, 3. On black activism on drug issues, see Roberts, "'Rehabilitation' as Boundary Object."

39. "Dope Death Shocks Girl's Friends Here," *Chicago Tribune*, February 2, 1970, 7.

40. "Pretty, Bright Young Militant Dead at 17," *Morgantown Post*, February 4, 1970, 9-A.

41. "Pretty Antoinette is Morgue Case No. 483," *Salinas Journal*, February 2, 1970, 1–2.

42. Barbara Campbell, "West Side Marchers Seek More Addict Facilities," *New York Times*, March 24, 1970, 30. The first African American woman hired through a *New York Times* reporter trainee program, Campbell was later nominated for the Pulitzer Prize for reporting on problems facing the black community.

43. Joseph O. Haff, "Drugs Kill Youth; 34th Here in 1970," *New York Times*, February 22, 1970. According to the article, overdose was the leading cause of death in New York City for people ages 14 to 35.

44. As Kohler-Hausman pointed out in "The 'Attila the Hun Law,'" targeting enforcement on "pushers" effectively criminalized the poor who became involved in the drug trade.

45. Thomas A. Johnson, "A 'Heroin Epidemic' Hits Schools," *New York Times*, February 16, 1970, 1, 32.

46. Ibid.

47. Benita Jones, "Addicts Rehabilitation Center: A Haven in Harlem," https://www.youtube.com/watch?v=60tELA2FX60. Allen opposed methadone maintenance and was sought out by reporters seeking to "balance" methadone enthusiasts.

48. George Gent, "Stars' Drug Deaths Stir Rock Scene," *New York Times*, November 3, 1970.

49. Rickie Solinger, *Wake Up Little Susie: Single Pregnancy and Race before Roe v. Wade* (New York: Routledge, 2000).

50. Johnson, "A 'Heroin Epidemic' Hits Schools."

51. Ibid.

52. Ibid.

53. "Child Addicts Depict Horrors of Drug Abuse to Pupils Who Won Anti-Narcotics Poster Contest Here," *New York Times*, March 22, 1970. The "child addicts" resided at Odyssey House, directed by Judianne Densen-Gerber, then-wife of medical examiner Michael Baden, who made it a policy to gain custody of all children treated in her program as a condition of participation.

54. F. E. Camps, "The Pathological Aspects of Drug Dependence," in Wilson, *The Pharmacological and Epidemiological Aspects of Adolescent Drug Dependence*, 235–237.

55. Helpern, "Causes of Death from Drugs of Dependence."

56. Ibid., 226.

57. Ibid., 224.

58. Ibid., 225–226.

59. Markham directed the White House Conference on Narcotics and Drug Abuse in 1962; was executive director of John F. Kennedy's Presidential Advisory Commission on Narcotic and Drug Abuse in 1963; and was a member of the US delegation to the United Nations Commission on Narcotics and Drugs.

60. Dean F. Markham, "Epidemiological Aspects of Adolescent Drug Dependence in the United States," in Wilson, *The Pharmacological and Epidemiological Aspects of Adolescent Drug Dependence*, 187–196, at 189.

61. "Heroin Hits the Young," *Time*, March 16, 1970, 11–15. For further analysis of media coverage of drug issues in this era, including addiction to prescription drugs, see Martin Halliwell, *Voices of Mental Health: Medicine, Politics, and American Culture, 1970–2000* (New Brunswick, NJ: Rutgers University Press, 2017), 72–96.

62. Helpern, *Autopsy*, 76.

63. Ibid.

64. Jan D. Hasbrouck, "The Antagonism of Morphine Anesthesia," *Anesthesia and Analgesia* 50, no. 6 (1971): 954–959.

65. Approximately 1 mg/60 kg or 0.017 mg/kg.

66. Monkeys exhibited convulsions at 60 mg/kg for thirty days; dogs experienced hind limb weakness after 4 mg/kg intravenously for fourteen days. Reproductive studies in rats at 100 and 500 times the proposed human dose failed to show significant effects on reproductive performance or offspring. See the summary of the basis for approval of NDA 16–636, June 3, 1969, 24.

67. Patients were given mixtures of oxymorphone 1.5 mg and naloxone 0.05 to 0.4 mg/cc in 2,111 subjects; 778 were pre- and postoperative patients; 1,333 were pregnant and in labor.

68. Max Fink, "EEG and Human Psychopharmacology," *Annual Review of Pharmacology* 9 (1969): 241–258.

69. Electropharmacologists explored whether the hallucinogenic effects of psychotomimetic agents might be based in electrophysiology, developing as they did so new technologies for studying brain function. As Mary A. B. Brazier put it in an early review, it was a "matter of astonishment that so many of the brain's secrets escape across the wall of the skull to electrodes fixed to the scalp of man. That they do so is testimony to the fact that the brain's electrical activity is a most sensitive indicator of its function" (Mary A. B. Brazier, ed., "Computer Techniques in EEG Analysis," in *Electroencephalography and Clinical Neurophysiology*, suppl. 20 [Amsterdam: Elsevier, 1961], 15, quoted in Fink, "EEG and Human Psychopharmacology," 242).

70. Max Fink, interview by Jonathan Cole, American College of Neuropsychopharmacology, San Juan, Puerto Rico, December 11, 1995.

71. Ibid., 12. See also Max Fink, "Pharmacoelectroencephalography: A Note on Its History," *Neuropsychobiology* 12, no. 2–3 (1984): 173–178; Max Fink, "Remembering the Lost Neuroscience of Pharmaco-EEG," *Acta Psychiatrica Scandinavica* 121, no. 3 (2010): 161–173.

72. See Max Fink, Arthur Zaks, Robert Sharof, Arnoldo Mora, Alfred Bruner, Stephen Levit, and Alfred M. Freedman, "Naloxone in Heroin Dependence," *Clinical Pharmacology and Therapeutics* 9, no. 5 (1968): 568–577; Max Fink, Arthur Zaks, Jan Volavka, and Jiri Roubicek, "Opiates and Antagonists," in *Narcotic Drugs*, ed. Doris H. Clouet (New York: Plenum, 1971), 452–467; Arthur Zaks, Alfred Bruner, Max Fink, and Alfred M. Freedman, "Diacetylmorphine (Heroin) in Studies of Opiate Dependence," *Diseases of the Nervous System* 30 (1969): 89–92; Jan Volavka, Arthur Zaks, Jiri Roubichek, and Max Fink, "Electrograph Effects of Heroin and Naloxone in Man," *Neuropharmacology* 9 (1970): 587–593.

73. Volavka et al., "Electrograph Effects," 593.

74. Title of an important collection of over 100 hours of tape-recorded, first-person interviews with people at the Henry Street Settlement, which operated a mental hygiene clinic in Manhattan's Lower East Side. The interviews were collected by Ralph Tefferteller and written up by Jeremy Larner as *The Addict in the Street* (New York: Grove Press, 1964).

75. George R. Gay and Darryl S. Inaba, "Treating Acute Heroin and Methadone Toxicity," *Anesthesia and Analgesia* 55, no. 4 (1976): 607–610. Writing about naloxone's clinical use when it should have been in wide use but was not, Gay and Inaba cited the original papers of Harold Blumberg, Hyman B. Dayton, and Peter S. Wolf, "Narcotic Antagonist Activity of Naloxone," *Federation Proceedings (FASEB)* 24, no. 2 (1965): 676, and Harold Blumberg, Hyman B. Dayton, and Peter S. Wolf, "Counteraction of Narcotic Antagonist Analgesics by the Narcotic Antagonist Naloxone," *Proceedings of the Society on Experimental Biology and Medicine* 123 (1966): 755–758. Gay and Inaba's article was first published by Medical Economics, Inc., but is currently unavailable in its original form. However, the editors of *Anesthesia and Analgesia* selected this article for a series on current research published in adjoining disciplines that was deemed particularly relevant for anesthesiologists.

76. Gay and Inaba, "Treating Acute Heroin and Methadone Toxicity."

77. Ibid. Dopram's counterindications were briefly discussed.

78. Ibid.

79. Ibid. Their detoxification regimen consisted of various drugs used to manage withdrawal from the perspective of four different "symptom complexes." They were, in short, thinking pharmacologically in ways almost as sophisticated as their patient population!

80. Ibid.

81. Ibid.

5   UNNATURAL ACCIDENTS

1. Peter Itzen, "Who Is Responsible in Winter? Traffic Accidents, the Fight against Hazardous Weather and the Role of Law in a History of Risks," *Historical Social Research* 41, no. 1 (2016): 154–175.

2. Thomas Schlich, "Trauma Surgery and Traffic Policy in Germany in the 1930s: A Case Study in the Co-Evolution of Modern Surgery and Society," *Bulletin of the History of Medicine* 90 (2006): 73–94.

3. Itzen, "Who Is Responsible in Winter?," 158.

4. Scholars of biopolitics, "postgenomics," and "biological citizenship" have traced how mental health status, serostatus, genetic profiles, and biogenetic markers have crept into the calculus of the moral worth of persons, their legal and political standing, and citizenship status. See Adriana Petryna, *Life Exposed: Biological Citizens after Chernobyl* (Princeton, NJ: Princeton University Press, 2002); Jenny Reardon, *The Postgenomic Condition: Ethics, Justice, and Knowledge after the Genome* (Chicago: University of Chicago Press, 2017); Nikolas Rose, *The Politics of Life Itself: Biomedicine, Power, and Subjectivity in the Twenty-First Century* (Princeton, NJ: Princeton University Press, 2006). Suzanne Fraser and Kylie Valentine's *Substance and Substitution: Methadone Subjects in Liberal Societies* (Basingstoke, UK: Palgrave Macmillan, 2008), addresses how methadone maintenance therapy constitutes its subjects in relation to categories of biological personhood as well as social identities.

5. Charles Perrow, *Normal Accidents: Living with High-Risk Technologies* (Princeton, NJ: Princeton University Press, 1999).

6. Peter Itzen, "Death on the Streets: Preliminary Thoughts on a Project on the History of Traffic Accidents in Germany, 1920–1980" (paper presented at the Freiburg Institute of Advanced Study [FRIAS], Freiburg, Germany, May 29, 2014).

7. As of September 9, 2016, this was reported in England, according to Ian Hamilton and Mark Monaghan: http://theconversation.com/drugs-fatalities-overtake-car -fatalities-for-the-first-time-64972. There were 1,989 total drug-related deaths in England as opposed to 1,732 traffic fatalities.

8. Johannes M. Just, Markus Bleckwenn, Rieke Schnakenberg, Philipp Skatuall, and Klaus Weckbecker, "Drug-Related Celebrity Deaths: A Cross-Sectional Study,"

Substance Abuse Treatment, Prevention, and Policy 11 (2016): 40; Kelly Ray Knight, "Women on the Edge: Opioids, Benzodiazepines and the Social Anxieties Surrounding Women's Reproduction in the US Opioid Epidemic," *Contemporary Drug Problems* 44, no. 4 (2017): 301–320; Rachel L. Shaw, Claire Whitehead, and David C. Giles, "Crack Down on Celebrity Junkies: Does Media Coverage of Celebrity Drug Use Pose a Risk to Young People?," *Health, Risk, and Society* 12, no. 6 (2010): 575–589; Michelle Wood, "Media's Positive and Negative Frames in Reporting Celebrity Deaths from Illegal Drug Overdoses versus Prescription Medication Overdoses" (Unpublished master's thesis, University of Kansas, 2011).

9. Robert E. Suter, "A History of Emergency Medicine in the United States," *World Journal of Emergency Medicine* 3, no. 1 (2012): 5–10.

10. Similarly, the history of Automated External Defibrillators was one of fits and starts, relying on studies of the use of electricity to defibrillate by Negovsky, who was working in the former Soviet Union and is known as the "father of 'reanimatology,'" along with Hooker and Kouwenhouen, who in studying electrical accidents showed that it was possible for defibrillation to occur in intact, live humans. Jonas A. Cooper, Joel D. Cooper, and Joshua M. Cooper, "Cardiopulmonary Resuscitation: History, Current Practice, and Future Direction," *Contemporary Reviews in Cardiovascular Medicine* 114 (2006): 2839–2849.

11. Stefan Timmermans, *Sudden Death and the Myth of CPR* (Philadelphia: Temple University Press, 1999), 48–49.

12. Ibid., 47n50, 219.

13. Ibid., 46–49.

14. Ibid., 87–89. "CPR for all" is Timmermans's provocative chapter title. "The strategy for universal CPR rests partly on deception and carries serious risks. In absolute numbers, it is possible that more lives are saved with the current emergency system than without the emergency infrastructure, but the relative numbers of lives saved remains disappointingly low." He attributed this state of affairs to unrealistic expectations in meeting the realities of sudden death. Naloxone was introduced to shift that calculus; it is rarely ineffective if administered in a timely fashion. Naloxone is less implicated in anoxia, brain damage, and death caused by other resuscitation methods. However, as more US states liberalized naloxone access, more news stories about brain damage from anoxia appeared.

15. According to the American Heart Association (2010), the current protocol is Chest compressions-Airway-Breathing (CAB), rather than the initial Airway-Breathing-Chest compressions (ABC).

16. David S. Leighninger, "Contributions of Claude Beck," in *Advances in Cardiopulmonary Resuscitation*, ed. Peter Elam and James O. Elam (New York: Springer-Verlag, 1975), 259–262.

17. At the time of writing, intranasal naloxone was far more expensive than injectable forms.

18. Most histories of emergency medicine mention the television shows "Emergency" and "ER."

19. Nancy L. Caroline, *Emergency Care in the Streets* (Boston: Little, Brown, 1979), 350. In addition, and in concert with Safar, Caroline also authored the first training standards for emergency medical technicians.

20. Ibid., 350.

21. Ibid., 351.

22. Ibid.

23. Ibid.

24. The history of Freedom House appeared in the introduction to the first edition of *Emergency Care in the Streets*, x. The "nation's first pilot project of national guidelines for ambulance design and equipment and ambulance attendant's training" ended the same year Caroline's book came out. Professionalization of the emergency medical services (EMS) system in Pennsylvania turned it largely white; few of the pioneering Freedom House African American paramedics made it into municipal ambulance service. One who did was John Moon, who as head of Pittsburgh City EMS joined with other stakeholders in the task force that first brought naloxone to Prevention Point Pittsburgh's needle exchange in 2005. See Alice Bell, interview by Nancy D. Campbell, Pittsburgh, Pennsylvania, May 3, 2017, 4; Ron Johnson, interview by Nancy D. Campbell, Pittsburgh, Pennsylvania, May 3, 2017; Alex Bennett, interview by Nancy D. Campbell, New York City, October 4, 2013, 27.

25. Peter Safar and Pat Sands, *University of Pittsburgh Department of Anesthesiology and Critical Care Medicine: The First 15 Years, 1961–1976* (Pittsburgh: University of Pittsburgh, 1976); Andrew T. Simpson, "Transporting Lazarus: Physicians, the State, and the Creation of the Modern Paramedic and Ambulance, 1955–73," *Journal of the History of Medicine and Allied Sciences* 68, no. 2 (2013): 163–197; Naomi Braine, Caroline J. Acker, Cullen Goldblatt, Huso Yi, Samuel Friedman, and Don C. DesJarlais, "Neighborhood History as a Factor Shaping Syringe Distribution Networks among Drug Users at a U.S. Syringe Exchange," *Social Networks* 30, no. 3 (2008): 235–246.

26. Ron Johnson, interview by Campbell, May 3, 2017, 3.

27. Ibid, 4.

28. Thomas Kent, "New Test Drug Robbing Heroin, Opiates of 'High,'" *Manchester Journal-Inquirer*, November 23, 1972, 4.

29. Jack Fishman, E. F. Hahn, B. I. Norton, A. Ronal, and Frances F. Foldes, "Preparation and Evaluation of a Sustained Naloxone Delivery System in Rats," *Pharmacology* 13 (1975): 513–519; Hyman B. Dayton and Harold Blumberg, "Prolongation of Narcotic Antagonist Action in Animals with Naloxone Pamoate and Naltrexone Pamoate Suspensions in Oil," *Federation Proceedings* 32 (1973): 693; Jack Fishman, Howard Roffwarg, and Leon Hellmann, "Disposition of Naloxone-7,8–3H in Normal and Narcotic-Dependent Men," *Journal of Pharmacology and Experimental Therapeutics* 187 (1973): 575–580.

30. Arthur Zaks, Thelma Jones, Max Fink, and Alfred M. Freedman, "Naloxone Treatment of Opiate Dependence: A Progress Report," *Journal of the American Medical Association* 215, no. 13 (1971): 2108–2110. If naloxone was to be made suitable for

"prophylaxis of opiate dependence," a long-acting formulation was necessary. See also Alfred M. Freedman, Arthur Zaks, Richard Resnick, and Max Fink, "Blockade with Methadone, Cyclazocine, and Naloxone," *International Journal of the Addictions* 5, no. 3 (1970): 507–515; Albert A. Kurland, Thomas E. Hanlon, and O. Lee McCabe, "Naloxone and the Narcotic Abuser: A Controlled Study of Partial Blockade," *International Journal of the Addictions* 9, no. 5 (1974): 663–672. These studies relied on heroin supplied to the researchers by the Bureau of Narcotics and Dangerous Drugs.

31. "Another Heroin Cure Gets a Shot in the Arm," *Science News* 102, no. 3 (July 15, 1972), 38.

32. Ibid.

33. Ibid.

34. Ibid.

35. Abraham Wikler, "Dynamics of Drug Dependence: Implications of a Conditioning Theory for Research and Treatment," *Archives of General Psychiatry* 28, no. 5 (1973): 611–616; Abraham Wikler and Frank T. Pescor, "Classical Conditioning of a Morphine Abstinence Phenomenon, Reinforcement of Opioid-Drinking Behavior and 'Relapse' in Morphine-Addicted Rats," *Psychopharmacologia* 10, no. 3 (1967): 255–284; Charles P. O'Brien, Thomas Testa, Thomas J. O'Brien, and Robert Greenstein, "Conditioned Narcotic Withdrawal in Humans," *Science* 195, no. 4282 (1977): 1000–1002; A. Thomas McLellan, Anna Rose Childress, Ronald Ehrman, Charles P. O'Brien, and Steven Pashko, "Extinguishing Conditioned Responses during Opiate Dependence Treatment Turning Laboratory Findings into Clinical Procedures," *Journal of Substance Abuse Treatment* 3, no. 1 (1986): 33–40.

36. "New Drug Holds Out Hope for Addicts," *Morgantown Post*, May 25, 1972, 4C.

37. Regina Whittington, "Overdose Records Show Problem Serious Here," *High Point Enterprise*, August 19, 1973.

38. Jordan Green, "Heroin Addiction: A Tour of Hell," *Triad City Beat*, January 21, 2015.

39. Bill Duncan, "Drugs—Overdoses Kill More than Traffic Accidents," *Independent Press-Telegram*, September 26, 1971, A-1; Evans Witt, "Painkiller Darvon May Be Further Restricted," *Chillicothe Constitution-Tribune*, March 3, 1976, 4.

40. See "Chelmsford Residents Opposing SHARE," and Nick Caragenas, "Three Die from Overdoses in Area in Past Two Weeks," *Lowell Sunday Sun*, January 28, 1973, B3.

41. Pam Jahnke, "Recent Cases Here Highlight Drug Overdosing by Juveniles," *Hillsdale Daily News*, April 20, 1979, 2.

42. Propoxyphene and related products were not removed from the US market until 2010 despite Public Citizen campaigns in 1978 and 2006. Deaths from overdose of the drug came to the attention of the North Carolina State Medical Examiner's Office; see Page Hudson, Michael Barringer, and Arthur J. McBay, "Fatal Poisoning with Propoxyphene: Report from 100 Consecutive Cases," *Southern Medical Journal* 70, no. 8 (1977): 938–942.

43. This social science unit produced careful accounts of local variations within ecological pockets of heroin use, arguing that national prevalence rates led to a distorted picture. These sociologists thus criticized overreliance on knowledge about opioid addiction that was generated at the federal narcotic farms.

44. Irving Lukoff, *The Epidemiology of Heroin and Other Drugs,* ed. Joan Dunne Rittenhouse, NIDA Research Monograph 16, 1977, 196.

45. Roberts, "'Rehabilitation' as Boundary Object," 148.

46. Ibid., 165.

47. Lee N. Robins and George T. Murphy, "Drug Use in a Normal Population of Young Negro Men," *American Journal of Public Health* 57, no. 9 (1967): 1580–1596; Dan Waldorf, "Life without Heroin: Some Social Adjustments during Long-Term Periods of Voluntary Abstention," *Social Problems* 18, no. 2 (1970): 228–243; Dan Waldorf and Patrick Biernacki, "Natural Recovery from Heroin Addiction: A Review of the Incidence Literature," *Journal of Drug Issues* 9, no. 2 (1979): 281–289.

48. David N. Nurco, Arthur J. Bonito, Monroe Lerner, and Mitchell B. Balter, "Studying Addicts over Time: Methodology and Preliminary Findings," *American Journal of Drug and Alcohol Abuse* 2, no. 2 (1975): 183–196; Robins, *Follow-Up of Vietnam Drug Users*; Lee N. Robins, *The Vietnam Drug User Returns*, Special Action Office Monograph, Series A, no. 2 (Washington, DC: US GPO, 1974); Lee N. Robins, Darlene H. Davis, and David N. Nurco, "How Permanent Was Vietnam Drug Addiction?," *American Journal of Public Health* 64 (1974): 38–43.

49. According to historian Jerry Kuzmarov's book *The Myth of the Addicted Army: Vietnam and the Modern War on Drugs* (Amherst: University of Massachusetts Press, 2009), the Nixon White House created the myth to justify expanding the War on Drugs. On Nixon's drug policy activities, see Mical Raz, "Treating Addiction or Reducing Crime? Methadone Maintenance and Drug Policy under the Nixon Administration," *Journal of Policy History* 29, no. 1 (2017): 58–86.

50. Quoting Audrey Halliday, director of a methadone program in San Diego, California.

51. Jack Webb, "Illegal Methadone Kills Hundreds of Americans." Syndicated by the Copley News Service, the story went national in the summer of 1974. See the *Capital-News* (Jefferson City, MO), August 28, 1974, and the *Daily Gazette* (Sterling, IL), July 19, 1974.

52. *Drugs and Death*, ed. Patricia Ferguson, Thomas Lennox, and Dan J. Lettieri (Rockville, MD: National Institute on Drug Abuse, 1974, DHEW Publication No. [ADM] 75-188), https://archive.org/stream/drugsdeathnonmed00docu/drugsdeath nonmed00docu_djvu.txt.

53. Center for Substance Abuse Treatment (CSAT), *Methadone-Associated Mortality: Report of a National Assessment*, CSAT Publication No. 28-03 (Rockville, MD: Center for Substance Abuse Treatment, Substance Abuse and Mental Health Services Administration, SAMHSA Publication No. 04-3904, 2003), 11–12. Appendix 4 contained a valuable table of national data indicating all known cases of "Past Methadone-Associated Mortality" up to 2002 on pages 52–55.

54. Ibid.

55. Edward J. Cone and Bruce A. Goldberger, "Pharmacokinetics and Tissue Distribution of Opiates," in *Drug Abuse in the Decade of the Brain*, ed. Gabriel G. Nahas and Thomas F. Burks (Amsterdam, The Netherlands: IOS Press, 1997), 71–77.

56. See Justin Wade Hubbard, "'You Can't Fool the P Test': American Science, the Department of Defense, and the Unlikely Invention of the War on Drugs, 1945–1980" (PhD diss., Vanderbilt University, 2019). The gas-liquid chromatograph became the gold standard device for urine testing. Hubbard explored the 1965 establishment of a Registry of Tissue Reactions to Drugs (RTRD) to study adverse drug reactions, including overdose, at an Armed Forces Institute of Pathology laboratory under direction of Leo Goldbaum. The RTRD acquired more than 2,700 tissue samples from cases of suspected drug poisoning. A 1969 report on thirty cases of "sudden death in narcotic addicts" described the difficulty of determining cause of death form tissue samples given the very low concentrations of morphine in them. See Edward Johnston, Leo Goldbaum, and Richard Whelton, "Investigation of Sudden Death in Addicts: With Emphasis on the Toxicologic Findings in Thirty Cases," *Medical Annals of the District of Columbia* 38, no. 7 (1969): 375–380.

57. Randall C. Baselt, Donna J. Allison, James A. Wright, James R. Scannell, and Boyd G. Stephens, "Acute Heroin Fatalities in San Francisco: Demographic and Toxicologic Characteristics," *Western Journal of Medicine* 122 (1975): 455–458.

58. Ibid., 458.

59. Ibid.

60. Louis A. Gottschalk, "Problems in Data Acquisition: Deaths," in Richards and Blevens, *The Epidemiology of Drug Use*, 98–122.

61. Ibid., 112.

62. Ibid., 118.

63. Ibid., 122.

64. Michael Alexander, "Indicators of Drug Abuse—Hepatitis," in *The Epidemiology of Drug Use*, ed. Louise G. Richards and Louise B. Blevens, NIDA Research Monograph 10, 1970, 123–125; see also Mark H. Greene, Stuart L. Nightingale, and Robert L. DuPont, "Evolving Patterns of Drug Use," *Annals of Internal Medicine* 83 (1975): 402–411; Lee Minichiello, *An Examination of Trends in Intravenous Drug Use Reflected by Hepatitis and DAWN Reporting Systems* (Washington, DC: Report to the Drug Enforcement Administration by the Institute for Defense Analysis, 1974); R. A. Garibaldi, B. Hanson, and M. B. Gregg, "Impact of Illicit Drug-Associated Hepatitis on Viral Hepatitis Morbidity Reporting in the United States," *Journal of Infectious Diseases* 126 (1972): 288–293.

65. Ibid., 124.

66. Discussion, in *The Epidemiology of Drug Use*, 126.

67. Ibid., 127.

68. Ibid., 128. Minichiello also produced internal reports: "Supply, Distribution, and Usage Patterns of Drugs of Abuse" and "Indicators of Intravenous Drug Use in the United States."

69. Mark H. Greene, "Applications of Indicator Data (Epidemics)," in Richards and Blevens, *The Epidemiology of Drug Use*, 206–218, quotation at 206.

70. Ibid.

71. Ibid., 210.

72. Ibid.

73. Ibid., 212.

74. Ibid., 218.

75. Sidney Schnoll, interview by Nancy D. Campbell, San Juan, Puerto Rico, June 19, 2008, 26.

6 ADOPTING HARM REDUCTION

1. Heather Edney and Emily E. Ager, "User-Driven Overdose Prevention," *Harm Reduction Communication* 14 (2002): 10–12.

2. Phil Brown, Stephen Zavestoski, Sabrina McCormick, Brian Mayer, Rachel Morello-Frosch, and Rebecca Gasior, "Embodied Health Movements: Uncharted Territory in Social Movement Research," *Sociology of Health and Illness* 26 (2004): 1–31; Phil Brown, Stephen Zavestoski, Sabrina McCormick, Brian Mayer, Rachel Morello-Frosch, and Rebecca Gasior Altman, "Embodied Health Movements: New Approaches in Health," *Sociology of Health and Illness* 26 (2004): 50–80.

3. Forum Droghe, *Preventing Opioid Overdose Deaths: A Research on the Italian THN Model* (2016), 22.

4. Forum Droghe (FD) relied on funding from OSF, Eclectica, and harm reduction–oriented Public Addiction Departments (Azienda Sanitaria Locale (ASL) 2 and 3 in Turin and ASL 1 in Naples). Participation was sought from organizations inclusive of people who use drugs (PWUDs) such as the Italian Network for Harm Reduction, the National Coordination for Residential Centers/Communities (CNCA), Isola di Arran, and IndifferenceBusters. Authored by Susanna Ronconi (FD), Franca Beccaria (Eclectica), Antonella Camposeragna (FD), Angelo Giglio (Addiction Department, ASL ex Torino 2), Paolo Jarre (Addiction Department, ASL Torino 3), Sara Rolando (Eclectica), Stefano Vecchio (Addiction Department, ASL Napoli 1), and Grazia Zuffa (FD), with contributions from Paolo Nencini, the report was translated into English by Elizabeth O'Neill.

5. John Strang, Shane Darke, Wayne Hall, Michael Farrell, and Robert Ali, "Heroin Overdose: The Case for Take-Home Naloxone," *British Medical Journal* 312 (1996): 1435–1436; Bruno Simini, "Naloxone Supplied to Italian Heroin Addicts," *The Lancet* 352 (1998): 967.

6. Forum Droghe, *Preventing Opioid Overdose Deaths*, 54.

7. Ibid.

8. Ibid.

9. Stephen Hinchliffe, Nick Bingham, John Allen, and Simon Carter, *Pathological Lives: Disease, Space, and Biopolitics* (Hoboken, NJ: John Wiley and Sons, 2016).

10. Marina Davoli, Carlo A. Perucci, Francesco Forastiere, Pat Doyle, Elisabetta Rapiti, Mauro Zacarelli, and Damiano D. Abeni, "Risk Factors for Overdose Mortality: A Case-Control Study within a Cohort of Intravenous Drug Users," *International Journal of Epidemiology* 22, no. 2 (1993): 273–277; Antonio Preti, Paola Miotto, and Monica De Coppi, "Deaths by Unintentional Illicit Drug Overdose in Italy, 1984–2000," *Drug and Alcohol Dependence* 66, no. 3 (2002): 275–282.

11. Forum Droghe, *Preventing Opioid Overdose Deaths*.

12. Vololona Rabeharisoa and Michel Callon, "Patients and Scientists in French Muscular Dystrophy Research," in *States of Knowledge: The Co-Production of Science and Social Order*, ed. Sheila Jasanoff (New York: Routledge, 2004), 142–160, quotation at 142.

13. Ibid., 157.

14. Ibid., 159.

15. Ibid., suggesting that "reflexiveness is at the core of organizations that are capable of mobilizing resources for redefining their own identity, by entering into sustained dialogue with their environment, and inventing and implementing new forms of co-production."

16. Sheila Jasanoff, "Science and Democracy," in *The Handbook of Science and Technology Studies*, 4th ed., ed. Ulrike Felt, Rayvon Fouche, Clark A. Miller, and Laurel Smith-Doerr (Cambridge, MA: MIT Press, 2017), 259–288, quotation at 272.

17. I pose these in the spirit of feminist standpoint epistemology, which has made the social locations of knower and known central for knowledge projects. Given that harm reduction is a source for new knowledges, feminist standpoint epistemology and intersectional analysis are crucial. See Patricia Hill Collins and Sirma Bilge, *Intersectionality: Key Concepts* (Cambridge: Polity Press, 2016).

18. For a most evocative account of people and place, see Garcia, *The Pastoral Clinic*.

19. Philip Fiuty, telephone interview by Nancy D. Campbell, November 10, 2013, 2.

20. Ibid.

21. Ibid.

22. Ibid., 5.

23. Ibid., 3.

24. Ibid., 3–4.

25. Ibid., 5.

26. Ibid., 4.

27. Ibid., 7.

28. Stefan Timmermans and Marc Berg, *The Gold Standard: The Challenge of Evidence-Based Medicine and the Standardization of Health Care* (Philadelphia: Temple University Press, 2003).

29. Vololona Rabeharisoa, Tiago Moreirab, and Madeleine Akricha, "Evidence-Based Activism: Patients', Users' and Activists' Groups in Knowledge Society," *BioSocieties* 9, no. 2 (2014): 111–128.

30. A voluntary sector organization from 1971 to 2017, Lifeline delivered a more holistic treatment approach to what was then called "risk reduction."

31. Portions were published as John Strang, "Drug Use and Harm Reduction: Responding to the Challenge," in *Psychoactive Drugs and Harm Reduction: From Faith to Science*, ed. Nick Heather, Alex Wodak, Ethan Nadelmann, and Pat O'Hare (London: Whurr Publishers, 1993), 3–20.

32. John Strang and Michael Farrell, "Harm Minimisation for Drug Users: When Second Best May Be Best First," *British Medical Journal* 304 (1992): 1128–1129, quotation at 1128.

33. In making such arguments, Strang joined Russell Newcombe, "The Reduction of 'Drug-Related' Harm: A Conceptual Framework for Theory, Practice and Research," in *The Reduction of Drug-Related Harm*, ed. Patrick O'Hare, Russell Newcombe, Alan Mathews, Ernst Buning, and Ernest Drucker (London: Routledge, 1992), 1–14.

34. Strang and Farrell, "Harm Minimisation for Drug Users," 1128.

35. Danielle Giffort, "The Rip Van Winkle Science: Credibility Struggles in the Psychedelic Renaissance" (PhD diss., University of Illinois at Chicago, 2015), showed how scientists who studied psychedelics were portrayed as transgressive and therefore not respectable. "Impure" psychonaut Timothy Leary contrasted to the "sober" positivism pervading Western conceptions of science. Today's psychedelic researchers face the daunting task of integrating impure and sober elements. Harm reductionists face similar dynamics as evidence-based policy contexts change the stakes and promote ever-more sober science.

36. John Strang, "Drug Use and Harm Reduction: Responding to the Challenge," in *Psychoactive Drugs and Harm Reduction: From Faith to Science*, ed. Nick Heather, Alex Wodak, Ethan A. Nadelmann, and Pat O'Hare (London: Whurr Publishers, 1993), 3–20.

37. Strang and Farrell, "Harm Minimisation for Drug Users," 1128.

38. Strang, "Drug Use and Harm Reduction: Responding to the Challenge," 17.

39. John Strang, interview by Nancy D. Campbell, London, March 7, 2017, 8.

40. Strang and Farrell, "Harm Minimisation for Drug Users"; Strang, "Drug Use and Harm Reduction: Responding to the Challenge."

41. Strang et al., "Heroin Overdose: The Case for Take-Home Naloxone." In a personal communication on August 27, 2018, Strang pointed out that the term "home-based naloxone" appeared in the body of the article and speculated that editorial staff had put "take-home naloxone" in the title.

42. Ibid.

43. Founded as part of the Australian government's late 1980s drug strategy, the National Drug and Alcohol Research Centre is a WHO Collaborating Centre in the Treatment and Prevention of Drug and Alcohol. The Sydney symposium was jointly convened with Central Sydney Area Drug and Alcohol Services on August 14–15, 1997.

44. Martin Donoghoe, Andrew Ball, and Alan Lopez, "Opioid Overdose: An International Perspective," in *Proceedings of an International Opioid Overdose Symposium,*

ed. Wayne Hall (Sydney: National Drug and Alcohol Research Centre, University of New South Wales, 1998), 9–26, quotation at 13.

45. D. English, C. J. D. Holman, E. Milne, M. G. Winter, G. K. Hulse, J. P. Codde, B. Cori, V. Dawes, N. De Klerk, J. J. Knuiman, G. F. Lewin, and G. A. Ryan, *The Quantification of Drug Caused Morbidity and Mortality in Australia, 1995 edition* (Canberra: Commonwealth Department of Human Services and Health, 1995).

46. Jason M. White, "The Pharmacology of Opioid Overdose," in Hall, *Proceedings of an International Opioid Overdose Symposium*, 63–72, quotation at 63. See also Jason M. White and Rodney J. Irvine, "Mechanisms of Fatal Opioid Overdose," *Addiction* 95 (1999): 961–972. Investigators used naloxone to elucidate the role of opioid peptides in normal respiration. J. E. Shook, W. D. Watkins, and E. M. Camporesi, "Differential Roles of Opioid Receptors in Respiration, Respiratory Disease and Opiate-Induced Respiratory Depression," *American Review of Respiratory Disease* 142 (1990): 895–909.

47. Duflou, "Classification of Opioid Deaths," 78, quoting Milton Helpern and Yong-Myun Rho, "Deaths from Narcotism in New York City," *New York State Journal of Medicine* 66 (1966): 2391–2408, and Milton Helpern and Yong-Myun Rho, "Deaths from Narcotism: Incidence, Circumstances and Postmortem Findings," *Journal of Forensic Science* 11 (1966): 1–16.

48. Johan Duflou, "Classification of Opioid Deaths—A Forensic Pathologist's Perspective," in Hall, *Proceedings of an International Opioid Overdose Symposium*, 77–83, quotation at 80; see also Deborah Zador, Sandra Sunjic, and Shane Darke, "Heroin-Related Deaths in New South Wales, 1992: Toxicological Findings and Circumstances," *Medical Journal of Australia* 164 (1996): 204–207.

49. Anne O'Loughlin and Ingrid van Beek, "Advances in the Management of Drug Overdose in an Outreach Setting," in Hall, *Proceedings of an International Opioid Overdose Symposium*, 84–86. Two nurses from the Kirketon Road Centre (KRC) described a naloxone program implemented in 1992, in which a medical officer prescribed naloxone via standing order to outreach RNs [registered nurses] who were trained to administer intramuscular naloxone. "KRC supports the continuation of this extended role of outreach RNs. Consideration should be given to extending the authority to administer naloxone to other outreach workers and even to IDUs themselves."

50. Simon Lenton, Tim Stockwell, and Robert Ali, "Heroin Overdose Workshop: Summary and Recommendations," in Hall, *Proceedings of an International Opioid Overdose Symposium*, 122.

51. Zador, Sunjic, and Darke, "Heroin-Related Deaths in New South Wales, 1992."

52. Randall C. Baselt and Robert H. Cravey, *Disposition of Toxic Drugs and Chemicals in Man*, 4th ed. (Foster City, CA: Chemical Toxicology Institute, 1995); Helpern and Rho, "Deaths from Narcotism in New York City"; Steven B. Karch, *The Pathology of Drug Abuse*, 2nd ed. (Boca Raton, FL: CRC Press, 1996).

53. Geoffrey Hunt and Kristin Evans. "'The Great Unmentionable': Exploring the Pleasures and Benefits of Ecstasy from the Perspectives of Drug Users," *Drugs* (Abingdon, UK) 15, no. 4 (2008): 329–349.

54. C. Robert Schuster, interview by Nancy D. Campbell, San Juan, Puerto Rico, June 14, 2004, 40–42.

55. Cathy J. Cohen, *The Boundaries of Blackness: AIDS and the Breakdown of Black Politics* (Chicago: University of Chicago Press, 1999); Ann Cvetkovich, *An Archive of Feelings: Trauma, Sexuality, and Lesbian Public Cultures* (Durham, NC: Duke University Press, 2003); Steve Epstein, *Impure Science: AIDS, Activism, and the Politics of Knowledge* (Berkeley: University of California Press, 1996).

56. Don C. DesJarlais, interview by Nancy D. Campbell and Joseph F. Spillane, Orlando, Florida, June 16, 2008, 22–23.

57. Ibid., 28–29.

58. Ibid., 30–31.

59. Ibid.

60. See Warwick Anderson, "The New York Needle Trial: The Politics of Public Health in the Age of AIDS," *American Journal of Public Health* 81, no. 11 (1991): 1506–1517, which described the political standoff over "experimental needle exchange" billed as a clinical trial between local politicians and New York City Health Department officials from 1985 to 1991. The health department's failure to establish political and cultural authority over the trial preceded the ACT UP needle exchange described in this chapter. The episode demonstrated the limits of medicalization rather than its triumph.

61. Richard Elovich, interview by Sarah Schulman for the ACT UP Oral History Project, interview no. 072, Brooklyn, New York, May 14, 2007.

62. Richard Elovich, interview by Nancy D. Campbell, New York City, April 29, 2015.

63. Remarks about noncompliance were summarized in Anthony S. Fauci, A. M. Macher, and D. L. Longo, "NIH Conference. Acquired Immunodeficiency Syndrome: Epidemiologic, Clinical, Immunologic, and Therapeutic Considerations," *Annals of Internal Medicine* 100, no. 1 (1984): 92–106; Anthony S. Fauci, "The Human Immunodeficiency Virus: Infectivity and Mechanisms of Pathogenesis," *Science* 239, no. 4840 (1988): 617–622. Fauci has directed the National Institute of Allergy and Infectious Diseases since 1984. Familiar to the AIDS movement, he was arguably a "convert" to movement views—see Epstein, *Impure Science*, 235–237.

64. Elovich, interview by Campbell, April 29, 2015.

65. Elovich, interview by Schulman, May 14, 2007, 38.

66. Ernest Drucker, "Epidemic in the War Zone: AIDS and Community Survival in New York City," *International Journal of Health Services* 20, no. 4 (1990): 601–615; Ernest Drucker, Peter Lurie, Philip Alcabes, and Alex Wodak, "Measuring Harm Reduction: The Effects of Needle and Syringe Exchange Programs and Methadone Maintenance on the Ecology of HIV," *AIDS* 12 (1998, Supplement A): S217–S230; Peter Lurie and Ernest Drucker, "An Opportunity Lost: HIV Infections Associated with Lack of a National Needle-Exchange Programme in the USA," *The Lancet* 34,

no. 9 (1997): 604–608; Ernest Drucker, "Advocacy Research in Harm Reduction Drug Policies," *Journal of Social Issues* 69, no. 4 (2013): 684–693.

67. Elovich, interview by Campbell, April 29, 2015, 9.

68. James Inciardi, interview by Nancy D. Campbell and Joseph F. Spillane, Scottsdale, Arizona, June 2006.

69. Don C. DesJarlais, "Editorial: Harm Reduction: A Framework for Incorporating Science into Drug Policy," *American Journal of Public Health* 85, no. 1 (1995): 10–12.

70. DesJarlais, interview by Campbell and Spillane, June 16, 2008, 42–44. See also David Vlahov, Don C. DesJarlais, Eric Goosby, Paula C. Hollinger, Peter G. Lurie, Michael D. Shriver, and Steffanie A. Strathdee, "Needle Exchange Programs for the Prevention of Human Immunodeficiency Virus Infection: Epidemiology and Policy," *American Journal of Epidemiology* 154, no. 12 (2001): S70–S77.

71. Holly Hagan, Don C. DesJarlais, David Purchase, Terry Reid, and Samuel R. Friedman, "The Tacoma Syringe Exchange," *Journal of Addictive Diseases* 10, no. 4 (1991): 81–88.

72. Buyer's and grower's clubs evolved to support compassionate use of medical marijuana (largely by HIV/AIDS patients); see Wendy Chapkis and Richard J. Webb, *Dying to Get High: Marijuana as Medicine* (New York: New York University Press, 2008); Dennis Hevesi, "Dave Purchase Dies at 73; Led Early Needle Exchange," *New York Times*, January 27, 2013. NASEN survived; see https://nasen.org/about/.

73. San Francisco AIDS Foundation, "History of Health: Needle Exchange in San Francisco," 2018, http://www.sfaf.org/client-services/syringe-access/history-of-needle-exchange.html.

74. DesJarlais, interview by Campbell and Spillane, June 16, 2008, 42–44. See Roberts, "Rehabilitation' as Boundary Object," and his forthcoming work on antidrug activism and harm reduction by communities of color in New York City.

75. Joyce Purnick, "Koch Bars Easing of Syringe Sales in AIDS Fight," *New York Times*, October 4, 1985, B5.

76. This history is based on the narrative and materials at http://www.leshrc.org/page/history.

77. Samuel Maull, "Judge Acquits ACT UP Defendants on Needle Exchange Charges," Associated Press, June 25, 1991.

78. The New York State Health Commissioner waived the part of the public health law prohibiting the sale, possession, and distribution of syringes for the purpose of preventing HIV transmission. Syringe exchanges could operate within the state to obtain, possess, and distribute hypodermic syringes and needles without prescription after 1992. That year LESNEP opened as an official, aboveground program. Allan Clear served as executive director.

79. Elovich, interview by Schulman, May 14, 2007, 46.

80. Douglas Crimp, *Melancholia and Moralism: Essays on AIDS and Queer Politics* (Cambridge, MA: MIT Press, 2002); Cvetkovich, *An Archive of Feelings*; David France,

How to Survive a Plague: The Story of How Activists and Scientists Tamed AIDS (New York: Vintage, 2017).

81. Rod Sorge, "I'm Sick," *junkphood: The Book of Death*, vol. 1 (Santa Cruz, CA: Santa Cruz Needle Exchange, 1997), 9.

82. Daniel Wolfe, interview by Nancy D. Campbell, New York City, October 6, 2013, 1.

83. Ibid., 2.

84. As this book neared completion, Stancliff retired and Kimberly Sue became medical director of the Harm Reduction Coalition.

85. Sharon Stancliff, telephone interview by Nancy D. Campbell, November 15, 2013, 8–9.

86. This history relies on the VOCAL-NY website at http://www.vocal-ny.org/?p=281.

87. See the Sawbuck Productions interview with Elizabeth Owens at https://vimeo.com/70970815.

7 "ANY POSITIVE CHANGE"

1. Allan Clear, interview by Sarah Schulman for the ACT UP Oral History Project, interview no. 178, New York City, March 1, 2015, 20–21. Clear directed the Lower East Side Harm Reduction Center and was the first executive director of the Harm Reduction Coalition in New York City. The Harm Reduction Coalition (HRC) is a national organization that promoted naloxone access at municipal, state, and national levels.

2. The latter worked for the San Francisco Department of Public Health.

3. Jennifer Gonnerman, untitled, *Village Voice*, September 9, 1997. Along with artist Brooke Lober, Edney and other collaborators created these collages using found materials.

4. *junkphood* itself has been the topic of academic writing about the role of zines in social movements. See Neil Wieloch, "Collective Mobilization and Identity from the Underground: The Deployment of 'Oppositional Capital' in the Harm Reduction Movement," *Sociological Quarterly* 43, no. 1 (2002): 45–72, which situates SCNE in the context of New Social Movements.

5. The anonymous story "Paramedic Tells All" appeared in *junkphood: The Book of Death*, vol. 1 (Santa Cruz, CA: Santa Cruz Needle Exchange, 1997), 12–16.

6. See Nancy D. Campbell, "Visual Art and Writing," in *The Oxford Companion to Women's Writing*, ed. Cathy N. Davidson and Linda Wagner-Martin (Oxford: Oxford University Press, 1995), 898–899.

7. Heather Edney, "User v. Addict/Abuser," *Harm Reduction Communication* 1 (1995): 6.

8. Two signal events were formation of Queer Nation in New York City and the Queer Theory conference convened by Teresa de Lauretis at UCSC; see Teresa de

Lauretis, "Queer Theory: Lesbian and Gay Sexualities an Introduction," *differences: A Journal of Feminist Cultural Studies* 3, no. 2 (1991): iii–xviii; and David Halperin, "The Normalization of Queer Theory," *Journal of Homosexuality* 45, no. 2/3/4 (2003): 339–343. "Queer" was in the Santa Cruz air, at first paired with lesbian, gay, and feminist theory in ways from which it later sheared. De Lauretis repurposed the term from the street, applying it to "queers of all sexes." De Lauretis guest edited a special issue of the feminist journal *differences*.

9. This latter category integrated the new politics of "social movement theory" with older leftist politics, divisions quite apparent in another early-1990s UCSC conference on social movements; see Marcy Darnovsky, Barbara Epstein, and Richard Flacks, *Cultural Politics and Social Movements* (Philadelphia: Temple University Press, 1995).

10. Wieloch, "Collective Mobilization," 67.

11. See Kristen C. Ochoa, Peter J. Davidson, Jennifer L. Evans, Judith A. Hahn, Kimberly Page-Shafer, and Andrew R. Moss, "Heroin Overdose among Young Injection Drug Users in San Francisco," *Drug and Alcohol Dependence* 80, no. 3 (2005): 297–302; the ten-year follow-up, Jennifer L. Evans, Judith I. Tsui, Judith A. Hahn, Peter J. Davidson, Pam J. Lum, and Kimberly Page, "Mortality among Young Injection Drug Users in San Francisco: A 10-Year Follow-Up of the UFO Study," *American Journal of Epidemiology* 175, no. 4 (2012): 302–308; Christopher Rowe, Glenn-Milo Santos, Eric Vittinghoff, Eliza Wheeler, Peter J. Davidson, and Phillip O. Coffin, "Predictors of Participant Engagement and Naloxone Utilization in a Community-Based Naloxone Distribution Program," *Addiction* 110, no. 8 (2015): 1301–1310; and Christopher Rowe, Glenn-Milo Santos, Eric Vittinghoff, Eliza Wheeler, Peter J. Davidson, and Phillip O. Coffin, "Neighborhood-Level and Spatial Characteristics Associated with Lay Naloxone Reversal Events and Opioid Overdose Deaths," *Journal of Urban Health* 93, no. 1 (2016): 117–130.

12. Kristen C. Ochoa, Judith Hahn, Pam Lum, Kimberly Page-Shafer, Rachel McLean, and Andrew Moss, "Overdose Common among Young Street Injection Drug Users," paper presented at the 127th Annual Meeting of the American Public Health Association, Chicago, November 10, 1999; and Kristen C. Ochoa, Judith A. Hahn, Karen H. Seal, and Andrew R. Moss, "Overdosing among Young Injection Drug Users in San Francisco," *Addictive Behavior* 26, no. 3 (May–June 2001): 453–460. Figures also appeared in *junkphood Presents the UFO Study*, a special issue of the zine published in September 1997.

13. Ibid., 173.

14. Ibid., 185. See Peter J. Davidson, Kristen C. Ochoa, Judith A. Hahn, J. L. Evans, and Andrew R. Moss, "Witnessing Heroin-Related Overdoses: The Experiences of Young Injectors in San Francisco," *Addiction* 97, no. 12 (2002): 1511–1516; and Peter J. Davidson and Kimberly Page, "Research Participation as Work: Comparing the Perspectives of Researchers and Economically Marginalized Populations," *American Journal of Public Health* 102, no. 7 (2012): 1254–1259.

15. Alex Kral, telephone interview by Nancy D. Campbell, February 7, 2014, 5.

16. Karen H. Seal, Alex H. Kral, Lauren Gee, Lisa D. Moore, Ricky N. Bluthenthal, Jennifer Lorvick, and Brian R. Edlin, "Predictors and Prevention of Non-Fatal Overdose among Street-Recruited Injection Heroin Users in the San Francisco Bay Area, 1998–1999," *American Journal of Public Health* 91, no. 11 (2001): 1842–1846.

17. Kral, interview by Campbell, February 7, 2014, 6.

18. Dates imprecise; in "Drug User's Tools of the Trade: Naloxone and Narcan," SFNE cofounder Ro Giuliano stated that the organization had been conducting "overdose management and prevention" programs for a year and a half with under-thirties in the Haight; see *Harm Reduction Communication* 9 (Fall 1999): 18–19.

19. Kral, interview by Campbell, February 7, 2014, 6–7.

20. Ibid., 12.

21. Ibid., 13.

22. Ibid., 13.

23. Seal et al., "Predictors and Prevention of Nonfatal Overdose." See also the Sawbuck Productions interview with DOPE Project Manager Eliza Wheeler at https://vimeo.com/71281393.

24. Kral, interview by Campbell, February 7, 2014, 10.

25. Ibid., 5–6; Peter J. Davidson, Rachel L. McLean, Alex Kral, A. A. Gleghorn, Brian R. Edlin, and Andrew R. Moss, "Fatal Heroin-Related Overdose in San Francisco, 1997–2000: A Case for Targeted Intervention," *Journal of Urban Health* 80, no. 2 (2003): 261–273.

26. My thanks to an anonymous reviewer, who assuredly took part in this assemblage, for pointing me to this language and compelling me to investigate the critical contributions of West Coast overdose prevention activism, particularly in the Bay Area and Santa Cruz (where I lived when a doctoral student from 1990 to 1995).

27. See *"junkphood* explained" at http://heatheredney.com/junkphood-2/.

28. Philippe Pignarre and Isabelle Stengers, *Capitalist Sorcery: Breaking the Spell*, trans. Andrew Goffey (Basingstoke, UK: Palgrave Macmillan, 2011), 120.

29. Dan Bigg, interview by Nancy D. Campbell, Chicago, Illinois, November 21, 2013, 3.

30. The story is told in full at http://harmreduction.org/issues/overdose-prevention/tools-best-practices/naloxone-program-case-studies/chicago-recovery-alliance/.

31. As with any larger-than-life figure, there are many sources of information on Bigg, who is typically credited with pioneering overdose prevention with naloxone. I have mainly relied on my own lengthy interviews with him and those who worked with him, although in the wake of his own confirmed overdose death on August 21, 2018, tributes and remembrances have poured forth from multiple sources within the international harm reduction movement.

32. Sarz Maxwell, interview by Nancy D. Campbell, Seattle, Washington, January 3, 2014, 5.

33. Ibid.

34. Ibid., 6.

35. Bigg, interview by Campbell, November 21, 2013, 20.

36. Maxwell, interview by Campbell, January 3, 2013, 6.

37. Ibid., 5.

38. Anonymous, "Overdose!," *Harm Reduction Communication*, (Fall 1999): 9.

39. Later published as Sarz Maxwell, Dan Bigg, Karen Stanczykiewicz, and Suzanne Carlberg-Racich, "Prescribing Naloxone to Actively Injecting Heroin Users: A Program to Reduce Heroin Overdose Deaths," *Journal of Addictive Diseases* 25, no. 3 (2006): 89–96.

40. Maxwell, interview by Campbell, January 3, 2013, 7.

41. Ibid.

42. Ibid., 21–22.

43. Kristen Ochoa, Rachel McLean, Heather Edney-Meschery, Dante Brimer, and Andrew Moss, "The Challenges and Rewards of Collaborative Research: The UFO Study," in *Preventing AIDS: Community-Science Collaborations*, ed. R. Dennis Shelby, Benjamin Bowser, Shiraz Mishra, and Cathy Reback (New York: Routledge, 2004), 163–190, at 184. See http://adai.washington.edu/training/conf/agenda.htm. Founded by Ethan Nadelmann in 1994, the Lindesmith Center was funded by George Soros's Open Society Foundation (OSF). Renamed the Drug Policy Alliance, it incorporated the Drug Policy Foundation, a centrist reform organization founded by American University professor Arnold Trebach, author of *The Heroin Solution* (his account appears in the second edition of *The Great Drug War*). OSF supported global naloxone distribution in post-Soviet countries, employing activist consultants such as Matt Curtis, Richard Elovich, Stephen Malloy, Roxanne Saucier, and Daniel Wolfe to institute overdose prevention to Afghanistan, Estonia, Vietnam, the Russian Federation, and other countries. Another important funder of domestic US naloxone and syringe distribution was the Comer Family Foundation; an early photograph of many protagonists mentioned in this chapter appears at http://www.comerfamilyfoundation.org/articles/two-decades-of-positive-change-a-brief-history-of-the-harm-reduction-coalition.

44. Pat O'Hare, "Merseyside, the First Harm Reduction Conferences, and the Early History of Harm Reduction," *International Journal of Drug Policy* 18 (2007): 141–144; John Ashton and Howard Seymour, "Public Health and the Origins of the Mersey Model of Harm Reduction," *International Journal of Drug Policy* 21 (2010): 94–96.

45. Bigg, interview by Campbell, November 21, 2013, 20.

46. John Strang, Beverly Powis, David Best, Lisa Vingoe, Paul Griffiths, Colin Taylor, Sarah Welch, and Michael Gossop, "Preventing Opiate Overdose Fatalities with Take-Home Naloxone: Pre-Launch Study of Possible Impact and Acceptability," *Addiction* 94, no. 2 (1999): 199–204.

47. John Strang, interview by Nancy D. Campbell, London, March 7, 2017, 10.

48. One of the pilots was on Jersey, one of the Channel Islands.

49. Detailed in Rosie Mundt-Leach, Siobhan Jackson, Francis Keaney, and Alun Morinan, "Citizen Medics," *Druglink* (January/February 2010): 24–25.

50. Ibid., 11–12.

51. Maxwell et al., "Prescribing Naloxone to Actively Injecting Heroin Users"; Dan Bigg, "Data on Take Home Naloxone Are Unclear But Not Condemnatory," *British Medical Journal* 324, no. 7338 (2002): 678.

52. The 2006 site visit report was prepared by the Scottish group who visited CRA as a model for the naloxone pilot in Glasgow.

53. Only one reversal was reported unsuccessful and that incident involved multiple drugs, including alcohol.

54. Sawbuck Productions is Greg Scott's film company; the film may be viewed at http://www.sawbuckproductions.org/overdose-prevention-naloxone.

55. The full protocol may be accessed at https://harmreduction.org/issues/overdose -prevention/tools-best-practices/naloxone-program-case-studies/chicago-recovery -alliance/.

56. Susan G. Sherman, Donald S. Gann, Gregory Scott, Suzanne Carlberg, Dan Bigg, and Robert Heimer, "A Qualitative Study of Overdose Responses among Chicago IDUs," *Harm Reduction Journal* 5, no. 2 (2008): 2.

57. Maxwell et al., "Prescribing Naloxone to Actively Injecting Heroin Users," 92.

58. The classic reference is David Sudnow, *Passing On: The Social Organization of Dying* (Englewood Cliffs, NJ: Prentice-Hall, 1967), 104.

59. Bigg, interview by Campbell, November 21, 2013, 17.

60. It is not my purpose to repeat that exercise. Scott Burris, Leo Beletsky, and colleagues did the Herculean job of documenting changes in law and policy. See Burris, Edlin, and Norland, "Legal Aspects of Providing Naloxone to Heroin Users in the United States"; Burris et al., "Stopping an Invisible Epidemic"; and Phillip O. Coffin et al., "Preliminary Evidence of Health Care Provider Support for Naloxone Prescription as Overdose Fatality Prevention Strategy in New York City," *Journal of Urban Health* 80, no. 2 (2003): 288–290.

61. Phillip O. Coffin, Benjamin P. Linas, Stephanie H. Factor, and David Vlahov, "New York City Pharmacists' Attitudes toward Sale of Needles/Syringes to Injection Drug Users before Implementation of Law Expanding Syringe Access," *Journal of Urban Health* 77, no. 4 (2000): 781–793; Barbara Gerbert, Bryan T. Maguire, Thomas Bleecker, Thomas J. Coates, and Stephen J. McPhee, "Primary Care Physicians and AIDS: Attitudinal and Structural Barriers to Care," *Journal of the American Medical Association* 266, no. 20 (1991): 2837–2842.

62. See Tessie Castillo, "A Space for Grief and Growth: The 12th National Harm Reduction Conference," in *The Fix: Addiction and Recovery, Straight Up*, October 29, 2018, https://www.thefix.com/space-grief-and-growth-12th-national-harm-reduction -conference.

8   HARM REDUCTION RESEARCH AND SOCIAL JUSTICE

1. Dan Kennedy's blog, *Media Nation*, analyzes these photographs, which reveal working people shooting up in the Boston Commons under the headline "Heroin Takes a Life" at https://dankennedy.net/2005/08/26/death-in-garden/.

2. Maya Doe-Simkins, telephone interview by Nancy D. Campbell, Chicago, Illinois, September 27, 2013, 2.

3. Ibid., 12.

4. Ibid.

5. When Allan Clear stepped down in 2016, Monique Tula became executive director of the HRC. That year the US harm reduction movement shifted from one largely led by white men to one in which for the first time women of color held prominent roles at HRC (see https://harmreduction.org/about-us/our-staff/monique-tula/) and Drug Policy Alliance, where Maria McFarland Sanchez-Moreno took the helm (see https://www.thefix.com/two-women-color-leading-fight-end-drug-war). Activists of color have been central shapers of the politics of drug treatment, particularly in New York City, but their role in harm reduction has not been highlighted; see Roberts, "'Rehabilitation' as Boundary Object" and forthcoming work on harm reduction in communities of color.

6. The PowerPoint show was dated 2012 and produced by people well known in harm reduction circles. Langis made a case for the moral worth of drug users in the local newspaper, the *Gloucester Daily Times*, on September 11, 2012: "When firefighters rush into a burning house," says Gary Langis, "they don't stop to ask what kind of lives they're saving, they don't make a judgment, is this life or that life worth it." "No," he adds. "A life is always worth saving."

7. Controversy erupted in 2016 on the pages of *Addiction* between many of the interviewees whose experiences informed this book. National Addiction Centre researchers John Strang, Rebecca McDonald, Basak Tas, and Ed Day raised the question, "Clinical provision of improvised nasal naloxone without experimental testing and without regulatory approval: imaginative shortcut or dangerous bypass of essential safety procedures?" Not wishing to rely on intranasal naloxone until proof of efficacy was established, Strang and McDonald published another letter on March 14, 2016, titled "New approved nasal naloxone welcome, but unlicensed improvised naloxone spray kits remain a concern: proper scientific study must accompany innovation." Although intranasal naloxone was not approved in the United Kingdom, it was used in the Scottish Highlands/Inverness naloxone pilot starting in 2007. Two weeks later, McDonald and Strang published "Are Take-Home Naloxone Programmes Effective? Systematic Review Utilizing Application of the Bradford Hill Criteria" (*Addiction* 111, no. 7 [2016]: 1177–1187). Two groups of US protagonists then published volleys in the skirmish. Phillip O. Coffin, Josiah Rich, Michael Dailey, Sharon Stancliff, and Leo Beletsky published "While We Dither, People Continue to Die from Overdose: Comments on 'Clinical Provision of Improvised Nasal Naloxone without Experimental Testing and without Regulatory Approval: Imaginative Shortcut or Dangerous Bypass of Essential Safety Procedures?'" (*Addiction*

111, no. 10 [2016]: 1880–1881); and Maya Doe-Simkins, Caleb Banta-Green, Corey S. Davis, Traci C. Green, and Alexander Y. Walley published "Clinical Provision of Improvised Nasal Naloxone without Experimental Testing and without Regulatory Approval: Imaginative Shortcut or Dangerous Bypass of Essential Safety Procedures?" (*Addiction* 111, no. 10 [2016]: 1879–1880). On July 14, 2016, a "Naloxone Virtual Issue" appeared at https://onlinelibrary.wiley.com/doi/toc/10.1111/(ISSN)1360-0443 .naloxone_virtual_issue. While seemingly technical, the controversy mapped onto differing approaches to the social organization of naloxone distribution. Throughout this time there was no licensed supplier of preassembled intranasal kits.

8. Alexander Walley, telephone interview by Nancy D. Campbell, November 5, 2013, 1.

9. Ibid., 2.

10. Ibid.

11. Ludwik Fleck, *Genesis and Development of a Scientific Fact*, trans. Frederick Bradley and Thaddeus J. Trenn (Chicago: University of Chicago Press, 1979).

12. Walley, interview by Campbell, November 5, 2013, 2.

13. Anonymous researchers, interviews by Nancy D. Campbell, 2013–2014.

14. This was the case in the women's and feminist health movement; on gender-specific drug and alcohol treatment, see Nancy D. Campbell and Elizabeth Ettorre, *Gendering Addiction: The Politics of Drug Treatment in a Neurochemical World* (Basingstoke, UK: Palgrave Macmillan, 2011). On reproductive rights, see Paula A. Treichler, "Feminism, Medicine, and the Meaning of Childbirth," in *Body/Politics: Women and the Discourses of Science*, ed. Mary Jacobus, Evelyn Fox Keller, and Sally Shuttleworth (New York: Routledge, 1990), 113–138.

15. Walley, interview by Campbell, November 5, 2013, 3.

16. For example, see Eliza Wheeler, "Surviving Fentanyl: Resisting Racialized Drug Profiling," *Medium*, July 28, 2017, https://medium.com/@wheeler_12737/surviving -fentanyl-2d404511a285.

17. In May 2006, to identify NPF-related deaths in six state and local jurisdictions, the CDC implemented an ad hoc case finding and surveillance system, later managed by the DEA. The system identified 1,013 NPF-related deaths occurring between April 4, 2005 and March 28, 2007. Fentanyl was not new; it had been implicated in overdose deaths since the 1970s when it first became available to heroin users. Actual historical levels of fentanyl's contribution to "heroin overdoses" is unknown and likely unknowable. CDC, "Nonpharmaceutical Fentanyl-Related Deaths— Multiple States, April 2005–March 2007," *Morbidity and Mortality Weekly Report* 57, no. 29 (2008): 793–796.

18. Alice Bell, interview by Nancy D. Campbell, Pittsburgh, Pennsylvania, May 3, 2017.

19. Ibid., 8–10.

20. Ibid., 9.

21. Ibid.

22. Discovering that the initial acronym they used, NOPE, was already in use, the group gradually morphed into a Google Group under the name OSNN, its current appellation. The group—which generously provided ongoing documentation of the diffusion of naloxone access and overdose prevention movement for the author—shares resources and lively conversations on topics of mutual interest.

23. Bell, interview by Campbell, May 3, 2017, 11–12.

24. Another fentanyl-fueled cycle of overdose deaths was underway as this book went to press.

25. By this time, Karen H. Seal, Robert Thawley, Lauren Gee, Joshua Bamberger, Alex H. Kral, Dan Ciccarone, Moher Downing, and Brian R. Edlin, "Naloxone Distribution and Cardiopulmonary Resuscitation Training for Injection Drug Users to Prevent Heroin Overdose Death: A Pilot Intervention Study," *Journal of Urban Health* 82, no. 2 (2005): 303–311, had been published.

26. The hearing was titled "Role of Naloxone in Opioid Overdose Fatality Prevention." It was convened by the FDA Center for Drug Evaluation and Research in conjunction with the CDC and the NIDA on April 12, 2012.

27. Bell, interview by Campbell, May 3, 2017, 15–16.

28. US Department of Health and Human Services. Food and Drug Administration. Center for Drug Evaluation and Research, "Role of Naloxone in Opioid Overdose Fatality Prevention." April 12, 2012, 262–264, 271, and 283.

29. Long a physician supportive of the consumer and patient rights movements, Lurie directed Public Citizen's Health Research Group from 2000 to 2009; worked at the FDA as Commissioner for Public Health Strategies and Analysis; and left the FDA in 2017 to direct the Center for Science in the Public Interest. See the September 19, 2016, press release "FDA Launches Competition to Spur Innovative Technologies to Help Reduce Opioid Overdose Deaths," at https://www.fda.gov/NewsEvents /Newsroom/PressAnnouncements/ucm520945.htm.

30. Davis and Carr, "Legal Changes to Increase Access to Naloxone."

31. The free toolkit may be accessed at https://store.samhsa.gov/product/Opioid -Overdose-Prevention-Toolkit-Updated-2016/All-New-Products/SMA16-4742. The statement is from a hearing on "Targeting Overdose-Reversing Drugs" at https:// www.drugabuse.gov/about-nida/legislative-activities/testimony-to-congress/2017 /federal-response-to-opioid-crisis. SAMHSA took on the role of funder of naloxone for overdose prevention, providing $11 million in Grants to Prevent Prescription Drug/Opioid Overdose-Related Deaths to twelve states in 2017. Grantees used funds to train first responders how to administer, purchase, and distribute naloxone. In September 2017, SAMHSA awarded funding for another $46 million over five years to grantees in twenty-two states to provide resources to first responders and treatment providers working directly with populations at highest risk for opioid overdose. This scale of funding could not have been envisioned before the state-by-state efforts to change naloxone access laws, paraphernalia laws, and Good Samaritan

legislation, all of which were tracked in real time by Burris, Beletsky, and their research team.

32. From *junkphood: The Book of Death*, vol. 2 (Santa Cruz, CA: Santa Cruz Needle Exchange, 1997). One or more of these "methods" might have coincidentally worked if an individual had witnessed someone on the "continuum of overdosing," who had not gone over so far that central nervous system stimulation no longer worked.

33. Sandro Galea, Jennifer Ahern, Ken Tardiff, Andy Leon, Phillip O. Coffin, Karen Derr, and David Vlahov, "Racial/Ethnic Disparities in Overdose Mortality Trends in New York City, 1990–1998," *Journal of Urban Health* 80, no. 2 (2003): 201–211.

34. Ibid., 202.

35. Ibid., 206.

36. Ibid., 208.

37. Rabeharisoa, Moreirab, and Akricha, "Evidence-Based Activism," 128.

38. Useful discussion of this point in relation to crime appears in Michelle Brown, *The Culture of Punishment: Prison, Society, and Spectacle* (New York: New York University Press, 2009).

39. Scott Frickel, Sahra Gibbon, Jeff Howard, Joanna Kempner, Gwen Ottinger, and David J. Hess, "Undone Science: Charting Social Movement and Civil Society Challenges to Research Agenda Setting," *Science, Technology, and Human Values* 35, no. 4 (2009): 444–473; David J. Hess, "Undone Science and Social Movements: A Review and Typology," in *Routledge International Handbook of Ignorance Studies*, ed. Matthias Gross and Linsey McGoey (New York: Routledge, 2015), 141–154.

40. Nabarun Dasgupta, telephone interview by Nancy D. Campbell, October 11, 2013, 2–3.

41. Bell, interview by Campbell, May 3, 2017, 32.

42. Walley, interview by Campbell, November 5, 2013, 3.

43. Seal et al., "Naloxone Distribution and Cardiopulmonary Resuscitation Training for Injection Drug Users"; Kerstin Dettmer, Bill Saunders, and John Strang, "Take Home Naloxone and the Prevention of Deaths from Opiate Overdose: Two Pilot Schemes," *British Medical Journal* 322 (2001): 895–896.

44. Walley, interview by Campbell, November 5, 2013, 3.

45. Maya Doe-Simkins, Alexander Walley, Andy Epstein, and Peter Moyer, "Saved by the Nose: Bystander-Administered Intranasal Naloxone for Opioid Overdose," *American Journal of Public Health* 99, no. 5 (2009): 788–791.

46. Walley, interview by Campbell, November 5, 2013, 3.

47. Andrew McAuley, interview by Nancy D. Campbell, Glasgow, Scotland, February 28, 2017, 8–9.

48. Andrew McAuley, George Lindsay, Maureen Woods, and Derek Louttit, "Responsible Management and Use of a Personal Take-Home Naloxone Supply: A Pilot Project," *Drugs: Education Prevention and Policy* 17, no. 4 (2010): 388–399. The study aimed

to "assess if Scottish drug users, their family and friends could be trained in critical incident management and the safe and effective administration of naloxone. The project also sought to monitor whether drug users can manage their own personal take-home naloxone supply and use it appropriately in an emergency opiate overdose situation."

49. Andrew McAuley, Janet Bouttell, Lee Barnsdale, Daniel Mackay, Jim Lewsey, Carole Hunter, and Mark Robinson, "Evaluating the Impact of a National Naloxone Programme on Ambulance Attendance at Overdose Incidents: A Controlled Time-Series Analysis," *Addiction* 12, no. 2 (2017): 301–308. Again, this was a study that was conceptualized in response to criticisms.

50. McAuley, interview by Campbell, February 28, 2017, 11–12.

51. Ibid., 12.

52. Ibid.

53. Hess, *Undone Science*, 27.

54. Scott Frickel and Abby Kinchy, "Lost in Space: Geographies of Ignorance in Science and Technology Studies," in *Routledge Handbook of Ignorance*, ed. Matthias Gröss and Linsey McGoey (New York: Routledge, 2015), 174–182.

55. Hess suggested that the goal of counterpublics in industrial transition movements is to change the distribution system by which people access technologies ("rather than the design of technologies themselves"). Such movements supposedly generate less "epistemic conflict"—unless and until they are framed as "giving handouts to the poor." Poor people's movements have evidenced a pattern in which radical demands are met with service delivery; as "clients," participants are absorbed into the very system they once opposed. In *Alternative Pathways*, Hess examined access movements ranging from hunger to housing that relied on public-private partnerships to deliver material goods and services. His categories do not quite fit the naloxone access advocacy movement, which does not simply seek universal access to specific types of material goods, and which is complex in terms of opposition to and cooperation with the pharmaceutical industry—which supplies naloxone as well as the opioids it antagonizes.

56. Opioid substitution therapy, methadone maintenance treatment, and medication-assisted treatment are all recognized evidence-based treatments.

57. Annette Dale-Perrara, interview by Nancy D. Campbell, London, March 6, 2017, 6–7, referring to a 2006 report from the Centre for Social Justice titled *Breakdown Britain: Addicted Britain*. This report was one of a series of Conservative Party responses to the National Treatment Agency for Substance Abuse.

58. Sharon Traweek, *Beamtimes and Lifetimes: The World of High Energy Physicists* (Cambridge, MA: Harvard University Press, 1992).

59. Mark Faul, Peter G. Lurie, Jeremiah M. Kinsman, Michael W. Dailey, Charmaine Crabaugh, and Scott M. Sasser, "Multiple Naloxone Administrations among Emergency Medical Service Providers Is Increasing," *Prehospital Emergency Care* 21, no. 4 (2017): 411–419. See also Sandro Contenta, "He Saved 17 People Who OD'd—But Police Want to Jail Him," *Toronto Star*, April 30, 2017.

60. Scott Frickel and Neil Gross, "A General Theory of Scientific/Intellectual Movements," *American Sociological Review* 70, no. 2 (2005): 204–232.

61. New levels of scrutiny are one side effect of inclusion; see Steven Epstein, "The New Attack on Sexuality Research: Morality and the Politics of Knowledge Production," *Sexuality Research and Social Policy* 3 (2006): 1–12.

62. Adele E. Clarke, *Disciplining Reproduction: Modernity, American Life Sciences, and the Problems of Sex* (Berkeley: University of California Press, 1998); Nancy D. Campbell and Susan J. Shaw, "Incitements to Discourse: Illicit Drugs, Harm Reduction and the Production of Ethnographic Subjects," *Cultural Anthropology* 23, no. 4 (2008): 688–717; Nicole Vitellone, *Social Science of the Syringe: A Sociology of Injecting Drug Use* (New York: Routledge, 2017).

63. Isabelle Stengers, "History through the Middle: Between Macro and Mesopolitics; An Interview with Isabelle Stengers," interview by Brian Massumi, *INFLeXions* no. 3 (2009): 1–16.

64. Karen Barad, *Meeting the Universe Halfway* (Durham, NC: Duke University Press, 2007).

65. Stengers, "History through the Middle."

66. Brown et al., "Embodied Health Movements: Uncharted Territory in Social Movement Research"; Brown et al., "Embodied Health Movements: New Approaches to Social Movements in Health."

67. Robert N. Proctor and Londa Schiebinger, *Agnotology: The Making and Unmaking of Ignorance* (Palo Alto, CA: Stanford University Press, 2008); Nancy Tuana, "The Speculum of Ignorance: The Women's Health Movement and the Epistemologies of Ignorance," *Hypatia* 23, no. 3 (2006): 1–19.

68. Michelle Murphy, *Seizing the Means of Reproduction: Entanglements of Feminism, Health, and Technoscience* (Durham, NC: Duke University Press, 2012), 31.

69. Joanna Kempner, "The Chilling Effect: How Do Researchers React to Controversy," *PLOS Medicine* 6 (2009): e1000014.

70. Joanna Kempner, Jon F. Merz, and Charles L. Bosk, "Forbidden Knowledge: Public Controversy and the Production of Nonknowledge," *Sociological Forum* 26, no. 3 (2011): 475–500, at 488.

71. Murphy, *Seizing the Means of Reproduction.*

72. Martin McCusker, George Porter, and Reuben Cole, interview by Nancy D. Campbell, London, March 9, 2017.

73. Judith Yates, interview by Nancy D. Campbell, London, March 8, 2017.

74. Jamie Bridges and Kirstie Douse, interview by Nancy D. Campbell, London, March 6, 2017; see also Ashton and Seymour, "Public Health and the Origins of the Mersey Model."

75. Austin Smith and Jason Wallace, interview by Nancy D. Campbell, Glasgow, Scotland, February 28, 2017.

9   RESUSCITATING SOCIETY

1. I. Peirce James, "The London Heroin Epidemic of the 1960's," *Medico-Legal Journal* 39, no. 1 (1971): 17–26.

2. Virginia Berridge, "The 'British System' and Its History: Myth and Reality," in *Heroin Addiction and the British System*, vol. 1, ed. Michael Gossop and John Strang (London: Routledge, 2005), 7–16; Alex Mold, *Heroin: The Treatment of Addiction in Twentieth-Century Britain* (DeKalb: Northern Illinois University Press, 2008). For what became of the system post-1965, see Alex Mold and Virginia Berridge, *Voluntary Action and Illegal Drugs: Health and Society in Britain since the 1960s* (Basingstoke, UK: Palgrave Macmillan, 2010).

3. Mold, *Heroin*, 29–36. This analysis built on the sociological insights of Gerry Stimson and Edna Oppenheimer, *Heroin Addiction: Treatment and Control in Britain* (London: Tavistock, 1982). The second Brain report lent itself to these sociologists' interpretation that after the 1960s, social control became the aim of British drug policy. The report presented the "phenomena of habituation, dependence, and addiction" as products of "social, medical, and psychological factors," but added the more localized element of social acceptance for drug-taking in such settings as Soho clubs.

4. Her Majesty's Government, Ministry of Health, *Drug Addiction: The Second Report of the Interdepartmental Committee on Drug Addiction* (also referred to as the Brain report, named after its chair, Walter Russell Brain) (London: Stationery Office, 1965).

5. Ibid.

6. James, "The London Heroin Epidemic of the 1960's."

7. The Drug Dependence Units drifted towards more "medical" approaches over their lifetimes. Some such as the Blenheim Project allowed injection on the premises or supplied clean injecting equipment; see Mold, *Heroin*, 39. This was a major difference between the United Kingdom and the United States, which moved toward criminalizing paraphernalia and more consequentially adopted incarceration as drug policy. See Michelle Alexander, *The New Jim Crow: Mass Incarceration in the Age of Colorblindness* (New York: New Press, 2010); and Elizabeth Hinton, *From the War on Poverty to the War on Crime* (Cambridge, MA: Harvard University Press, 2016).

8. I. Peirce James, "Suicide and Mortality amongst Heroin Addicts in Britain," *British Journal of the Addictions* 62 (1967): 391–398.

9. Peter Reuter and Alex Stevens, *An Analysis of UK Drug Policy* (London: UK Drug Policy Commission, 2007), 25–27.

10. Lord Privy Council, *Tackling Drug Misuse: A Summary of the Government's Strategy* (London: HMSO, 1985); Lord President of the Council and Leader of the House of Commons, *Tackling Drugs Together: A Strategy for England 1995–98* (London: HMSO, 1995).

11. Her Majesty's Government, President of the Council, *Tackling Drugs to Build a Better Britain: The Government's Ten-Year Strategy for Tackling Drugs Misuse* (London: Stationery Office, 1998).

12. Ibid., 3.

13. Sarah G. Mars, *The Politics of Addiction: Medical Conflict and Drug Dependence in England since the 1960s* (Basingstoke, UK: Palgrave Macmillan, 2012), 22–24.

14. Ibid., 24.

15. Steven Hayle, "The Politics of Harm Reduction: Comparing the Historical Development of Needle Exchange Policy in Canada and the UK between 1985 and 1995," *Social History of Alcohol and Drugs* 33 (2018): 81–103.

16. Mars, *The Politics of Addiction*, 16. In January 1986, the ACMD announced that priority should be given to preventing HIV. By fall 1986, needle and syringe programs were in place at Peterborough Drug Dependence Unit and the Mersey Regional Drug Training and Information Centre. These Liverpool efforts occurred nearly simultaneously as those in Edinburgh; see Jeannette Carr and Steve Dalton, "Syringe Exchange: The Liverpool Experience," *DrugLink* (May/June 1988): 12–14.

17. Constance Nathanson, *Disease Prevention as Social Change: The State, Society, and Public Health in the United States, France, Great Britain, and Canada* (New York: Russell Sage Foundation, 2007), 183.

18. Roy Robertson, interview by Nancy D. Campbell, Edinburgh, Scotland, March 1, 2017, 3.

19. Nathanson, *Disease Prevention as Social Change*, 181.

20. Ibid., 184.

21. J. Roy Robertson, Aidan B. V. Bucknall, Philip Welsby, J. J. Roberts, J. M. Inglis, J. F. Peutherer, and Raymond P. Brettle, "Epidemic of AIDS-Related Virus (HTLV-III/LAV) Infection among Intravenous Drug Abusers," *British Medical Journal* 292 (1986): 527–529.

22. Robertson, interview by Campbell, March 1, 2017, 7.

23. Frances X. Cline, "Via Addict Needles, AIDS Spreads in Edinburgh," *New York Times*, January 4, 1987.

24. Two infectious disease specialists, Claire McCarthy and Philip D. Welsby, conducted a study showing that Edinburgh University students who had read or seen *Trainspotting* incorrectly believed that HIV/AIDS prevalence was higher in Edinburgh than in other European cities to which HIV had been introduced much earlier than it first appeared in Edinburgh (1982–1983). Lamenting the predominance of "parochial news," McCarthy and Welsby set out to undo Edinburgh's "undeserved and inappropriate accreditation" as the "AIDS Capital of Europe" on the pages of the *Scottish Medical Journal*. See also Liz Matthews, "Begbie and Co. Blamed for 'AIDS Capital' Reputation," *The Scotsman*, March 23, 2003.

25. See https://www.youtube.com/watch?v=YtH8MDilprc&list=PL194165CB4A2E3AB6 &index=2. Social entrepreneur Jonathan Derricott made this film in addition to those discussed in chapter 8, along with the National Treatment Agency campaign "Harm Reduction Works." In 2002 Derricott and former NHS nurse Andrew Preston cofounded Exchange Supplies, which distributes clean injection supplies such as

citric acid packets that reduce heroin injectors' reliance on nonsterile lemon juice. "As we intended to break the law, it seemed wrong to ask people to take responsibility for that … so the only other option at the time was to be a company." According to Derricott. "Within a couple of years, with our help, the law was being systematically broken all over the UK, with the full agreement of the police. As new agreements were negotiated, we posted them on our website which in turn influenced other police forces to agree to allow the supply of citric. Of course the fact that it did work has been as much down to the enthusiasm and willingness of drug workers to push boundaries, and stand up for the rights of users, as it was to anything that we did."

26. Robertson, interview by Campbell, March 1, 2017.

27. Ibid., 10.

28. Macleod et al., "The Edinburgh Addiction Cohort." The initial cohort was 814, of whom 227 were deceased and 139 declined participation or were disqualified because they had no history of injection; the remainder were interviewed and matched to controls.

29. J. Roy Robertson, Peter J. M. Ronald, Gillian M. Raab, Amanda J. Ross, and Tamiza Parpia, "Deaths, HIV Infection, Abstinence, and Other Outcomes in a Cohort of Injecting Drug Users Followed Up for 10 Years," *British Medical Journal* 309, no. 6951 (1994): 369–372.

30. C. A. Skidmore, J. Roy Robertson, and G. Savage, "Mortality and Increasing Drug Use in Edinburgh: Implications for HIV Epidemic," *Scottish Medical Journal* 35, no. 4 (1990): 100–102.

31. Robertson in Martin Plant et al., eds., *Alcohol and Drugs: The Scottish Experience* (Edinburgh: Edinburgh University Press, 1992), 152–161.

32. Robertson, interview by Campbell, March 1, 2017, 11.

33. Ibid.

34. John Macleod, Lorraine Copeland, Matthew Hickman, James McKenzie, Jo Kimber, Daniela De Angelis, and J. Roy Robertson, "The Edinburgh Addiction Cohort: Recruitment and Follow-Up of a Primary Care Based Sample of Injection Drug Users and Non Drug-Injecting Controls," *BMC Public Health* 10 (2010): 101–114; Jo Kimber, Lorraine Copeland, Matthew Hickman, John Macleod, James McKenzie, Daniela De Angelis, and J. Roy Robertson, "Survival and Cessation in Injecting Opiate Users: Prospective Observational Study of Outcomes and Effect of Opiate Substitute Treatment," *British Medical Journal* 340 (2010): 3172.

35. See J. Roy Robertson, "Misadventure in Muirhouse. HIV Infection: A Modern Plague and Persisting Public Health Problem," *Journal of the Royal College of Physicians of Edinburgh* 47, no. 1 (2017): 88–93, on the 2015 reemergence of HIV in Glasgow as the result of "policy failure."

36. Susanne MacGregor, "'Tackling Drugs Together': Ten Years On," *Drugs: Education, Prevention and Policy* 13, no. 5 (2006): 393–398, quotation at 398.

37. Ibid., 401–402; Her Majesty's Government, President of the Council, *Tackling Drugs to Build a Better Britain: The Government's Ten-Year Strategy for Tackling Drugs*

Misuse (London: Stationery Office, 1998), 1, 6. See also K. Duke and Susanne Mac-Gregor, *Tackling Drugs Locally* (London: HMSO, 1997); Michael Farrell and John Strang, "Britain's New Strategy for Tackling Drugs Misuse: Shows a Welcome Emphasis on Evidence," *British Medical Journal* 316, no. 7142 (1998): 1399–1400.

38. Clare Gerada, interview by Nancy D. Campbell, London, March 6, 2017. See also Clare Gerada, "The GP and the Drug Misuser in the New NHS: A New 'British System,'" in *Heroin Addiction and the British System: Treatment and Policy Responses*, vol. 2, ed. John Strang and Michael Gossop (London: Routledge, 2005), 68–80. Drug treatment in the United Kingdom had previously been dominated by addiction psychiatrists.

39. This general ethos pervaded the space; see Michael Gossop, "The Treatment Mapping Survey: A Descriptive Study of Drug and Alcohol Treatment Responses in 23 Countries," *Drug and Alcohol Dependence* 39 (1995): 7–14; Michael Gossop, *Treating Drug Misuse Problems: Evidence of Effectiveness* (London: National Treatment Agency for Substance Misuse, 1995); MacGregor, "'Tackling Drugs Together': Ten Years On" and Susanne MacGregor, "'Tackling Drugs Together' and the Establishment of the Principle That 'Treatment Works,'" *Drugs: Education, Prevention and Policy* 13, no. 5 (2006): 399–408.

40. Advisory Council on the Misuse of Drugs, *Reducing Drug Related Deaths* (London: The Stationery Office, 2000), 59.

41. Ibid., 95.

42. My thanks to Sheila Bird for clarifying the significance of this distinction in personal communication by email on January 17, 2019.

43. National Treatment Agency for Substance Misuse, *The Naloxone Programme: Investigation of the Wider Use of Naloxone in the Prevention of Overdose Deaths in Pre-Hospital Care* (London: National Addiction Centre Institute of Psychiatry, June 2007).

44. National Treatment Agency for Substance Misuse, *The NTA Overdose and Naloxone Training Programme for Families and Carers* (London, 2011), 3, https://www.drugsandalcohol.ie/15680/1/NTA_naloxonereport2011.pdf.

45. Dale-Perrara, interview by Campbell, March 6, 2017, 3.

46. Strang, interview by Campbell, March 7, 2017, 1–2.

47. Ibid., 3.

48. Ibid.

49. See the letter coauthored by Bird and McAuley, "Scotland's National Naloxone Programme," *The Lancet* 393, no. 10169 (January 26, 2019): 316–318, which provides six-year outcomes for the Scottish National Naloxone Program discussed in chapter 10 and suggests that England was distributing over twenty thousand naloxone kits each year by 2018.

50. *DrugLink* 2000, 3.

51. Advisory Council on the Misuse of Drugs, *Reducing Drug Related Deaths*, 79.

52. Ibid.

53. Ibid., 80–81. See Rebecca McDonald, Nancy D. Campbell, and John Strang, "Twenty Years of Take-Home Naloxone for the Prevention of Overdose Deaths from Heroin and Other Opioids—Conception and Maturation," *Drug and Alcohol Dependence* 178 (2017): 176–187.

54. Ibid., 81.

55. Dale-Perrara, interview by Campbell, March 6, 2017, 3.

56. Ibid., 13.

57. Marilyn Strathern, ed., *Audit Cultures: Anthropological Studies in Accountability, Ethics, and the Academy* (New York: Routledge, 2000); Nikolas Rose, *Powers of Freedom: Reframing Political Thought* (Cambridge: Cambridge University Press, 1999); Cris Shore and Susan Wright, eds., *Anthropology of Policy: Critical Perspectives on Governance and Power* (London: Routledge, 1997); Cris Shore, Susan Wright, and Davide Pero, eds., *Policy Worlds: Anthropology and the Analysis of Contemporary Power* (New York: Berghahn, 2010).

58. Dale-Perrara, interview by Campbell, March 6, 2017, 13–14.

59. National Treatment Agency for Substance Misuse (NTA), *Treatment Effectiveness Strategy* (n.p.: NTA, 2005–2006).

60. Centre for Social Justice, *Breakdown Britain: Addicted Britain* (London: Centre for Social Justice, 2006), 20. All CSJ reports are here: https://www.centreforsocialjustice.org.uk/about/story.

61. Ibid., 10–11.

62. Ibid., 78–79.

63. Virginia Berridge, "The Rise, Fall, and Revival of Recovery in Drug Policy," *The Lancet* 379, no. 9810 (2012): 22–23.

64. Rosie Mundt-Leach, interview by Nancy D. Campbell, London, March 7, 2017, 9–10.

65. Virginia Berridge, *AIDS in the UK: The Making of Policy, 1981–1994* (Oxford: Oxford University Press, 2012).

66. William L. White, "A Recovery Revolution in Philadelphia," *Counselor* 8, no. 5 (2007): 34–38; William L. White, *Recovery Management and Recovery-Oriented Systems of Care: Scientific Rationale and Promising Practices* (n.p.: Northeast Addiction Technology Transfer Center, the Great Lakes Addiction Technology Transfer Center, and the Philadelphia Department of Behavioral Health Service, 2008); see also Center for Substance Abuse Treatment, *Guiding Principles and Elements of Recovery-Oriented Systems of Care: What Do We Know from the Research?* (Washington, DC: US Department of Health and Human Sciences, 2009), http://www.samhsa.gov; Epstein, *Inclusion*.

67. Berridge, "The Rise, Fall, and Revival of Recovery in Drug Policy"; see also David Best and Alexandra B. Laudet, *The Potential of Recovery Capital* (London: Royal Society for the Arts, 2010).

68. Strang, interview by Campbell, March 7, 2017, 31.

69. John Strang, Michael Kelleher, David Best, Soraya Mayet, and Victoria Manning, "Emergency Naloxone for Heroin Overdose: Should It Be Available Over-the-Counter?," *British Medical Journal* 333 (2006): 614–615; United Kingdom, Medicines and Healthcare Products Regulatory Agency, Proposal for Amendments to the Prescription Only Medicine (Human Use) Order 1997.

70. John Strang, personal communication to the author, June 4, 2017.

71. Yates, interview by Campbell, March 8, 2017, 1–2.

72. The NTA pointed out the need for evaluations of pilot studies in England in a report dated June 2007 prepared by the National Addiction Centre headed by Strang. Up to that point, there had been no evaluations of pilot studies in England.

73. Yates, interview by Campbell, March 8, 2017, 2–3.

74. Tehseen Noorani, "Service User Involvement, Authority and the 'Expert-by-Experience' in Mental Health," *Journal of Political Power* 6 (2013): 49–68. On equivalent movements in the United States, see Nancy Tomes, "From Outsiders to Insiders."

75. Dale-Perrara, interview by Campbell, March 6, 2017, 4–5.

76. Bird chaired the Royal Statistical Society's Working Party on "Performance Monitoring in the Public Services." See "Performance Indicators: Good, Bad, and Ugly."

77. John Strang, Victoria Manning, Soraya Mayet, David Best, Emily Titherington, Laura Santana, Elizabeth Offor, and Claudia Semmler, "Overdose Training and Take-Home Naloxone for Opiate Users: Prospective Cohort Study of Impact on Knowledge and Attitudes and Subsequent Management Of Overdoses," *Addiction* 103 (2008): 1648–1657; Soraya Mayet, Victoria Manning, Anna Williams, Jessica Loaring, and John Strang, "Impact of Training for Healthcare Professionals on How to Manage an Opioid Overdose: Effective, But Dissemination Is Challenging," *International Journal of Drug Policy* 22 (2011): 9–15; Mundt-Leach et al., "Citizen Medics."

78. Mundt-Leach, interview by Campbell, March7, 2017, 2.

79. Beresford South London Patient Accounts, unpublished document.

80. "Therapeutic commitments" are examined in Andrew McAuley, Alison Munro, and Avril Taylor, "'Once I'd Done It Once, It Was Like Writing Your Name': Lived Experience of Take-Home Naloxone Administration by People Who Inject Drugs," *International Journal of Drug Policy* 58 (2018): 46–54.

81. Beresford South London Patient Accounts, unpublished document.

82. Ibid.

83. Ibid.

84. Anonymous service user, interview by Nancy D. Campbell, London, March 9, 2017, 4.

85. Ibid.

86. See the Aurora Project website at http://www.auroraprojectlambeth.org.uk/our-work.html.

87. McCusker, Porter, and Cole, interview by Campbell, March 9, 2017, 33.

88. Ibid., 16–17.

89. Trevor H. Bennett and Katy Holloway, "The Impact of Take-Home Naloxone Distribution and Training on Opiate Overdose Knowledge and Response: An Evaluation of the THN Project in Wales," *Drugs: Education, Prevention and Policy* 19, no. 4 (2012): 320–328; see also Substance Misuse Programme of Public Health Wales, *Harm Reduction Database—Wales: Take-Home Naloxone, 2015–16* (Cardiff: Public Health Wales, 2016).

90. Michael Kelleher, P. Keown, Colin O'Gara, F. Keaney, Michael Farrell, and John Strang, "Dying for Heroin: The Increasing Opioid-Related Mortality in the Republic of Ireland, 1980–1999," *European Journal of Public Health* 15 (2005): 589–592.

91. Chris Rintoul, telephone interview with Nancy D. Campbell, June 13, 2017, 10–11.

92. Gillian W. Shorter and Tim Bingham, *Service Review: Take-Home Naloxone Programme in Northern Ireland: Consultation with Service Users and Service Providers* (Belfast: Public Health Agency, 2016).

93. Ibid.

94. Ibid., 11.

95. Ibid, 7–8.

96. David Best, Romina Gaston, Andrew McAuley, Duncan Hill, and Jennifer Kelly, "The Stuttering Story of Naloxone Distribution," *Drink and Drug News*, December 1, 2008, 12–13.

97. Ibid., 13.

10 "GROWING ARMS AND LEGS"

1. One of the world's largest global suppliers of opioid alkaloids, MacFarlan and Smith was established in west Edinburgh in the early nineteenth century. Kenneth Bentley began to synthesize powerful analgesics there in the 1960s, continuing the line of inquiry that yielded buprenorphine at Reckitt and Coleman in Hull; see John Lewis, "In Pursuit of the Holy Grail," CPDD Nathan B. Eddy Award Acceptance Speech, June 1998. McFarlan and Smith remained in Gorgie, Edinburgh, later becoming sole legal grower of opium poppies in the United Kingdom.

2. Robertson, interview by Campbell, March 1, 2017, 15.

3. Ibid.

4. Irvine Welsh, *Skagboys* (London: Random House, 2012), 359.

5. Ria Din, "Discussion of the Origins of *The Road to Recovery*—the Scottish Government's Recovery Policy for Drugs," *Journal of Groups in Addiction & Recovery* 6 (2011): 38–48.

6. See Gerry McCartney, Chik Collins, David Walsh, and David Batty, *Accounting for Scotland's Excess Mortality: Towards a Synthesis* (Glasgow: Glasgow Centre for Population Health, 2011), 65–75.

7. Ibid. The Scotland and Glasgow Effects have been taken to indicate a "new, and troubling, mortality pattern" paralleling that of post-Soviet Eastern Europe.

8. Reuter and Stevens, *An Analysis of UK Drug Policy*.

9. Ibid.; Ziggy MacDonald, Louise Tinsley, James Collingwood, Pip Jamieson, and Stephen Pudney, *Measuring the Harm from Illegal Drugs Using the Drug Harm Index* (London: Home Office Online Report 24, 2005); Ziggy MacDonald, "The Answer Is 42: Life, the Universe, and the Drug Harm Index," *DrugLink* (2006), https://www .drugwise.org.uk/druglink-article-2006-the-answer-is-42-life-the-universe-and-the -drug-harm-index-by-ziggy-macdonald/; Russell Newcombe, "A Review of the UK Drug Strategy PSA Targets and Drug Harm Index," July 2006, http://www.dldocs.stir .ac.uk/documents/psa.pdf.

10. Reuter and Stevens, *An Analysis of UK Drug Policy*, 47.

11. Paul Slovic and Daniel Vastfjall, "The More Who Die, the Less We Care: Psychic Numbing and Genocide," in *Numbers and Nerves: Information, Emotion, and Meaning in a World of Data*, ed. Scott Slovic and Paul Slovic (Corvallis: Oregon State University Press, 2015), 27–41.

12. Daniel Vastfjall, Paul Slovic, and Marcus Mayorga, "Pseudoinefficacy and the Arithmetic of Compassion," in Slovic and Slovic, *Numbers and Nerves*, 42–52; Deborah A. Small, George Loewenstein, and Paul Slovic, "Sympathy and Callousness: The Impact of Deliberative Thought on Donations to Identifiable and Statistical Victims," *Organizational Behavior and Human Decision Processes* 102 (2007): 143–153.

13. Anthropology of suffering and subjectivity provides inspiration; see Joao Biehl and Adriana Petryna, eds., *When People Come First: Critical Studies in Global Health* (Princeton, NJ: Princeton University Press, 2013); Joao Biehl, Byron J. Good, and Arthur Kleinman, *Subjectivity: Ethnographic Investigations* (Berkeley: University of California Press, 2007); Arthur Kleinman, Veena Das, and Margaret M. Lock, eds., *Social Suffering* (Berkeley: University of California Press, 1997); Veena Das et al., *Remaking a World: Violence, Social Suffering and Recovery* (Berkeley: University of California Press, 2001).

14. Din, "Discussion of the Origins of *The Road to Recovery*."

15. Ibid.

16. Campbell, *Discovering Addiction*.

17. Jeffrey D. Roth and David Best, *Addiction and Recovery in the UK* (London: Routledge, 2012), xiii. In his foreword to *Addiction and Recovery in the UK*, "recovery writer" William L. White traced the once-influential epidemiological theory of the shift from acute to chronic disease as leading killer in the developed world. See Abdel R. Omran, "The Epidemiologic Transition: A Theory of the Epidemiology of Population Change," *The Milbank Quarterly* 83, no. 4 (2005): 731–757. In addition to offering a determinist account of social and economic development, the "epidemiological transition" narrative misunderstands the role of drugs and alcohol in mortality. Proponents of the theory manifest wishful thinking that "society" conforms to the orderly patterns and "natural" laws of a linear, deterministic progression implied by modernization theory. Responding to Michael Marmot's influential account of

"excess deaths" in post-Soviet countries summarized in *The Status Syndrome: How Social Standing Affects Our Health and Longevity* (London: Bloomsbury, 2004), Eugene Raikhel writes in *Governing Habits: Treating Alcoholism in the Post-Soviet Clinic* (Ithaca, NY: Cornell University Press, 2016, 43) that "far from being 'natural,' [these deaths] were the product of 'shock therapy' privatization schemes promoted by many neoliberal economists during the 1990s." See also Michelle A. Parsons, *Dying Unneeded: The Cultural Context of the Russian Mortality Crisis* (Nashville, TN: Vanderbilt University Press, 2014).

18. Ashton and Seymour, "Public Health and the Origins of the Mersey Model"; John Ashton and Howard Seymour, *The New Public Health: The Liverpool Experience* (Philadelphia: Open University Press, 1988); O'Hare, "Merseyside, the First Harm Reduction Conferences, and the Early History of Harm Reduction," *International Journal of Drug Policy* 18, no. 2 (2007): 141–144.

19. Gerada, interview by Campbell, March 6, 2017; Robertson, interview by Campbell, March 1, 2017; Saket Priyadarshi, interview by Nancy D. Campbell, Glasgow, Scotland, February 27, 2017; Yates, interview by Campbell, March 8, 2017.

20. David Best and Grace Ball, "Recovery and Public Policy: Driving the Strategy by Raising Political Awareness," *Journal of Groups in Addiction and Recovery* 6 (2011): 7–19, at 10; on the Scottish recovery agenda, see Roseanna Cunningham, "Recovery in Scotland—Playing to Strengths," *Drugs: Education, Prevention and Policy* 19, no. 4 (2012): 291–293; Ian Wardle, "Five Years of Recovery, December 2005 to December 2010: From Challenge to Orthodoxy," *Drugs: Education, Prevention and Policy* 19, no. 4 (2012): 294–298; Home Office, *Drugs: Protecting Families and Communities— 2008–2018* (London, 2008); Her Majesty's Government, *Drug Strategy 2010: Reducing Demand, Restricting Supply, Building Recovery: Supporting People to Live a Drug-Free Life* (London: Stationery Office, 2010).

21. Scottish Government, *The Road to Recovery: A New Approach to Tackling Scotland's Drug Problem* (Edinburgh, May 29, 2008).

22. David Best, Andrew Rome, Kirstie A. Hanning, William L. White, Michael Gossop, Avril Taylor, and Andy Perkins, *Research for Recovery: A Review of the Drugs Evidence Base* (Edinburgh: Scottish Government, 2010).

23. Din, "A Discussion of the Origins of *The Road to Recovery*," 42.

24. Her Majesty's Government, *Drug Strategy 2010*, 2; Best and Ball, "Recovery and Public Policy," 9.

25. The National Treatment Agency conducted a UK-wide survey designed to document the scope of the unevenness of needle exchange. It was carried out in partnership with the Welsh Substance Misuse Policy Development Team and the Northern Ireland Department of Health, Social Services, and Public Safety. Funded by the Scottish Executive under the Drug Misuse Research Program, the Scottish arm of the study occurred between January 15 and September 30, 2005.

26. Deborah Zador, Brian Kidd, Sharon J. Hutchinson, Avril Taylor, Matt Hickman, Tom Fahey, Andrew Rome, and Alex Baldacchino, *National Investigation in Drug Related Deaths in Scotland, 2003* (Edinburgh, 2005), https://www2.gov.scot/Publications/2005 /08/03161745/17507.

27. Deborah Zador, Andrew Rome, Sharon J. Hutchinson, Matthew Hickman, Alec Baldacchino, Tom Fahey, Avril Taylor, and Brian Kidd, "Differences between Injectors and Non-Injectors, and a High Prevalence of Benzodiazepines among Drug-Related Deaths in Scotland 2003," *Addiction Research and Theory* 15, no. 6 (2007): 651–662.

28. "Making Choice Real," *Drink and Drugs News*, February 8, 2017, https://drinkand drugsnews.com/making-choice-real/.

29. Shaun R. Seaman, Raymond P. Brettle, and Sheila M. Gore, "Pre-AIDS Mortality in the Edinburgh City Hospital HIV Cohort," *Statistics in Medicine* 16, no. 21 (1997): 2459–2474; Sheila M. Bird and Sharon J. Hutchinson, "Male Drugs-Related Deaths in the Fortnight after Release from Prison: Scotland, 1996–99," *Addiction* 98 (2003): 185–190; Sheila M. Bird, Sharon J. Hutchinson, and David J. Goldberg, "Drug-Related Deaths by Region, Sex, and Age Group per 100 Injecting Drug Users in Scotland, 2000–01," *The Lancet* 362, no. 9388 (2003): 941–944. The last was the first to highlight a lower DRD rate for female injectors. This study adhered to the ACMD 2000 recommendation discussed in the previous chapter: The usual death-rate calculus of 100,000 persons from the general population was scuttled in favor of reporting the rate in 100 problem drug users, a practice adopted to make overdose deaths more discernible.

30. Zador et al., "Differences between Injectors and Non-Injectors."

31. Amanda Laird, interview by Nancy D. Campbell, Glasgow, Scotland, February 27, 2017, 3.

32. Strang et al., "Emergency Naloxone for Heroin Overdose."

33. Carole Hunter, interview by Nancy D. Campbell, Glasgow, Scotland, February 27, 2017, 17.

34. Anonymous general practitioner, interview by Nancy D. Campbell, London, March 6, 2017.

35. Ibid.

36. Ibid.

37. Joanne Neale and John Strang, "Naloxone: Does Over-Antagonism Matter? Evidence of Iatrogenic Harm after Emergency Treatment of Heroin/Opioid Overdose," *Addiction* 110 (2015): 1644–1652. This article was based on a retrospective analysis of data that Neale collected originally for a Glasgow study by Neil McKeganey. Strang encouraged Neale to dust off the data and look further at drug user attitudes towards naloxone and experiences with it prior to the launch of the SNNP.

38. While Neale could never have anticipated that Scotland would launch a National Naloxone Program a decade after her ethnographic and qualitative work, her study offers compelling reasons why it is important to multiply perspectives from which knowledge about overdose and overdose prevention is produced.

39. Joanne Neale, "Experiences of Illicit Drug Overdose: An Ethnographic Study of Emergency Hospital Attendances," *Contemporary Drug Problems* 26 (Fall 1999): 505–530, at 517.

40. Ibid., 519.

41. Ibid.

42. Smith and Wallace, interview by Campbell, February 28, 2017, 2.

43. Malloy, interview by Campbell, January 23, 2019, 1.

44. Smith and Wallace, interview by Campbell, February 28, 2017, 3.

45. Neale, "Experiences of Illicit Drug Overdose," 525.

46. Priyadarshi, interview by Campbell, February 27, 2017, 3–4.

47. Hunter, interview by Campbell, February 27, 2017, 16.

48. Robertson, interview by Campbell, March 1, 2017.

49. Chicago Site Visit Report, 2006, 4. I am grateful to Carole Hunter for supplying this unpublished document, which was central to my reconstruction of events in Scotland and Chicago. All quotations in this paragraph are from the report.

50. Ibid., 8.

51. Ibid.

52. Preble and Casey, "Taking Care of Business."

53. Chicago Site Visit Report, 12. This situation was greatly changed by the time of writing, due to a parent support group, Live4Lali, which became a harm reduction naloxone distribution program in close conversation with Bigg until his death in 2018. One reason for documenting harm reduction history is because social movements are always evolving as new participants articulate new goals and work to expand the repertoire of tools and practices.

54. Bigg, interview by Campbell, November 21, 2013.

55. Chicago Site Visit Report, 19.

56. Ibid., 19–20. Each of these concerns was subsequently addressed by researchers—among them Coffin, Green, McAuley, and Strang—in the years to come.

57. Ibid., 20.

58. Ibid., 9.

59. Maxwell, interview by Campbell, January 3, 2014, 1.

60. Priyadarshi, interview by Campbell, February 27, 2017, 10.

61. Hunter, interview by Campbell, February 27, 2017, 2.

62. Priyadarshi, interview by Campbell, February 27, 2017, 10–11.

63. Sheila Bird, interview by Nancy D. Campbell, Loddington, UK, March 5, 2017, 10–11.

64. Hunter, interview by Campbell, February 27, 2017, 12.

65. Ibid.

66. Priyadarshi, interview by Campbell, February 27, 2017, 12.

67. As one of the main advocates for THN, the SDF has built a vast peer education and research network that was coordinated at the time of my interviews by Jason Wallace. Initially the National Naloxone Coordinator was Steven Malloy, followed by Kirsten Horsburgh.

68. Scottish Drugs Forum (SDF), "Take Home Naloxone: Reducing Drug Deaths" (Occasional Paper 2, August 2010), 3.

69. Hunter, interview by Campbell, February 27, 2017, 8.

70. Andrew McAuley, David Best, Avril Taylor, Carole Hunter, and Roy Robertson, "From Evidence to Policy: The Scottish National Naloxone Programme," *Drugs: Education, Prevention, and Policy* 19, no. 5 (2012): 309–319.

71. McAuley, interview by Campbell, February 28, 2017, 4.

72. See the SDF's more schematic timeline at http://www.sdf.org.uk/what-we-do /reducing-harm/take-home-naloxone/naloxone-training-scotland-timeline/.

73. Ibid., 2.

74. Ibid., 3.

75. Ibid., 4–5.

76. Priyadarshi, interview by Campbell, February 27, 2017.

77. McAuley, interview by Campbell, February 28, 2017; see also McAuley et al., "Responsible Management and Use of a Personal Take-Home Naloxone Supply."

78. See Lisa Ross, "The Scottish Highland Overdose Prevention Program Goes to Prison," Overdose Prevention Alliance, February 28, 2013, http://overdoseprevention .blogspot.de/2013/02/the-scottish-highland-overdose.html.

79. McAuley, interview by Campbell, February 28, 2017, 5.

80. Bird, interview by Campbell, March 5, 2017, 10.

81. Each pilot encountered at least one situation in which naloxone failed or was not on hand to be administered by someone who had been trained. The vast majority of actions undertaken yielded a good outcome in that the pilots bolstered the personal self-confidence of those trained to administer naloxone—and helped make the policy case to Scottish ministers. See McAuley et al., "From Evidence to Policy," 313.

82. Chaired by Hunter, the SLWG included Stephen Heller-Murphy, Addiction Policy Development Manager, Scottish Prison Service; Andrew McAuley, Public Health Adviser, Substance Misuse/Alcohol, NHS Health Scotland; Sam Perry, Emergency Medicine Consultant, Western Infirmary, Glasgow; and Lisa Ross, Clinical Harm Reduction Nurse Specialist, NHS Highlands. NFDRD annual reports are archived on the Scottish government website. The annual report for 2009–2010 is available at http://www.gov.scot/Publications/2010/07/30140320/0.

83. Hunter, interview by Campbell, February 27, 2017, 10–11.

84. Ibid.

85. Adults in England and Scotland experience specific forms of bereavement, according to Lorna Templeton et al., "Bereavement Following a Fatal Overdose: The Experiences of Adults in England and Scotland," *Drugs: Education, Prevention and Policy* 24, no. 1 (2017): 58–66.

86. Laird, interview by Campbell, February 27, 2017, 12–13.

87. Amanda Laird was responsible for reporting figures on how much naloxone was distributed through the SNNP to the NHS Information Services Division. Since October

2015 she has converted groups such as NHS harm reduction nurses who were operating under the old PGD to the new framework. At her invitation I attended two events in March 2017 to observe how these legal mechanisms work in practice. The chapter title, "Growing Arms and Legs," is Laird's expression, used to illustrate the developmental process through which naloxone distribution passed in Scotland as it matured.

88. Ibid.

89. Andrew McAuley, Lorna Aucott, and Catriona Matheson, "Exploring the Life-Saving Potential of Naloxone: A Systematic Review and Descriptive Meta-Analysis of Take Home Naloxone (THN) Programmes for Opioid Users," *International Journal of Drug Policy* 26 (2015): 1183–1188.

90. Hunter, interview by Campbell, February 27, 2017, 13.

91. At the time of the interviews Kirsten Horsburgh was National Naloxone Coordinator and Jason Wallace was the National Naloxone Peer Training and Support Officer. Horsburgh has since become Strategy Coordinator for Drug Death Prevention; Wallace is Senior Development Officer for Volunteering and Engagement.

92. Nathanson, *Disease Prevention as Social Change*, 611.

93. Hunter, interview by Campbell, February 27, 2017, 13. In summer 2017, Sheila Bird organized an international letter-writing campaign when it appeared official reporting might be truncated.

94. Sheila M. Bird and Andrew McAuley, "International Comparators Catching up Fast on Scotland's National Naloxone Programme," *The Lancet* 393, no. 10169 (January 26, 2019): 316–318, citing Eliza Wheeler, T. Stephen Jones, M. K. Gilbert, and Peter J. Davidson, "Opioid Overdose Prevention Programs Providing Naloxone to Laypersons—United States," *Morbidity and Mortality Weekly Report* 64, no. 23 (2015): 631–635. They argue that US programs distributed "barely six times as many naloxone kits" as would be necessary to affect the overdose death rate. They estimate the number should be twenty times the number of opioid-related deaths in any given locale.

95. Smith and Wallace, interview by Campbell, February 28, 2017, 13.

96. Campbell, *Using Women*, 38–54.

97. Smith and Wallace, interview by Campbell, February 28, 2017, 13.

98. Ibid., 17.

99. Ibid.

100. Alan M. Wilson, "Mystery Shopping: Using Deception to Measure Service Performance," *Psychology and Marketing* 18 (2001): 721–734; Alan M. Wilson, "The Role of Mystery Shopping in the Measurement of Service Performance," *Managing Service Quality: An International Journal* 8, no. 6 (1998): 414–420.

101. Smith and Wallace, interview by Campbell, February 28, 2017, 4.

102. Neale and Strang, "Naloxone."

103. Sherman et al., "A Qualitative Study of Overdose Responses among Chicago IDUs"; Lankenau et al., "Injection Drug Users Trained by Overdose Prevention Programs."

104. McAuley, interview by Campbell, February 28, 2017, 13.

105. Vincanne Adams, "Evidence-Based Global Public Health: Subjects, Profits, Erasures," in *When People Come First: Critical Studies in Global Health*, ed. João Biehl and Adriana Petryna (Princeton, NJ: Princeton University Press, 2013), 54–90; Michael Agar, "How the Drug Field Turned My Beard Grey," *International Journal of Drug Policy* 13, no. 4 (2002): 249–258; Philippe Bourgois, "Anthropology and Epidemiology on Drugs: The Challenges of Cross-Methodological and Theoretical Dialogue," *International Journal of Drug Policy* 13, no. 4 (2002): 259–269.

106. McAuley, interview by Campbell, February 28, 2017, 13.

107. See Kari Lancaster, Carla Treloar, and Alison Ritter, "'Naloxone Works': The Politics of Knowledge in 'Evidence-Based' Drug Policy," *Health: An Interdisciplinary Journal for the Social Study of Health, Illness, and Medicine* 21, no. 3 (2017): 278–294; Dorothy E. Smith, "The Relations of Ruling: A Feminist Inquiry," *Studies in Cultures, Organizations and Societies* 2, no. 2 (1996): 171–190.

108. Lancaster, Treloar, and Ritter, "Naloxone Works," 12.

11 EVIDENCE FROM PILLAR TO POST

1. Harry M. Marks, *The Progress of Experiment: Science and Therapeutic Reform in the United States, 1900–1980* (Cambridge: Cambridge University Press, 1997); Meldrum, *Departures from the Design*; Yoshioka, "Use of Randomisation."

2. Fredric Jameson, *The Seeds of Time* (New York: Columbia University Press, 1996), 15.

3. Timmermans and Berg, *The Gold Standard*, 22.

4. Ibid. Timmermans and Berg approach the politics of standards from three tenets: "situated knowledges," "blurred agency," and "emergent politics." The first of these is most important to my study. Donna Haraway's 1988 account of "situated knowledges" is crucial for rendering discernible the multiple literacies and trajectories involved in all social interactions, including those occurring in science and scientific research contexts.

5. Thomas Mathar and Yvonne J. F. M. Jansen, eds., *Health Promotion and Prevention Programmes in Practice: How Patients' Health Practices Are Rationalized, Reconceptualized, and Reorganized* (Bielefeld, Germany: Verlag, 2010), 14–15; Alan R. Peterson and Deborah Lupton, *The New Public Health: Discourses, Knowledges, Strategies* (Thousand Oaks, CA: Sage, 1996).

6. Downtown Eastside Vancouver, British Columbia, Canada is a profound example of this; see Susan Boyd, Donald MacPherson, and Bud Osborn, *Raise Shit! Social Action Saving Lives* (Halifax, NS: Fernwood Publishing, 2009); Travis Lupick, *Fighting for Space: How a Group of Drug Users Transformed One City's Struggle with Addiction* (Vancouver, BC: Arsenal Pulp Press, 2017); Gabor Maté, *In the Realm of Hungry Ghosts: Close Encounters with Addiction* (Toronto: Vintage Canada, 2009); Jarrett Zigon, *A War on People: Drug User Politics and a New Ethics of Community* (Oakland: University of California Press, 2019).

7. Judith Yates, interview by Nancy D. Campbell, March 8, 2017, 14–15.

8. Ibid., 15.

9. Ibid., 15–16.

10. Outright rejection of naloxone was rare in England. One site of rejection was Liverpool, paradoxically considered an origin of harm reduction, and site of the first International Harm Reduction Conference. On naloxone availability as a "post-code lottery," see *BBC Newsnight*, January 10, 2017, https://www.youtube.com /watch?v=e28xYxsse1c.

11. Lindsay Prior and Mick Bloor, "Why People Die: Social Representations of Death and Its Causes," *Science as Culture* 3, no. 3 (1993): 346–374.

12. Ibid., 363.

13. Ibid., 370.

14. Ibid., 368.

15. Ibid., 371.

16. Anna Williams and John Strang, in "Opioid Overdose Deaths: Risks and Clusterings in Time and Context" (in *Preventing Opioid Overdose Deaths with Take-Home Naloxone*, ed. John Strang and Rebecca McDonald [Luxembourg: Publications Office of the European Union, 2016], 37–40), wrote that such deaths were subject both to underreporting and "underascertainment" (37). Changing criteria in what counted as a DRD resulted in false positives and false negatives.

17. Jacqueline Wernimont, *Numbered Lives: Life and Death in Quantum Media* (Cambridge, MA: MIT Press, 2018), 40–41. Early modern "political arithmetic" provided the kernel of today's tabular technologies and "quantum media," which Wernimont places in the context of the practice of enumerating the "dead who matter"—which have varied across time in relation to colonial projects, actuarial practices, and the governmentalities of empire.

18. Office of Disease Prevention and Health Promotion, *National Action Plan for Adverse Drug Event Prevention* (Washington, DC: US Department of Health and Human Services, 2014).

19. Lively discussion of the epidemic of quantification as a social problem has also been taken up by those studying global illicit drug markets; see Peter Andreas, "The Politics of Measuring Illicit Flows and Policy Effectiveness," in *Sex, Drugs, and Body Counts: The Politics of Numbers in Global Crime and Conflict*, ed. Peter Andreas and Kelly M. Greenhill (Ithaca, NY: Cornell University Press, 2010), 23–45; Peter Reuter, "The Social Costs of the Demand for Quantification," *Journal of Policy Analysis and Management* 5 (1986): 807–812; Deborah Stone, *The Art of Political Decision Making* (New York: Norton, 1997).

20. The SNNP followed on from the pilots described in the previous chapter, which Bird did not have a hand in designing.

21. Shaun R. Seaman, Raymond P. Brettle, and Sheila M. Gore, "Mortality from Overdose among Injecting Drug Users Recently Released from Prison: Database Linkage Study," *British Medical Journal* 316 (1998): 426–428, found a high DRD rate in HIV-positive drug injectors after release from Edinburgh Prison during 1993 and 1994 in a

study comparing 332 male patients in Dr. Raymond Brettle's Edinburgh City Hospital HIV Cohort embedded alphabetically (to meet privacy concerns) with 332 other male patients attending the City Hospital for Infectious Diseases for other reasons, and those incarcerated in Saughton Prison.

22. Seaman, Brettle, and Gore, "Mortality from Overdose," documented increased risk after prison release, a finding validated by Bird and Hutchinson in "Male Drugs-Related Deaths."

23. Bird, interview by Campbell, March 5, 2017, 2.

24. The idea for a controlled trial of naloxone distribution to ascertain the costs, benefits, and ethics was also central to Shane Darke and Wayne Hall, "The Distribution of Naloxone to Heroin Users," *Addiction* 92, no. 9 (1997): 1195–1199.

25. Ibid. See also Sheila M. Bird, Colin M. Fischbacher, Lesley Graham, and Andrew Fraser, "Impact of Opioid Substitution Therapy for Scotland's Prisoners on Drug-Related Deaths Soon after Prisoner Release," *Addiction* 110, no. 10 (2015): 1617–1624.

26. Ibid.

27. Michael Farrell and John Marsden, "Acute Risk of Drug-Related Death among Newly Released Prisoners in England and Wales," *Addiction* 103 (2007): 251–255.

28. Campbell, *Using Women.*

29. Farrell and Marsden, "Acute Risk of DRD."

30. Ibid.; see also Nina G. Shah, Sarah L. Lathrop, R. Ross Reichard, and Michael G. Landen, "Unintentional Drug Overdose Death Trends in New Mexico, USA, 1990–2005: Combinations of Heroin, Cocaine, Prescription Opioids and Alcohol," *Addiction* 103 (2007): 126–136. Although not a prison-release population, this study found comparatively elevated risk of death among women. While 77 percent of New Mexico overdose deaths were male between 1990 and 2005, the rate increased 123 percent in males but 664 percent in females in tandem with increasing deaths from prescription opioids. Racial-ethnic populations showed differential increases, but none so great as the gender differential.

31. A meta-analysis found elevated risk for DRD shortly after prison release, but excluded two-thirds of the eighteen studies initially considered relevant due to lack of statistical power. See Elizabeth L. C. Merrall, Azar Kariminia, Ingrid A. Binswanger, Michael S. Hobbs, Michael Farrell, John Marsden, Sharon J. Hutchinson, and Sheila M. Bird, "Meta-analysis of DRDs Soon after Release from Prison," *Addiction* 105 (2010): 1545–1554; Rebecca M. Turner, Sheila M. Bird, and Julian P. T. Higgins, "The Impact of Study Size on Meta-analyses: Examination of Underpowered Studies in Cochrane Reviews," *PLoS ONE* 8, no. 3 (2013): e59202. Smaller studies may lack statistical power to demonstrate plausible effect size and challenge the "Cochrane mantra in favor of efficient-science" (Bird, personal communication via email, January 20, 2019). Yet they argue for inclusion of smaller studies despite the "overenthusiasm of researchers for the effectiveness of a new intervention, problems with recruitment to the study, and inaccurate sample size calculations," all of which are endemic in experimental research and yet also crucial for scientific advancement.

32. As Bird and Hutchinson, "Male Drugs-Related Deaths," outlined, "At £10 each, naloxone minijet dispensers would be clearly cost-effective if they deferred (even for a year) ten recently released DRDs of the forty otherwise expected among 8,000 adult injector releases, and hugely cost-effective if they deferred twenty such deaths."

33. Strang, interview by Campbell, March 7, 2017, 20.

34. Animal models of addiction have been used since the early twentieth century to establish the lethal dose and effects of opioid drugs, cocaine, cannabis, and other substances; such studies are of limited utility for studying overdose. See Campbell, *Discovering Addiction*.

35. Adams, "Evidence-Based Global Public Health," 68.

36. Strang, interview by Campbell, March 7, 2017, 20.

37. Ibid.

38. As indicated by my interview with Alexander Walley, this idea was not confined to the United Kingdom.

39. Recognized as a strong naloxone proponent, Strang was credited by many of those I interviewed as a source of personal and public support, the latter due to programmatic statements such as the so-called Orange Guidelines for clinicians. Strang and colleagues consult widely but nonexclusively with the WHO, the UN, the EMCDDA, and pharmaceutical companies attempting to improve naloxone formulations and delivery systems. He might be said to have left no stone unturned in the study of naloxone.

40. Bird, interview by Campbell, March 5, 2017, 5.

41. Closure is defined as a theoretical mechanism by which social processes shape technology. See Trevor J. Pinch and Wiebe E. Bijker, "The Social Construction of Facts and Artifacts: Or How the Sociology of Science and the Sociology of Technology Might Benefit Each Other," *Social Studies of Science* 14, no. 3 (1984): 399–441. The degree to which closure may be seen as determining has been contested by those who see it as indeterminate or, as in the naloxone case, even unsettling.

42. Bird, interview by Campbell, March 5, 2017, 6.

43. Strang, interview by Campbell, March 7, 2017, 25.

44. Ibid., 21–22.

45. Ibid.

46. Priyadarshi, interview by Campbell, February 27, 2017, 8.

47. Hunter, interview by Campbell, February 27, 2017, 7.

48. Ibid., 8.

49. Ibid.

50. Priyadarshi, interview by Campbell, February 27, 2017, 8–9.

51. Bird, interview by Campbell, March 5, 2017, 7.

52. Ibid., 8.

53. Ibid., 10.

54. Ibid., 11.

55. Strang, interview by Campbell, March 7, 2017, 20–21.

56. Ibid., 25. Wales was similarly excluded because its subnational naloxone program went live just after Scotland but with less documentation, a smaller population, and less public awareness.

57. McAuley, interview by Campbell, February 28, 2017, 7–8.

58. Bird, interview by Campbell, March 5, 2017, 11.

59. Ibid., 12.

60. Ibid., 12.

61. McAuley, interview by Campbell, February 28, 2017, 8.

62. Michael Schillmeier, *Eventful Bodies: The Cosmopolitics of Illness* (New York: Routledge, 2014), 11–12.

63. Ibid., 12.

64. Campbell, *Using Women*, 38.

65. Schillmeier, *Eventful Bodies*, 12.

66. Strang, interview by Campbell, March 7, 2017, 18.

67. Donna J. Haraway, *Modest Witness@Second Millenium:FemaleMan Meets Oncomouse* (New York: Routledge, 1997).

68. Bruno Latour and Steve Woolgar's early laboratory-based account was titled *Laboratory Life: The Construction of Scientific Facts* (Princeton, NJ: Princeton University Press, 1986), although there was said to be nothing particularly important about the conceptual or scientific contributions of the laboratories studied. Science and Technology Studies researchers sought good hosts that would allow observation and interactions on the basis of which they could publish accounts of laboratory activities. In the decades since these early studies, many "laboratory lives" have been produced, typically without justifications of the significance of particular laboratories.

69. Strang, interview by Campbell, March 7, 2017, 16–17. Working with respiratory physiologist Caroline Jolley, Strang and colleagues traced plummeting oxygen levels in the first five minutes after heroin injection, assuming that the "phenomena we were able to observe were the same as overdoses which lead to death except that, in our study sessions, the extent of overdose was much less severe. These individuals were taking a dose of heroin which had been individually titrated and was deemed clinically to be their appropriate regular maintenance dose."

70. At the US PHS Narcotic Farm laboratory, neurophysiological studies such as these were conducted; however, the object of study was the physiology of withdrawal, or what was then called the morphine abstinence syndrome, and the abuse liability of particular compounds. See Acker, *Creating the American Junkie*; Campbell, *Discovering Addiction*.

71. Campbell, *Discovering Addiction*.

72. Strang, interview by Campbell, March 7, 2017, 16–17.

73. John Strang, Teodora Groshkova, Ambros Uchtenhagen, Wim van den Brink, Christian Haasen, Martin T. Schechter, Nick Lintzeris, James Bell, Alessandro Pirona, Eugenia Oviedo-Joekes, Roland Simon, and Nicola Metrebian, "Heroin on Trial: Systematic Review and Meta-analysis of Randomised Trials of Diamorphine-Prescribing as Treatment for Refractory Heroin Addiction," *British Journal of Psychiatry* 207, no. 1 (2015): 5–14.

74. Strang, interview with Campbell, March 7, 2017, 17.

75. Sheila M. Bird, John Strang, Deborah Ashby, Julie Podmore, J. Roy Robertson, Sarah Welch, Angela M. Meade, and Mahesh K. B. Parmar, "External Data Required Timely Response by the Trial Steering-Data Monitoring Committee for the Naloxone Investigation (N-ALIVE) Pilot Trial," *Communications in Contemporary Clinical Trials* 5 (2017): 100–106; Mahesh K. B. Parmar, John Strang, Louise Choo, Angela M. Meade, and Shelia M. Bird, "Randomised Controlled Pilot Trial of Naloxone-on-Release," *Addiction* 112, no. 3 (2017): 502–515.

76. Strang, interview by Campbell, March 7, 2017, 16–17.

77. Priyadarshi, interview with Campbell, February 27, 2017, 6.

78. Peterson and Lupton, *The New Public Health*, 159.

79. Ibid., 161.

80. Priyadarshi, interview by Campbell, February 27, 2017, 7.

81. Smith and Wallace, interview by Campbell, February 28, 2017, 6.

82. Ibid.

83. Ibid.

84. Ibid., 9.

85. Ibid., 5.

86. Ibid., 6.

87. Ibid.

88. David Liddell and Biba Brand, *Developing a Model of User Involvement and Social Research in Scotland* (Amsterdam: Correlation Network, 2008).

89. Jorge Anker et al., "Drug Users and Spaces for Legitimate Action," in *Empowerment and Self-Organizations of Drug Users: Experiences and Lessons Learnt*, ed. Georg Broring and Eberhard Schatz (Amsterdam: Correlation Network, 2008), 17–38, quotation at 19.

90. Smith and Wallace, interview by Campbell, February 28, 2017, 4.

91. Ibid., 15.

92. Ibid., 17.

93. Adele E. Clarke, Carrie E. Friese, and Rachel S. Washburn, *Situational Analysis: Grounded Theory after the Interpretive Turn*, 2nd ed. (Thousand Oaks, CA: Sage, 2017).

94. Hugh McLaughlin, "What's in a Name? 'Client,' 'Patient,' 'Customer,' 'Consumer,' 'Expert by Experience,' 'Service User'—What's Next?," *British Journal of Social Work* 39 (2009): 1101–1117, at 1106–1109.

95. Nancy Tomes, "From Outsiders to Insiders: The Consumer-Survivor Movement and Its Impact on U.S. Mental Health Policy," in *Patients as Policy Actors: A Century of Changing Markets and Missions,* ed. Nancy Tomes, Beatrix Hoffman, Rachel Grob, and Mark Schlesinger (New Brunswick, NJ: Rutgers University Press, 2011), 113–131.

96. Lavinia Black, Ian Connelly, Mel Getty, Cassandra Hogan, Paul Lennon, Martin McCusker, Joanne Neale, and John Strang, "Poor Implementation of Naloxone Needs to Be Better Understood to Save Lives," *Addiction* 112, no. 5 (2017): 911–912.

97. Anonymous service user, interview by Nancy D. Campbell, London, March 9, 2017, 3–4.

98. I am grateful to Sheila Bird for drawing my attention to this, and for patiently educating me about the statistical designs of the studies referenced in this chapter.

99. Elizabeth Ettorre, "Revisioning Women and Drug Use: Gender, Sensitivity, Embodiment and Reducing Harm," *International Journal of Drug Policy* 15, no. 5–6 (2004): 327–335; Natasha Du Rose, "Marginalised Drug-Using Women's Pleasure and Agency," *Social History of Alcohol and Drugs* 31 (2017): 42–64.

100. Strang, interview by Campbell, March 7, 2017, 32.

12 OVERDOSE AND THE CULTURAL POLITICS OF REDEMPTION

1. Langdon Winner, "Do Artifacts Have Politics?," in special issue, "Modern Technology: Problem or Opportunity?," *Daedalus* 109, no. 1 (1980): 121–136, at 125.

2. Nancy D. Campbell, "Technologies of Suspicion: Coercion and Compassion in Post-disciplinary Surveillance Regimes," *Surveillance and Society* 2, no. 1 (2004): 78–92.

3. Winner, "Do Artifacts Have Politics?," 127.

4. Duane Nieves, interview by Nancy D. Campbell, Harrisburg, Pennsylvania, February 2, 2017. At the time, Nieves was Chief and Director of Field Operations for Holy Spirit EMS, a Geisinger affiliate near Harrisburg, the capitol city of Pennsylvania.

5. Bennett, interview by Campbell, October 4, 2013.

6. Rose and Abi-Rached, *Neuro.*

7. Inaccessible until the early 2000s, the documentary intercut home movies from Prince's childhood with 1970s footage: https://www.youtube.com/watch?v=o_M8Yaoz-PDE.

8. In a 2009 retrospective at https://www.youtube.com/watch?v=12jfiDdLo4E, Prince narrated his awareness of the stakes at the time of the incident: "I totally understood that and the ramifications of fucking it up were gihugic" [26:30]. The 1978 scenes alternated with the scenes from *Pulp Fiction* as Prince's retrospective voice-over continues to 29:04.

9. See Nancy D. Campbell, "Rewriting the 'Antidrug': The Ambivalence of the Pharmakon in the Cultural Discourse of Overdose," in *The Pharmakon,* ed. Hermann Herlinghaus (Heidelberg, Germany: Winter Verlag, 2018), 343–369.

10. Played by Ving Rhames, who was later cast as the drug dealer in *Pulp Fiction*.

11. Wolfe, interview by Campbell, October 6, 2013.

12. Hansen, "Assisted Technologies"; Helena Hansen and Jules Netherland, "Is the Prescription Opioid Epidemic a White Problem?," *American Journal of Public Health* 106, no. 12 (2016): 2127–2129; Jules Netherland and Helena B. Hansen, "The War on Drugs That Wasn't: Wasted Whiteness, 'Dirty Doctors,' and Race in Media Coverage of Prescription Opioid Misuse," *Culture Medicine and Psychiatry* 40, no. 4 (2016): 664–686.

13. Sonia Mendoza, Allyssa S. Rivera, and Helena B. Hansen, "Re-racialization of Addiction and the Redistribution of Blame in the White Opioid Epidemic," *Medical Anthropology Quarterly* 33, no. 2 (2018): 242–262.

14. Ginia Bellafante, "In the Bronx, Heroin Woes Never Went Away," *New York Times*, March 23, 2017, MB1, https://www.nytimes.com/2017/03/23/nyregion/bronx-heroin-opioid-crisis.html.

15. Anne Case and Angus Deaton, "Rising Morbidity and Mortality in Midlife among White Non-Hispanic Americans in the 21st Century," *Proceedings of the National Academy of Sciences of the United States of America* 112, no. 49 (2015): 15078–15083; Anne Case and Angus Deaton, "Mortality and Morbidity in the 21st Century," *Brookings Reports on Economic Activity* (Spring 2017): 397–467. "Deaths of despair" are a *combination* of deaths resulting from alcoholism, suicide, and opioid overdose in the white, non-college-educated, midlife US population. See also Christopher Ruhm, who in "Deaths of Despair or Drug Problems?" (National Bureau of Economic Research Working Paper 24188, January 2018) argued that the "drug environment" (price and availability) were responsible for mortality. Case and Deaton reply in "Deaths of Despair Redux: A Response to Christopher Ruhm," at https://www.princeton.edu/~accase/downloads/Case_and_Deaton_Comment_on_CJRuhm_Jan_2018.pdf.

16. Katharine Q. Seelye, "Obituaries Shed Euphemisms to Chronicle Toll of Heroin," *New York Times*, July 11, 2015.

17. Tony Walter, "New Mourners, Old Mourners: Online Memorial Culture as a Chapter in the History of Mourning," *New Review of Hypermedia and Multimedia* 21, nos. 1–2 (2015): 10–24; Sue-Anne Ware, "Anti-memorials and the Art of Forgetting: Critical Reflections on a Design Practice," *Public History Review* 1 (2008): 61–76.

18. Parsons, *Dying Unneeded*.

19. Ibid., 11.

20. Orlando Patterson, *Slavery and Social Death: A Comparative Study* (Cambridge, MA: Harvard University Press, 2018).

21. See Chris McGreal, *American Overdose: A Tragedy in Three Acts* (London: Guardian Faber Publishing, 2018).

22. Michael J. Zoorob and Jason L. Salemib, "Bowling Alone, Dying Together: The Role of Social Capital in Mitigating the Drug Overdose Epidemic in the United States," *Drug and Alcohol Dependence* 173 (2017): 1–9.

23. Ernest Moy, Macarena C. Garcia, Brigham Bastian, Lauren M. Rossen, Deborah D. Ingram, Mark Faul, Greta M. Massetti, Cheryll C. Thomas, Yuling Hong, Paula W. Yoon, and Michael F. Iademarco, "Leading Causes of Death in Nonmetropolitan and Metropolitan Areas—United States, 1999–2014," *Morbidity and Mortality Weekly Report* 66, no. SS-1 (2017): 1–8. This 2017 study found that annual age-adjusted death rates for heart disease, stroke, cancer, unintentional injury, and chronic lower respiratory disease were highest in nonmetropolitan areas from 1999 to 2014.

24. Michael Faul, Michael W. Dailey, David E. Sugerman, Scott M. Sasser, Benjamin Levy, and Len J. Paulozzi, "Disparity in Naloxone Administration by Emergency Medical Service Providers and the Burden of Drug Overdose in US Rural Communities," *American Journal of Public Health* 105, no. S3 (2015): e26–e32.

25. Sam Quinones, *Dreamland: The True Tale of America's Opiate Epidemic* (London: Bloomsbury, 2015).

26. Faul et al., "Multiple Naloxone Administrations."

27. Steven Allan Sumner, Melissa C. Mercado-Crespo, M. Bridget Spelke, Leonard Paulozzi, David E. Sugerman, Susan D. Hillis, and Christina Stanley, "Use of Naloxone by Emergency Medical Services during Opioid Drug Overdose Resuscitation Efforts," *Prehospital Emergency Care* 20, no. 2 (2016): 220–225.

28. Christopher M. Jones, Peter G. Lurie, and William M. Compton, "Increase in Naloxone Prescriptions Dispensed in US Retail Pharmacies since 2013," *American Journal of Public Health* 106, no. 4 (2016): 689–690.

29. Traci C. Green and Maya Doe-Simkins, "Opioid Overdose and Naloxone: The Antidote to an Epidemic?," *Drug and Alcohol Dependence* 163 (2016): 265–271.

30. Nancy Kreiger, *Epidemiology and the People's Health* (Oxford: Oxford University Press, 2013).

31. Darke, *The Life of the Heroin User*, 129, 132.

32. Chris Benedetto, interview by Nancy D. Campbell, Carlisle, Pennsylvania, February 2, 2017.

33. Joseph S. Seidel, "Emergency Medical Services and the Adolescent Patient," *Journal of Adolescent Health* 12, no. 2 (1991): 95–100.

34. Faul et al., "Multiple Naloxone Administrations." MNAs are most likely to occur when emergency responders arrive on a scene where naloxone was previously administered by parents or peers; the Northeast is one of few regions where naloxone is readily available due to the community saturation model adopted in Massachusetts.

35. Users gain social status from habit size and the prodigious quantities of opiates they can consume—just as do drinkers. "Handling one's liquor" is an ability generally attached to particular ways of doing masculinity.

36. Denise Holden, interview by Nancy D. Campbell, Harrisburg, Pennsylvania, January 30, 2017, 8.

37. Ibid., 14.

38. Ibid., 6.

39. Shilo Murphy (Jama), interview by Nancy D. Campbell, Seattle, Washington, January 3, 2014, 1.

40. Ibid., 7.

41. Ibid., 12.

42. See Phillip O. Coffin, Lara S. Coffin, Shilo Murphy, Lindsay M. Jenkins, and Matthew R. Golden, "Prevalence and Characteristics of Femoral Injection among Seattle-Area Injection Drug Users," *Journal of Urban Health* 89, no. 2 (2012): 365–372; Caleb Banta-Green, Leo Beletsky, Jennifer A. Schoeppe, Phillip O. Coffin, and Patricia C. Kuszler, "Police Officers' and Paramedics' Experiences with Overdose and Their Knowledge and Opinions of Washington State's Drug Overdose–Naloxone–Good Samaritan Law," *Journal of Urban Health* 90, no. 6 (2013): 1102–1111.

43. Murphy (Jama), interview by Campbell, January 3, 2014, 23.

44. Daniel Raymond, "Holding Space for the Unredeemed: Harm Reduction and Justice," *Medium*, September 26, 2016, https://medium.com/@danielraymond/holding -space-for-the-unredeemed-harm-reduction-and-justice-1d70ca675f25.

45. Hess, "Undone Science and Social Movements."

46. Nabarun Dasgupta, Leo Beletsky, and Daniel Ciccarone, "Opioid Crisis: No Easy Fix to Its Social and Economic Determinants," *American Journal of Public Health* 108, no. 2 (2018): 182–186.

47. Tessie Castillo, "'Narcan Party' Hysteria Puts a Value on Drug Users Lives," *The Fix*, November 5, 2017, https://www.thefix.com/narcan-party-hysteria-puts-value-drug -users-lives.

48. "Narcan Parties Becoming Disturbing Trend, Police Say," *Atlanta Journal-Constitution*, April 27, 2017, https://www.ajc.com/news/national/narcan-parties -becoming-disturbing-trend-police-say/2dEFxaNfqAO68mHD64cVsJ/. This story embedded a news link in which Katherine Amenta, a Channel 11 news anchor from WFMZ-TV, reported that "police in Pennsylvania say addicts and dealers are holding Narcan parties." Claiming dealers threw parties "where they sell heroin and Narcan as a package deal," Amenta reported that Pittsburgh police had used Narcan in 2,300 overdoses in 2016, nearly twice the number they had in 2015.

49. Tina Terry, "Narcan Parties," Wsoctv.com, August 4, 2017, https://www.boston 25news.com/news/narcan-parties-drug-users-overdosing-to-be-brought-back-to -life/581594427.

50. Nancy Fraser and Linda Gordon, "A Genealogy of Dependency: Tracing a Keyword of the US Welfare State," *Signs* 19, no. 2 (1994): 309–336; Sanford Schram, "In the Clinic: The Medicalization of Welfare," *Social Text* 62 (18, no. 1) (2000): 81–107. In the United States, the safety-net metaphor has stretched to encompass government social or health insurance programs on which the poor and uninsured have become "dependent" in much the same ways that "addicts" are "dependent" on drugs.

51. Craig Reinarman, "The Social Construction of Drug Scares," in *Constructions of Deviance: Social Power, Context, and Interaction*, ed. Patricia A. Adler and Peter Adler

(Belmont, CA: Wadsworth, 1994), 92–105; Edward M. Brecher and the editors of *Consumer Reports*, *Licit and Illicit Drugs: The Consumers Union Report on Narcotics, Stimulants, Depressants, Inhalants, Hallucinogens, and Marijuana—Including Caffeine, Nicotine, and Alcohol* (Boston: Little, Brown & Co., 1972), 321–334.

52. This topic was discussed in briefing materials prepared by Adapt Pharma, makers of Narcan, the brand name product best known in the United States, for an October 5, 2016, meeting of the Anesthetic and Analgesic Drug Products Advisory Committee and the Drug Safety and Risk Management Advisory Committee, "FDA Advisory Committee on the Most Appropriate Dose or Doses of Naloxone to Reverse the Effects of Life-Threatening Opioid Overdose in the Community Settings," September 2, 2016, https://www.fda.gov/downloads/advisorycommittees/committeesmeetingmaterials/drugs/anestheticandanalgesicdrugproductsadvisorycommittee/ucm522688.pdf.

53. Ibid., 16.

54. Ibid., 4.

55. Joanne Neale and John Strang, "Naloxone: Does Over-antagonism Matter? Evidence of Iatrogenic Harm after Emergency Treatment of Heroin/Opioid Overdose," *Addiction* 110 (2015): 1644–1652.

56. World Health Organization (WHO), *Adherence to Long-Term Therapies: Evidence for Action* (Geneva: WHO, 2003).

57. Patient package inserts are the "fine print" by which consumers receive information about legal drug effects in the United States. See Lorna Ronald, "Empower to Consume: Direct-to-Consumer Advertising, Free Speech, and Pharmaceutical Governance" (Unpublished diss., Rensselaer Polytechnic Institute, Troy, NY, 2006).

58. Jennifer L. Doleac and Anita Mukherjee, "The Moral Hazard of Lifesaving Innovations: Naloxone Access, Opioid Abuse, and Crime," March 31, 2019, SSRN, https://papers.ssrn.com/sol3/papers.cfm?abstract_id=3135264.

59. Brookings distanced itself from the economists in response to demands for retraction from public health scholars, addiction experts, and harm reductionists. Szalavitz posted Brookings's response on December 14, 2018: "Brookings supports the academic freedom of our resident and nonresident scholars, including their right to publish controversial research that adheres to our quality and independence standards. However, Brookings as an institution does not take positions on issues, nor do we endorse Doleac's response to the criticism and feedback she received. Public health experts, researchers, and advocates have made important and critical contributions to our society, and their efforts have saved countless lives by informing America's response to the opioid crisis." I am indebted to personal communication with Maia Szalavitz and her online article "An Influential Think Tank Suggested That Harm Reduction Doesn't Work," *Tonic*, December 13, 2018, https://tonic.vice.com/en_us/article/gy7ke9/brookings-institution-suggested-that-harm-reduction-doesnt-work.

60. Jing Xu, Corey S. Davis, Marisa Cruz, and Peter G. Lurie, "State Naloxone Access Laws Are Associated with an Increase in the Number of Naloxone Prescriptions Dispensed in Retail Pharmacies," *Drug and Alcohol Dependence* 189 (2018): 37–41; Davis and Carr, "State Legal Innovations."

61. Davis and Carr, "Legal Changes to Increase Access."

62. Doleac and colleagues acknowledged neither their own "priors," nor assumptions integral to econometrics. Most problematic for contributing useful knowledge on the topic, their study ended in 2015—more than a year before all states had changed their naloxone access laws. Thus many of the laws were not yet in effect.

63. Jennifer L. Doleac and Anita Mukherjee, "The Moral Hazard of Lifesaving Innovations: Naloxone Access, Opioid Abuse, and Crime," IZA DP No. 11489 (Bonn, Germany: IZA Institute of Labor Economics, April 2018), 2, https://www.iza.org/publications/dp/11489/the-moral-hazard-of-lifesaving-innovations-naloxone-access-opioid-abuse-and-crime.

64. The December 7, 2018, piece appeared on the Brookings blog with the title, "Research Roundup: What Does the Evidence Say about How to Fight the Opioid Epidemic?," https://www.brookings.edu/blog/up-front/2018/12/07/research-roundup-what-does-the-evidence-say-about-how-to-fight-the-opioid-epidemic/.

65. Ibid., 2.

66. Quoted in Maia Szalavitz, "An Influential Think Tank Suggested That Harm Reduction Doesn't Work," Vice, December 13, 2018, https://www.vice.com/en_us/article/gy7ke9/brookings-institution-suggested-that-harm-reduction-doesnt-work.

67. Doleac and Mukherjee, "The Moral Hazard of Lifesaving Innovations," 1.

68. David Rowell and Luke B. Connelly, "A History of the Term Moral Hazards," Journal of Risk and Insurance 79, no. 4 (2012): 1051–1075.

CONCLUSION

1. In "Does Naloxone Availability Increase Opioid Overdose? The Case for Skepticism," Health Affairs Blog, March 19, 2018, https://www.healthaffairs.org/do/10.1377/hblog20180316.599095/full/, Richard G. Frank, Keith Humphreys, and Harold A. Pollack criticized the Brookings study for its "failure to capture what has been driving naloxone use in the United States."

2. Rintoul, interview by Campbell, June 13, 2017, 3.

3. Malloy, interview by Campbell, January 23, 2019, 16.

4. Martha Bebinger, "Fentanyl-Linked Deaths: The U.S. Opioid Epidemic's Third Wave Begins," NPR Morning Edition, March 21, 2019, https://www.npr.org/sections/health-shots/2019/03/21/704557684/fentanyl-linked-deaths-the-u-s-opioid-epidemics-third-wave-begins. See Merianne Rose Spencer, Margaret Warner, Brigham A. Bastian, James P. Trinidad, and Holly Hedegaard, "Drugs Deaths Involving Fentanyl, 2011–2016," National Vital Statistics Reports 68, no. 3 (March 21, 2019): 1–16.

5. Green and Doe-Simkins, "Opioid Overdose and Naloxone," 269.

6. Tessie Castillo, "What's Next after Naloxone?," The Fix, January 11, 2017, https://www.thefix.com/whats-next-after-naloxone.

7. Scott Burris, "Where Next for Opioids and the Law? Despair, Harm Reduction, Lawsuits, and Regulatory Reform," Public Health Reports 133, no. 1 (2018): 29–33.

8. Martha Bebinger, "What Doesn't Kill You Can Maim: Unexpected Injuries from Opioids," *All Things Considered*, National Public Radio, April 13, 2017.

9. See Jarrett Zigon, "Attunement: Rethinking Responsibility," in *Competing Responsibilities: The Ethics and Politics of Contemporary Life*, ed. Susanna Trnka and Catherine Trundle (Durham, NC: Duke University Press, 2017), 49–68.

10. Nikolas Rose and Joelle M. Abi-Rached, *Neuro: The New Brain Sciences and the Management of the Mind* (Princeton, NJ: Princeton University Press, 2013).

11. Natasha Myers, *Rendering Life Molecular: Models, Modelers, and Excitable Matter* (Durham, NC: Duke University Press, 2015); Deboleena Roy, *Molecular Feminisms: Biology, Becomings, and Life in the Lab* (Seattle: University of Washington Press, 2018).

12. Zigon, "Attunement," 54.

13. Ibid., 56.

14. Ibid., 63–64.

15. Wolfe, interview by Campbell, October 6, 2013, 7.

16. Ibid.

17. Carl L. Hart, "People Are Not Dying Because of Opioids," *Scientific American* 317 (2017): 5, 11.

18. Ryan Hampton, "America's Crisis Isn't Opioids," *The Hill*, December 18, 2018, https://thehill.com/blogs/congress-blog/healthcare/421868-americas-crisis-isnt -opioids-its-ignorance.

19. The remark occurred in an interview with Hertel that was part of Greg Scott's forthcoming film *Reach for Me: Fighting to End the American Drug Overdose Epidemic* by Sawbuck Productions. A self-identified former drug user, Hertel is a harm reduction educator in Minneapolis, who found in OD prevention work, the "thing that we all want in our lives, the thing that makes us happy and makes everything worth it."

20. Pharmacists are playing an increased role in the world of naloxone coprescription in the United States; see Lim et al., "Prescribe to Prevent."

INDEX

Scottish Drugs Forum (SDF), 228, 230, 236, 238, 243–245, 271–274, 276–280, 288, 376n67, 377n72
Scottish government, 4, 224, 226, 228–230, 234, 236–238, 243, 261, 377n82
Scottish Highlands, 11, 237, 239–240, 360n7, 377nn81–82
Scottish National Naloxone Program (SNNP), 4, 19, 191, 220, 223, 224, 231, 234, 237–240, 242, 243–245, 249, 256, 263–266, 268–269, 272, 369n49, 375n37, 377n87, 378n91, 380n20
Scottish National Party (SNP), 224, 238, 243, 249
Scottish Parliament, 225, 238, 242
Scottish Prison Service (SPS), 236, 256–257, 260, 262, 263, 377n82
Seattle, Washington, 109, 166, 167, 186, 217, 296
Sedatives, 36, 86, 101, 334n22
Service users, 135, 197, 216, 218, 219, 221, 270, 272, 273, 275, 281–282, 310–311, 371n74
Sexuality, 84, 93, 152, 161, 192, 275, 314, 355n8
Siegel, Shepard, 15–16
Simpich, Bill, 156
Sinatra, Frank, 31, 36
Singapore, 46
Skagboys (Welsh), 224
Smith, Anna Nicole, 10
Smith, Austin, 274–276
Snyder, Solomon H., 43, 44
Sober science, 351n35
Social
  actors, 128, 135, 185
  death, 73, 84, 289, 300
  infrastructure, 109, 175, 307, 313
  justice, 19, 21, 105, 106, 171, 178, 179, 184, 186, 198, 248, 293, 294
  movements, 1, 3, 13, 20–21, 23, 123, 148, 152, 167, 175, 177, 180, 184, 185, 190, 192–193, 195–196, 243, 252, 275, 299, 312–314, 356n9, 376n53
  embodied health, 125–126
  evidence- or knowledge-based, 134–135, 140
  reflexive, 128–129
  organization, 21, 91, 125, 128, 135, 360n7
  relation(s), 73, 75, 126, 135, 161, 166, 267, 300
  situatedness, 17, 251, 293, 308, 379n4
Sorge, Rod, 143–146
Spanish-speaking people, 70, 79, 98
Staley, Layne, 10
Stancliff, Sharon, 145, 148, 178, 355n84
Statistical
  study or method, 14, 83, 174, 188, 206, 226, 252, 255, 257, 258, 261, 265, 303, 371n76, 381n31, 385n98
Stengers, Isabelle, 194–195
Stewart, Justice Potter, 66
Stigmatization, 6, 11, 88, 118, 122, 127, 134, 159, 163, 169, 185, 205, 223, 226, 233, 255, 256, 281, 288, 297, 298, 300
Stimson, Gerry, 203
Strang, John, 135–138, 162, 167, 207–210, 214–217, 230, 237–238, 246, 258–260, 262, 263, 267, 268, 281–282, 351n33, 351n41, 360n7, 371n72, 375n37, 376n56, 382n39, 383n69
Suboxone. See Buprenorphine
Substance Abuse and Mental Health Services Administration (SAMHSA), 177, 181, 347n53, 362n31
Sudnow, David, 289, 359n58
Suicide, 25, 27–29, 66, 83–85, 100, 225, 286, 322n4, 323n20, 386n15